3/94

URBAN LANDSCAPE DYNAMICS

Urban Landscape Dynamics

A multi-level innovation process

Edited by
ARMANDO MONTANARI
GERHARD CURDES
LESLIE FORSYTH

Avebury

Aldershot · Brookfield USA · Hong Kong · Singapore · Sydney

Published by
Avebury
Ashgate Publishing Limited
Gower House
Croft Road
Aldershot
Hants GU11 3HR
England

Ashgate Publishing Company
Old Post Road
Brookfield
Vermont 05036
USA

A CIP catalogue record for this book is available from the British Library

ISBN 1 85628 203 1

Printed and Bound in Great Britain by
Athenaeum Press Ltd, Newcastle upon Tyne.

CONTENTS

Part II. INNOVATION AND URBAN DEVELOPMENT:
SELECTED ITEMS

FIGURES

TABLES

PREFACE

The URBINNO Project (Innovation and Urban Development: the Role of Technological and Social Change) started officially in Vienna in March 1987 at a Seminar organized by the Interdisciplinary Institute for Regional Planning of the Vienna University of Economics. The idea for the Programme had originated about three years earlier at a seminar at the University of Munich in May 1984. Participants from previous multi-disciplinary comparative research projects conducted between the end of the Seventies and the start of the Eighties at institutes specializing in multinational scientific cooperation in the social sciences were invited to participate in this Seminar. These included people from the European Coordination Centre for Research and Documentation in Social Sciences (Vienna Centre) in Vienna, and the International Institute for Applied Systems Analysis (IIASA) in Laxenburg. In this context, special reference can be made to the 'Costs of Urban Growth (CURB)' project, the 'Human Settlement Systems' study and the project entitled 'Metropolitan Dynamics'.

Several preparatory meetings were held between 1984 and 1987. In the course of these meetings possible participants were identified, the research agenda defined, as well as the organization of the research within the URBINNO Project. The Volkswagen Foundation

1

in Hannover sponsored it and the European Centre for Social Welfare Training and Research in Vienna agreed to coordinate and be responsible for the administrative aspects of the Project. The first phase of the URBINNO Project ended in 1989. The results of this initial phase are being presented in a series of volumes published in the Urban Europe Series, which started publication in 1990.

Four Working Groups (WGs) with members from around 50 different national research groups operated within the framework of the URBINNO Project on the following themes: (WG I) Population and Social Structure; (WG II) Employment and Economy; (WG III) Political and Institutional Organization; (WG IV) Built-form Environment and Land Use. The Project was directed by a Steering Committee made up of the Coordinating Committee working in conjunction with the Chairpersons of the four Working Groups.

In this volume we are presenting the results of the research conducted on 'Built-form Environment and Land Use' (WG IV). The Working Group met in Vienna (March 1987), in Aachen (March 1988), in Bari (September 1988), in Aachen (March 1989) and in Chapel Hill (July 1989) where draft chapters were presented. In the course of these meetings the Group discussed and analyzed the results of the research. Further meetings were held in Dubrovnik (May 1990) and Vienna (November 1991) to check the editorial aspects, fill in whatever elements were missing and plan and organize the second stage of the Project. The preliminary results of the research conducted by WG IV were presented at the two Conferences organized as part of the programme of the URBINNO Project: the first in Bari (Italy) from September 22-26, 1988 on the theme 'Innovation and Urbanization: Economic, Demographic and Territorial Implications', and the second in Chapel Hill (USA) from July 25-30, 1989 on the theme: 'Comparative Regional Research: Setting the Post-1991 Agenda'. There were several other opportunities directly or indirectly related to the URBINNO Project - public occasions (conferences and seminars) and publications (papers and specialized journals) - for the members of WG IV to present the results of their studies, in addition to the meetings organized by the Project and the many informal occasions during which the possibility of continuing bilateral and multilateral scientific cooperation was discussed. It is only possible for us to present a synthetic volume on the work undertaken and this must per force be a kind of meditated compromise between the various cultural positions of the individual researchers. One of the most successful achievements of the Project is

2

that of having fostered lengthy and harmonious cooperation among researchers from different disciplines, cultures and countries.

In the first phase of the Project, WG IV was directed by Gerhard Curdes (Aachen University of Technology) and Armando Montanari (National Research Council of Italy). Other participants were John Allpass (a practicing architect in Copenhagen), Halina Dunin-Woyseth (University of Oslo), Vitor Matias Ferreira (ISCTE, Lisbon), George Giannopoulos (University of Thessaloniki), Nora Hörcher (HABITAT, Budapest), Hidenobu Jinnai (Hosei University, Tokyo), David W. Massey (University of Liverpool), and V. Braco Music (University of Ljubljana). Other participants belonged to national research groups and their names have been indicated among the authors of each chapter.

WG IV's research programme was conceived, developed and implemented by Gerhard Curdes and Armando Montanari. A pilot study designed by Curdes and supported by the Volkswagen Foundation was used as a guide for the case-studies. Although this volume is the outcome of constant consultation and cooperation between the three editors, Leslie Forsyth was responsible for editing Part One and Armando Montanari for Part Two. The generous commitment of all those participating in the working group under whatever title enabled us to obtain the fruitful results presented in this volume. Special thanks also to Sylvia Trnka and Helmut Wintersberger of the European Centre who coordinated the activities of the URBINNO Project, and to Roy Drewett and Uwe Schubert who directed it. Our gratitude to Barbara Dawes for the linguistic editing of Part II and to Antonio Marra for his competence and skill in processing the texts and preparing the camera-ready copy.

ARMANDO MONTANARI, Naples, September 1992

on behalf of the Series Editors: ROY DREWETT, Chairman
                                               ARMANDO MONTANARI
                                               UWE SCHUBERT

1. INTRODUCTION

This study concerns the consequences of innovation on the form and structure of urban areas in the last century. Although it is a purely historical study, our aim was to identify a scientific theory that could help us to predict the impact that innovative processes can be expected to have on the form of urban areas in the coming decades. Indeed historical studies are always future-oriented, given that the test of every scientific theory is the extent to which it is capable of making reliable predictions. Man has always tried to anticipate the future and shaped his cultural attitude around environmental features as they evolve. For this reason, the research selected a number of constants, in a sense rules, that have regulated the relationship between innovation and urban forms in the past. It was felt that, if suitably updated, they would presumably continue to exercise this function, through a 'mutation-selection' process. As already mentioned in the Preface, the research which forms the object of this publication is the result of a series of studies based on comparative analysis, like the CURB Project, begun in the 1970s, of which it has gathered all the fruits. In its specific nature, the URBINNO Project has involved wide-ranging multidisciplinary collaboration and it has thus been difficult to find single definitions for events and phenomena which social sciences often view in different ways. In this volume, therefore, we will avoid addressing issues and definitions which form the basis of the collaboration among the Project's 4 Work Groups and have already been analysed in other volumes in the same series.

The role of innovative processes over time: invention, innovation and dissemination

Considered over a sufficiently long timespan, there is a very clear relationship between the economic scenario at a given point in time and studies on the characteristics of urban area development. In fact, a correlation exists between economic growth and the interest in undertaking analyses of immediate situations, for the purpose of formulating policies that will lead to concrete and short-term results. By contrast, it is in times of economic hardship that greater attention is devoted to long-term cultural and social needs. Studying the past is not only important in itself; retrospective studies also lay the theoretical foundations for subsequent economic development.

The beginning of the Eighties was characterized by an economic slump, by environmental disruption, by the devaluation of landscape, and on a theoretical level, by a lack of scientific studies containing original formulations, capable of reawakening the necessary interest and cultural commitment. European cities in the past decade have reached and exceeded their highest growth record, non-built up extraurban areas have decreased significantly and should now be considered as a very valuable commodity not to be squandered. The situation is thus conducive to an attitude whereby greater importance is attached to historical studies, and therefore to the conservation, reutilization and redevelopment of existing urbanized areas. The implementation of studies that will focus attention on the phases of urban development, examined in their long-term interrelations in space and time, may foster greater interest in the historical and geographic context in which the urban landscape developed.

This continual evolution of the urban structure is not a phenomenon that can be expressed with a constant curve, but rather with a sinusoidal trend, with ups and downs, characterized by phases of development alternated with phases of recession. In this context, phases of recession are to be intended as instances in which an operational transfer occurs between mature and new enterprises, which are able to provide new cultural stimuli and therefore once again affect the evolution of the urban structure. Inventions, innovations, and the dissemination of the latest findings are the driving forces of this process.

However, it is necessary to specify exactly what is meant by invention, innovation and dissemination, and it is especially important to define the parameters characterizing the evolution of such

6

processes in space and time. If reference is made to the purely technological sphere, the evolution of transportation media could be taken as an example. Usually, it is sufficient to look up an invention in the encyclopedia to find out when it took place; for instance, one can look up the steam engine to find out who invented it and when.

It may be asserted that an innovation occurred when the first steam locomotive prototype came out of the factory, while its dissemination may be viewed as the period of time during which the steam locomotive ran before being outperformed by the subsequent innovation, which introduced the electric locomotive. It is important to define the elements that distinguish an innovation from subsequent developments and enhancements. Replacing a valve with another mechanism is likely to have been a great technological innovation for a steamship, while it may not have been considered as an actual innovation for the area in which the boat operated.

The terms 'innovation' and 'invention' are considered in this context in the light of all of their social spin-offs, and not on a technological basis alone. However, it may be useful to refer to the contribution of a German scholar, G. Mensch, who collected several statistical data on the inventions and innovations which have occurred over the past 300 years. This study, too, pointed out that the invention-innovation process in the technological sphere is not linear, but is instead characterized by ups and downs. Invention-innovation processes thus occurred in waves, the first having taken place in 1802, the second in 1857, the third in 1920, and the fourth in 1980.

At the end of the Eighties a new cycle of innovation started in the field of computers, electronics, telecommunications, new materials, biotechnologies and robotics. Possibly the most dramatic innovation of this period was the forceful emergence of the environmental challenge, which is destined to become a key issue in the next century. By then, cities will have to have found a way to ensure environmental compatibility in the production, impact and recycling of waste, understood in the broad sense. Will these innovations be such as to model the cities of the future? To what extent will an analysis of recent processes contribute to indicating trends and, therefore, solutions for the decades to come?

There has been a great deal of debate on these issues in recent years in various international conferences and many research programmes have been undertaken on urban history both before and after industrialization. Numerous other scientific works have also been produced, and articles published in scientific journals, dealing with geography, architecture, territorial planning and regional sciences. In the second half of the Seventies, several journals on urban history were published for the first time. Among these: the 'Journal of Urban History' (Beverly Hills, 1974), the 'Urban History Year

7

Book' (Leicester, 1974), the 'Urban History Review - Revue d'histoire urbaine', (Winnipeg, 1975), 'Storia Urbana' (Milano 1977), Storia della città ' (Milano 1976), and 'Urbi: Art, Histoire et Ethnologie des Villes' (Paris, 1979).

WG IV aimed at developing this process of analysis further. Urban history was the core of the research of the major multinational URBINNO Project. The overall end of the Urbinno Project was to define the role of innovation in the process of social and technological change of urban areas.

It was decided to study the effects of innovation on the urban form from the second half of the nineteenth century, with reference to the main economic cycles and historical events common to all European countries. We divided this period into four blocks: from the middle of the nineteenth century to the beginning of World War I; between the two world wars from the end of World War II to 1970; and from 1970 onwards. These blocks of historical periods were further subdivided in the studies presented in Parts One and Two of this volume.

The role of innovative processes in space: morphology and urban form

Over time, the contradictory relations between the environment and economic development have had a strong impact on urban morphology. The use of the term 'morphology' is deliberate for it conveys the notions expression and content, thus giving us a key for understanding the meaning of things which are normally only interpreted in their 'form'.

The term 'form' is generally accepted to mean figure or image. Since form is by its very nature separate from, and counterpoised to, matter, it could imply the intention of separating the object from its image. The term 'morphology' deriving as it does from the Greek 'morfè' (form) and 'logos' (speech) has the meaning of speech, reasoning; hence, the doctrine of form.

The URBINNO Project adopted a circumlocution, something equivalent to the Latin 'formarum modorumque descriptio', rather than the word 'morphology' because this latter term was introduced into the main European languages from the German 'morphologie' but without always keeping the same meaning. 'Morphology' could have been confusing for those involved in studying the results of a complex multinational project such as the URBINNO Project for the first time. In this article the term 'morphology' is understood in the same way as German scholars, among whom Goethe, used it in the nineteenth century. However, morphology is not intended as the doctrine of form in a static sense - 'Gestalt' - but as a process of formation - 'bildung' - of organisms.

Seven levels of analysis were established to (i) identify the impact of innovation on urban form, (ii) contribute to the comparative study, (iii) verify the empirical results with other WGs and (iv) plan further research activities. The definition of the various levels of study is sufficiently clear and follows the pattern customarily used in specialized literature. The concept of definition of the metropolitan area (Level 1) is a functional urban region containing a core-continuous built-up area (Level 2) and a ring. The main functional units of the metropolitan area have also been established. For example the core-centre, the urban settlements situated in the ring and the most important connections and links between the core and the ring centres (Level 3). Reference is made to the major urban districts within the core (Level 4) and to zones performing specific functions (Level 5). In Level 4 reference is made to large areas such as the centre of the city, or zones performing a particular function acting as a link or a filter between the centre and the periphery. These could be zones dominated by the presence of particular infrastructures, such as universities and hospitals. Level 5 groups areas that are identifiable by the activities undertaken within them (industrial areas, port areas, homogeneous residential areas) or by their geographical characteristics (urban parks, waterfronts, river and lake areas) or by the functions performed there over time (historical centre, transition zones, border areas). In order to focus on the detailed characteristics of urban structures, blocks of buildings were also considered. For example urban blocks in the middle of streets or squares (Level 6) and individual buildings (Level 7).

A scale of hierarchies for the identification of change and innovation was also established for each level. The work of cataloging has been carried out using the main existing studies, and does not claim to have taken any further such elaboration. In this respect it has been used as an authentic tool of analysis to make the task of elaborating and comparing the data easier. Very briefly the scale is as follows:

Level 1: Metropolitan area. Studies of changes in land use, major traffic flows and urban settlements. This scale makes an important reference to demographic changes, the location of work places and effects on mobility. Among the particularly significant indicators are: the identification and position of poles of development at the regional level, the location of the main technological innovations, elements around which functions have either amalgamated or separated, and cases in which the dissemination of some innovations has been slow.

Level 2: Core-Continuous Built-up Area. At this level we studied the impact that the dissemination of innovations of international, national or local worth, has had on urban form. Particular attention is paid to studying areas of transition, areas of stability, the functions

9

and features of road infrastructures (link roads, ring roads, etc.), areas bordering with other areas of a different morphological character, the regulatory environment and urban planning tools, their implementation, the aggregation and division of different functions, the location of major commercial and service areas.

Level 3: Functional Units of the Metropolitan Area. Focus on changes in population settlement patterns, the work place and infrastructures set up within the area for trade and services. Identification of areas of transition and stability, assessment of the impact of the introduction of new technologies on transport. Extent and characteristics of urban mobility and the changes in urban form resulting from the introduction of new building regulations and the emergence of a new urban planning culture.

Level 4: Districts. Identification of public or private bodies in charge of administration, with a major role in determining development processes with respect to economic trends involving the metropolitan area. Innovations in the technological, regulatory and cultural environment.

Level 5: Specific Areas. The size and characteristics of specific areas is closely related to the function they perform. This level is dominated by the presence of a specific type of activity, a management body and a user. It concerns the study of relations between development and change involving the three main actors and how the dissemination of innovations has occurred.

Level 6: Blocks. Study of changes concerning the use of territory, the size of blocks, the structure of built-up areas, types of building. Assessment of adaptability to change, capacity to undergo transformation and flexibility vis à vis innovations. Ratio between public and private areas.

Level 7: Plots and Buildings. Changes in the use, dimensions, structure and typology of each building. Adaptability to change, capacity to undergo transformation and rigidity vis à vis innovations.

Conclusions

The URBINNO Project showed very clearly that cities are favourable to innovation. Even relatively minor innovations influence city life and urban form. Cities seem particularly suited to disseminating new technologies at the local, national and international level. What proved more complex was to reach a definite conclusion with regard to the mechanisms that guide such processes and influence urban morphology.

The Fourth Working Group of the URBINNO Project tried to identify the relationship that exists between technological and cul-

Figure 1.1. *From a case-study to a thematic approach*

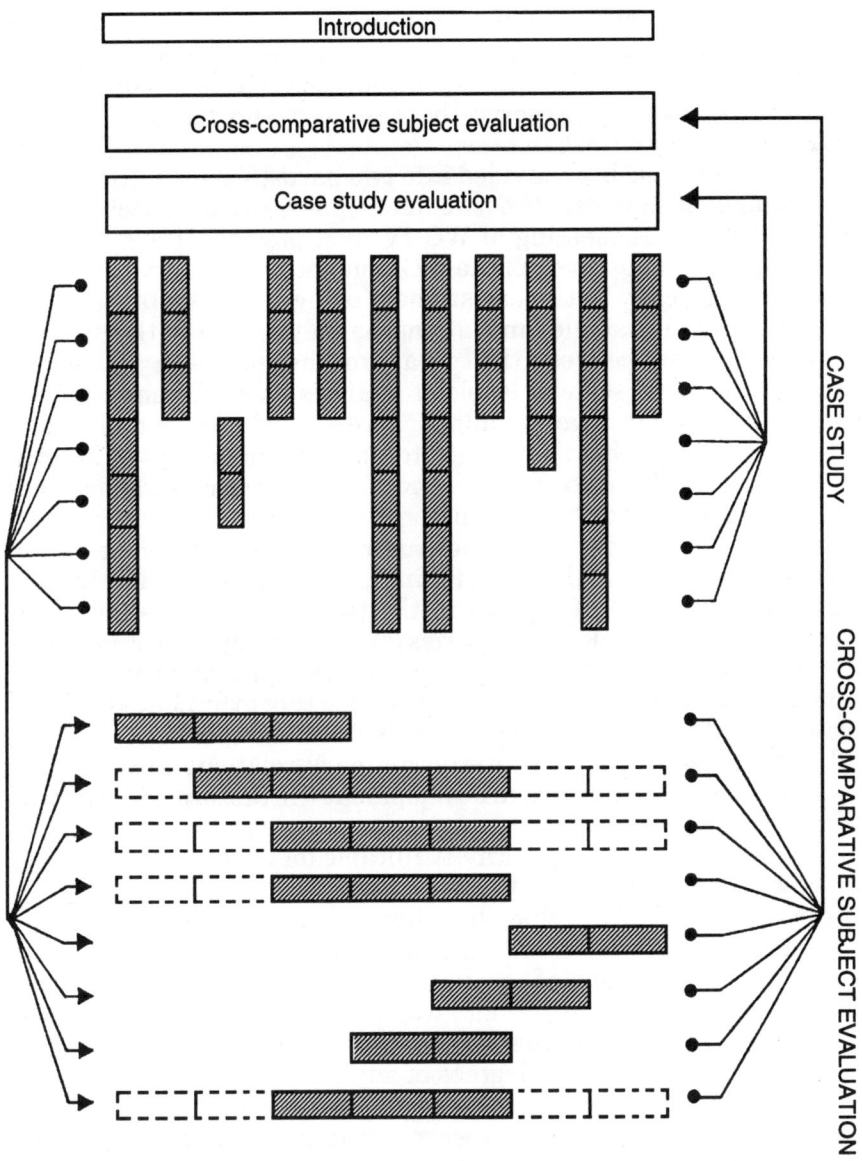

tural innovations, and the role played by innovation per se in determining urban form. Several attempts were made at establish-

11

ing whether innovations stimulate and bring about changes in urban form or whether they are simply tools for adapting the urban form to the cultural evolution of the society in which they occur. There is no one definite answer to this complex issue, and nor should there be at this stage. The results of WG IV reflect the variety of approaches of the members of the group. Despite the rigorous approach in editing and synthesizing the concepts expressed, each chapter of this volume has been allowed to preserve the approach and features of each individual group of researchers.

The volume has been divided into two parts (Figure 1.1.) in order to distinguish between the two methodological approaches that emerged from the meeting of WG IV in September 1988. On that occasion several members of the WG thought it necessary to test the results empirically with case-studies, using a horizontal analysis aimed at studying single innovations, as they appear in the existing literature. A link between the two approaches has been assured by keeping the same scale of levels of analysis as a reference in both cases. In Part One we examined a restricted number of urban structures and, within them, tried to examine the impact of innovation processes. In Part Two we look at the impact of individual innovation processes on urban form. The two parts are closely related: the results of the first were used to check those of the second and vice versa. Particularly useful in this respect is the fact that the same researchers participated in the two study phases and were therefore able to check their theories from two different perspectives.

WG IV chose a strategy of progressive selection of issues so as to limit the number of innovative processes to those that have played a primary and decisive role in changes in urban morphology. This will allow us to establish new interdisciplinary relations in which the study of innovation and urban development will be coupled with more targeted empirical studies.

The themes which appear most suitable for assessing innovatory processes are those related to the urban environment and especially the study of the impact that the following have had on the latter:
- the ways, processes and timescales of urban expansion;
- the transformation of production;
- the inhabitants' culture and way of life;
- the qualitative and quantitative nature of services;
- types of mobility and degrees of accessibility;
- organisation of built-on areas and conditions of habitation.

Obviously this list is not a research programme but it gives us a general idea of some of the issues which seem more suitable for explaining innovatory processes and are therefore able to provide a picture of the changes that will effect our towns and cities in the immediate future.

Part I
URBAN SETTLEMENT
INNOVATION:
EMPIRICAL EVIDENCE

2. INTRODUCTION

This part of the book presents a series of case-studies and a review of their contribution to the study of the relationship between innovation and urban development or, more specifically, between innovation built form, environment and land use. The general aim of the study was to investigate the role played by innovation in the physical/ spatial development of towns and cities, to try to establish if innovation has a fundamental effect on the form and function of a town and what factors are involved in process.

When considering the changes which take place in a town it can be problematic to isolate aspects which are directly attributable to innovation. In many cases the continuous evolutionary process of change has had a significantly greater effect than the occasional innovation. On the other hand, when the effects of certain major innovations are observed it can be seen that in a very short time they can produce dramatic changes which shape the process of urban development.

From the point of view of a study group seeking innovation, or the effects of innovation, within specific boundaries, there is a further question concerning the origin of the innovation. Did it occur within the realms of built form, environment and land use, or is it an innovation which has taken place in some other area, for example social or economic policy, and only the effects of the innovation manifest themselves within the morphological structure. The cause and effect nature of certain innovations introduces the idea of time as a variable.

In examining the role of innovation in this study three principle aspects were to be considered. The type of town, in an attempt to illustrate whether innovation has a universal effect or is restricted to specific types and sizes of settlements. The physical scale at which innovation has its effect and whether this is more significant at the macro or micro-level. The time-factor, to assess which phase of an innovation has the greatest effect on urban development, as well as whether an innovation was simply an expression of the range of solutions available at a given time or appeared because the threshold of a particular problem had been reached. The reasons for adopting this approach to a study of innovation were based on the premise that the urban form or morphology of a town reveals, in spatial terms, the political, social and economic changes and innovations of previous eras. In this respect we should be able to use morphology as a data-base to trace the important influences on urban development and the role played by innovation in the process.

This review is structured in four main sections. Firstly, there are some general comments about the case-study approach and the editorial process. Secondly, the original programme and methods are described to illustrate the intentions of the working group at the outset of the project. The third section discusses the contributions from the eight towns and cities which were involved in the study, Rome, Lisbon, Liverpool, Thessaloniki, Aachen, Bari, Kecskemét and Tromsø. The individual approaches and methods used in each case-study are discussed so that the similarities and differences are made clear. This is followed by an appraisal of these approaches with reference to the original programme. Finally, in the last section the nature and type of innovations which were revealed are considered. These may not be as comprehensive as was originally intended, however, when seen in conjunction with the great variety of methods used to reach them they provide an interesting conclusion to this stage of the study.

The case-study approach contains considerable advantages for a project which intends to produce a comparison, if a suitable range of samples can be found and investigative techniques and criteria are evolved to provide a common methodology during the individual stages of the project. This cooperation and coordination is important for the later, comparative stages of the work since it is only when some form of coordination has taken place that any type of comparison can be meaningful. However, the dispersed and de-centralised nature of cross-comparative, international research contains potential problems which should be addressed from the inception of the project.

During the initial phase of this work two particular problems were not fully resolved: the number and variety of suitable towns as examples for case-studies and the necessary commitment to a method and format which guaranteed a reasonable basis for comparison in subsequent phases. There was also the question of under-resourcing for some of the individual studies. The work of the individual case-studies was dependent on locally-raised finance, which made it difficult for any centrally determined methodology or standard to be imposed. The discrepancies among the projects can be seen from the extent to which they were able to fulfil the original programme. The amount of information which had to be examined in order to comprehensively document the various stages of development before considering change and innovation meant that some of the authors had to concentrate their attention on specific areas, whereas others were able to examine a wider spectrum.

The results of attempting to summarise large amounts of information also influenced the editorial process. A balance had to be struck over the length of individual contributions, taking account of the details presented in a work focusing on one level along with the comprehensiveness and consequent length of a work which has studied all levels. For the most part this was achieved through a process of summarising the results which, in some cases, causes an abbreviated, almost artificial style of presentation. The overall approach adopted during the editorial process was that of attempting to encourage comparison through an agreement over the levels of the programme which had been examined and conclusions which had been reached. Within that overall policy the content and emphasis of each study remains as in the original report.

Recommended programme and methods

The three main components of the original programme were the selection of the types of town to be studied, the time periods which were to be used and the levels of investigation which were suggested

17

as appropriate.

Town selection

The three types of towns originally to be studied were regional centres, national centres or regional metropolises and industrial cities. These types were selected on the grounds that change was likely to have occurred continuously in such centres over the last hundred years, therefore the likelihood of innovation appearing was higher. At the outset it was proposed that a large number of cities within each group should be examined, but it was also recognised that in some countries the metropolitan areas were limited in number and to expect more than one town per type would be unreasonable. However, it was also noted that the richness of examples in other countries could compensate for this. The initial programme for the first two years envisaged a maximum of three cities in each participating country.

Periods of investigation

The investigations were to examine a period of approximately 100 years, from 1880 to 1980, to cover the major changes caused by the industrial and post-industrial phases of development. Within this overall timescale there were suggestions for shorter phases which might merit investigation.

Up to 1900: the end of the second period of prosperity, effects of sanitation, development of mass transport systems, large scale homogeneous neighbourhoods.

1930-1938: between the wars, stagnation, economic crisis, new forms of buildings and settlements, effects of the motor car.

1945-1955: reconstruction, renovation, new patterns of buildings within the built up area.

1955-1970: the post-war growth period, systems planning approach, criticisms of the modern movement.

1970-1988: public space in the city, inner-city living, conservation, regeneration.

As with the selection of towns it was recognised that the periods of investigation would differ among the countries involved and participants were asked to formulate their own periods if investigation revealed that they were likely to be significantly different from those suggested.

Levels of investigation

Seven levels of investigation were suggested for the organisation of data, corresponding to different categories and hierarchies of problems. Of course it was also hoped that this would facilitate comparison between examples and eventually in the second stage with the work of the other working groups.

Level 1   metropolitan area
Level 2   core-continuous built-up area
Level 3   functional units of the metropolitan area; a. core-centre, b. ring-centre (small towns, villages), c. major axes
Level 4   districts (homogeneous morphological units)
Level 5   specific areas
Level 6   blocks or detached buildings
Level 7   plots and buildings

Investigations of levels one to four were to be more general with the remaining three levels limited to those areas which showed a very clear transformation or the influence of innovation. A process of sifting to be carried out through areas showing change to determine whether innovation was a part of that process. It was also suggested that investigations into all seven levels should or could only be carried out for regional centres since the others were likely to be too large for such an extensive study during the first phase of the work.

Individual methods and approaches

This section summarises the methods and approaches used in each study in an attempt to illustrate the internal logic used to define the extent of the work, the time and themes which were considered important, and the means by and categories under which innovation has been defined. Consideration of each individual case-study is followed by an appraisal using the original programme as a reference. The order in which the studies are considered is based on the three categories of towns which were used in the work. First are the capital cities of Lisbon and Rome. These are followed by Liverpool and Thessaloniki representing national/regional centres, not the industrial cities envisaged in the programme. Finally there are the four regional centres of Aachen, Bari, Kecskemét and Tromsø.

Lisbon

This study forms part of a larger research project dealing with the process of social and urban change in the city of Lisbon, which has consequently conditioned the methods and substantive content. The levels examined are one, two and three.

Different time periods are considered at each level within the context of the major changes in Lisbon's urbanisation process during the last century. These phases are analysed according to: mode of urbanisation, process of territorial structure and urban differentiation, and urban image. From this background elements of change are defined to analyse the metropolitan and core-continuous built-up areas. These consist of urban design and land use, built environment, infrastructure and transport, and urban instruments of planning. A brief description of the key factors within each period is given in the text supplemented by a matrix describing the changes which took place.

At level one, demographic growth and territorial expansion between 1890 and 1980 are discussed in four periods defined by the agglomeration of Lisbon's interdependence and dispersion, differentiation and density, suburbanisation process, and the structuring and expansion of the metropolis. Four time periods are selected within level two according to phases of expansion and densification, political control and urban planning, demographic intensification and suburban peripheral expansion, and the structuring of the metropolitan area. Within level three, seven time phases are distinguished in the development of a main axis of expansion. In each section the major changes to the built environment and infrastructure are noted, then adjustments to the morphological structure evaluated.

The conclusions are drawn in accordance with the definitions which guided the research, social and territorial achievements, and conditions and specifics of social innovation significant in urban change. Then transformations are analysed using the same categories which defined the elements of change at levels one and two: urban form and space structure, built environment and the transformation of land use, productive space and transport sub-structures, and urban instruments as forms of intervention. This analysis forms the background for a discussion about urban change and social innovation.

Rome

The historic centre of the city, a block and building within that area, corresponding to levels five, six and seven, are the focus of the Rome case-study.

The examination of the historic centre provides the context for the other levels through a definition of the geographic core, historical development and chronology, although each of the other levels provides its own periods within that framework. 1870 is taken as the starting point, the year of Rome's annexation to the Kingdom of Italy, with the subsequent century divided into five periods. Within each period the structure for investigation is the same. A description of the historical development is given highlighting the main events which had significant effects not only on built form but also for less obvious transformations in use and social structure. In this respect the important background information is illustrated through level five with the others providing more detailed consequences and highlights.

In the conclusion innovation is suggested for each time period at each level. In the case of the historic centre, is in general this categories of economic, political and social policy; however, at the other levels more detailed diagnoses are carried out within the framework suggested for the historic centre.

Liverpool

The investigation encompasses levels one to four: the metropolitan area of Merseyside, the core-continuous built-up area of Liverpool, the core-centre and a ring-centre, as well as the university/hospital precinct.

The time periods selected vary according to different factors at each level; however, there are some common periods particularly at the first three levels. Levels one and two set the context for the remainder of the study with a brief summary of the early development of the region and city before concentrating on the last hundred years. For levels three and four only the last century is considered. At all levels and within all time periods a consistent approach is adopted. A brief description of the key historical developments is followed by an outline of the morphological changes which occurred and then by suggestions as to the probable causes for these changes.

The conclusion attempts to isolate innovation within each time period and at each level. This takes the form of a matrix which locates an innovation within a phase of the economic cycle and suggests the development agent responsible for the implementation of the innovation. The innovations themselves are categorised according to Whitehand (1987) into four fields: construction (building materials, legislation, technology), function (methods of manufacture and trade), transport (as an important subdivision of function), and town planning (as a form of public intervention). The innovations isolated in this process are only those which have affected the urban fabric. There is no attempt to trace the wider social or economic innovation which may have been responsible for the process.

Thessaloniki

The Thessaloniki study examines the core-continuous built-up area, level two, and the core- centre, level three. In order to define the five sub-periods between 1880 and 1980 the authors use different structures of spatial organisation which were imposed by more general socio- economic rules. With these time periods as a background the changes which took place during each period at each level are studied using the themes of land use, building size and typology, road network, public transport and services, and points of attraction in the urban fabric.

Within level two a historical description of the evolution and development of the continuous built-up area is followed by a section highlighting the effects of town planning and another dealing with the specifics of population change. In the core-centre analysis the development of the area is described and the changes and transformations are discussed.

Conclusions are drawn for the core-centre as a whole, but using the themes of land use, population and transport for level two. Innovation is suggested as a list of individual events in chronological order without any attempt to define categories.

This is one of the two case studies which investigate all of the levels from one to seven. The time periods used vary among the levels although they generally follow the pattern adopted in level one. The themes of investigation also vary among the different levels but the topics of morphological change and urban innovation used in level two reappear. Levels six and seven employ a different style of examination.

At the level of the region the context for all other levels is created. Four major time periods relevant to Aachen and based on the main periods as defined by Henckel (1986) are adopted. These periods are then reviewed under categories of growth, urbanisation process, and urban structure change, the last category being further subdivided into extension, which is considered per time period, and densification, which is considered for the whole period of investigation. Main periods of innovation are highlighted and their geographic influences described.

The main themes of investigation in the core-continuous built-up area and the core-centre are morphological change and urban innovation. Morphological change is subdivided into enlargement, densification and development trends. These themes are then used to examine relevant time periods which display significant development phenomena before innovation is isolated.

Within the university precinct, level four, the morphological changes over the period 1870 to 1982 are described. In level five the method of analysis concerns morphological change as a consequence of urban innovation in a major east-west axis over most of the twentieth century. Level six describes the transformation of city blocks from different time periods through changes of use and building type and discusses the influence of dimension and form on stability. At the level of individual buildings and plots the criteria discussed are planning, investment and user demands.

In the conclusions the effects of innovation on sequences of form, use and function are explained. Innovation is categorised using six themes: production, transport, built form, infrastructure, green/ public space, and environment protection. The levels at which these innovations are to be found are given along with a short description of their effects and the dates at which they occurred.

Bari

The metropolitan area of Bari forms the basis of this case study. However, there is also a matrix outlining the urban development of the city or core-continuous built-up area. The time periods selected vary between the levels. For level one they act only as a background dictated by the availability of cartographic and census information covering the period from 1861 to 1981. Within the matrix of urban development for level two the periods result from ideas which were inherent to the expansion of the city.

The survey and analysis of the metropolitan area uses topics of investigation appropriate to regional analysis, growth and density of inhabited areas, construction of dwellings, economic indicators from the three industrial sectors, transport and utilities, and sites for industrial and tourist development. These are followed by a brief description of the regional structure in 1988. At level two the matrix illustrates five time periods from 1874 to 1977. These periods of expansion are evaluated in categories of ideals, legal tools and the resulting built form.

The conclusion lists the innovations which have taken place drawing a distinction between those which can be considered as local specific and those which are common to other regional contexts.

Kecskemét

The study of Kecskemét focuses on levels two, three and five, the core-continuous built-up area, the core-centre and specific areas immediately adjacent to the centre. Three characteristic stages in the recent history of the town are selected as the time periods for investigation: the period of intensive growth between 1880 and World War I, a period of stagnation until the town became the county seat in 1950, then a further period of development and growth.

Within the three levels different approaches are used to illustrate the major influences and the changes which they caused. The development of the continuous built-up area is described chronologically before a summary of changes in the morphological structure is given. In contrast, a thematic approach is used to illustrate the transformations which took place in the central core, using the categories of traffic, building, new modes of use and new modes of design. At level five the study returns to a descriptive approach enumerating the changes which took place in each time period within the neighbourhoods.

In the conclusion the results are drawn together into innovation types with their corresponding morphological responses. The categories used are social, economic and technical (transport, urban development, building type/construction/investment systems) with the morphological results listed chronologically in each category.

Tromsø

This is the second case study which examines all of the levels of the proposed programme, from the regional context through to individual plots and buildings. Due to the specific character of the area and the size of the town the levels of the core-centre and homogeneous morphological unit (three and four) are combined.

The time periods selected vary according to the level of investigation but, apart from the regional context in which transport is the significant determinant, they tend to be structured around the seven time periods established for the continuous built-up area by extensions to the town boundary. The exceptions are levels six and seven in which types of change are used as criteria. These are defined as new uses, change within plots and by plot increment, and plot amalgamation.

At all levels and for all time periods the approach of the study is consistent. A historical review establishes the general features of development and changes in urban form, then innovation is traced from these changes. In the conclusion innovations are brought together in a table. They are distinguished according to level but not time periods. Throughout the levels all innovations are categorised as 'hard' or 'soft', the former relating to technological progress while the latter is more concerned with ideals and lifestyle preferences.

Appraisal

Two differences emerged between the scope of the sample discussed in the original programme and the actual results. The first concerns the types of towns programme under study, which altered slightly to include national capitals, national/regional centres and regional centres, in response to the examples which were available. The second is more significant, that is the limited number of towns in the overall sample, instead of three from each participating country there are two from Italy and one from each of the other six. However, although this may affect comparability of the results, it does not diminish the value of an appraisal of the methods which the various authors have used in reaching their conclusions.

All of the case studies have three aspects in common. Each uses the framework of levels suggested in the programme, each addresses itself more or less to the last one hundred years, and all point in some way towards the stated goal of change and innovation. However, within these common aspects there are significant differences in the selection of levels for study, the way in which time is considered, the themes used to examine change and the way innovation is defined and isolated.

The selection of levels to be studied shows no clear pattern either over the whole range of studies or within the defined town types (Figure 2.1). The level most consistently examined is the core-continuous built-up area; only in one of the capital cities is this omitted. Next come levels one and three, which is understandable due to their common information and data overlaps with level two, but there is no evidence to suggest why specific towns have certain combinations of levels. It is within the regional centres that the two most comprehensive studies appear, Aachen and Tromsø, where all the levels have been addressed. The reasons for the selection of levels within case studies appear to be pragmatic: the most easily accessible and relevant information was available at the levels chosen. In this context it must be remembered that, in almost all cases, the individual research work was not funded and the time in which the work had to be produced was very limited, for some participants little more than a year.

Figure 2.1. *Levels examined in each case study*

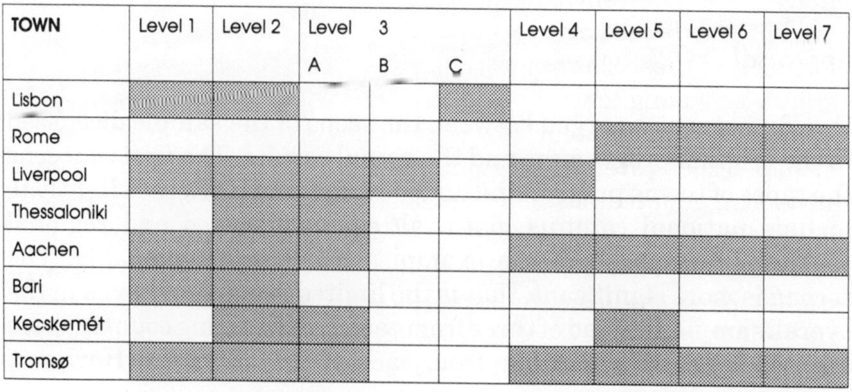

TOWN	Level 1	Level 2	Level 3 A	B	C	Level 4	Level 5	Level 6	Level 7
Lisbon	▓	▓			▓				
Rome							▓	▓	▓
Liverpool	▓	▓	▓		▓	▓			
Thessaloniki		▓							
Aachen	▓	▓	▓	▓	▓	▓	▓	▓	▓
Bari		▓	▓						
Kecskemét		▓					▓	▓	
Tromsø	▓	▓	▓	▓	▓	▓	▓	▓	▓

As previously mentioned all of the studies focus on the last one hundred years. However, there are minor differences evident in the starting and concluding years, not only among the case-studies but also within individual studies on different levels. This is usually relevant since the selection is due to some specific local event. In the two Italian studies both take unification as the starting point, but whereas for Bari national unity came in 1861 for Rome it came in 1870. The time periods used in each level vary significantly both horizontally across the whole sample and vertically within the various levels of single studies. This is to be expected since, although the project programme gave guidelines, the towns in the sample were subject to important influences at different times and in most cases it is these factors which have been selected to define sub-periods, as in the case of the fire which destroyed the centre of Thessaloniki in 1917.

In some studies, for example Lisbon, the logic behind the selection of sub-periods is given considerable attention and used as the starting point for discussing change, while in others the selection was governed by the availability of cartographic and census material nearest to the phases recommended in the original programme. A further distinction in the approaches to time involves those studies which employ different time periods for each level, for example Liverpool, and those which select the time periods for the overall development of the city or region and then apply these at each level of investigation, as in the case of Kecskemét.

In addition to the range of ways used to determine important time periods a variety of themes or categories of analysis is used to examine the effects of change. In some cases these are applied throughout the whole time period of a particular level, as with level one of the Bari study or the core-centre in the Kecskemét study. In other studies the established themes are used throughout all the levels of study. The selection of these themes or topics shows no consistency within the sample or among town types. Moreover, within individual studies it is not always made clear why particular themes have been selected. In some cases categories recommended in the original programme have been adopted but in others this is not the case.

The methods used to isolate and categorise innovation are as varied as those for distinguishing time and themes of analysis. All studies follow the basic steps suggested in the programme by seeking areas of change and then innovation and this is done in four main ways, for specific levels, within each time period, for each theme of analysis or overall for the whole study. What is difficult to assess from the examples is the ways used to distinguish whether impact was merely due to evolutionary change or the result of innovation. In examples where only one or two levels are studied, Bari and Thessaloniki, innovations are suggested without categorisation, whereas the Rome study isolates a relevant innovation within the separate time periods at each level. Innovations in the Tromsø study are traced for each level from the morphological change suggested in each time period. A similar process is used in the Aachen study, but this is taken a step further by a categorisation of themes before these are attributed to levels.

The categories used to trace and illustrate innovation form an expansive classification system. They range from those suggested in the programme and used in the Aachen study to those of Tromsø, which draw a simple, but very important, distinction between technological and societal innovation. Liverpool relies on categories existing in the relevant literature and attempts to extend them using economic and developmental elements. In the study of Lisbon the thoroughness of the evolution of the categories is extremely sophisticated, as is to be expected given the larger project of which it is a part. Two studies, Bari and Thessaloniki, concentrate on listing the actual innovations with no overall attempt to determine categories. General classifications of social, economic, political and technical are used by the Rome and Kecskemét studies and where necessary they are further broken down into more detail.

The variety of methods used in the case-studies reflects the different approaches of the various authors to the proposed programme. These approaches were influenced by diverse scientific understanding and the availability of personnel and other resources, but also, and perhaps more importantly, by the multifarious development of the individual towns in the sample.

Innovation

The limited number of case-studies available narrows the scope for comparison of the innovation process. A further restriction on the potential for comparison can be taken from the wide range of methods used to categorise change and innovation. However, while the limited number of examples means that there is no opportunity for national comparisons within one country, this does not negate the possibilities for a European analysis using the available studies, since there are at least two examples in each of the categories of town types selected. Similarly, while the range of methods used to determine the existence of innovation could restrict detailed comparisons among the studies it does not prevent discussion of the macro-level innovations which appear in almost all case-studies.

Taking account of the factors influencing any potential comparisons this conclusion tries to examine the innovations isolated in the case-studies from two opposing stances. Firstly, the innovations which appear most often over the range of the studies are discussed. Secondly, definitions which try to reflect the mechanisms of innovation are reviewed and these characteristics are then used to examine the innovations in the case-studies. This dual approach allows consideration of aspects of the case-studies which are capable of being compared and enables a discussion the outcome of the studies using general characteristics applicable to innovation.

The two innovations which appear most frequently in all the case studies and at all levels are those that can be grouped under the headings of transport and urban instruments. Innovations attributable to transport have had a tremendous effect on urban development throughout Europe during the last century, the railway, motorised road transport and the aeroplane. Each of these created an impetus for development while at the same time directing and affecting its form and nature through their demands for land and other resources. The pattern of the road networks emerges as being fairly consistent (Figure 2.2). What is apparent is that innovations in the field of transport were generally significant but more so at the larger scale and the fact that very few major distinctions are evident due to geographic location or the type of town.

Urban instruments represent a category of innovation which has had, in certain respects, a greater effect than transport since innovations in this category are observed at all levels of the case studies. In contrast to transport which tends to exert more influence at the larger scale levels of the metropolitan or core-continuous built-up area, urban instruments can be equally influential at the level of the neighbourhood or block. In the context of this study urban instruments relate to legislative tools and procedures which deal with the expropriation, development and control of land, buildings and the environment. For the most part, but not exclusively, this takes the form of planning legislation, development plans and the ideas and theories which guided physical planning over a given period.

Figure 2.2. *Development patterns of the road and rail networks*

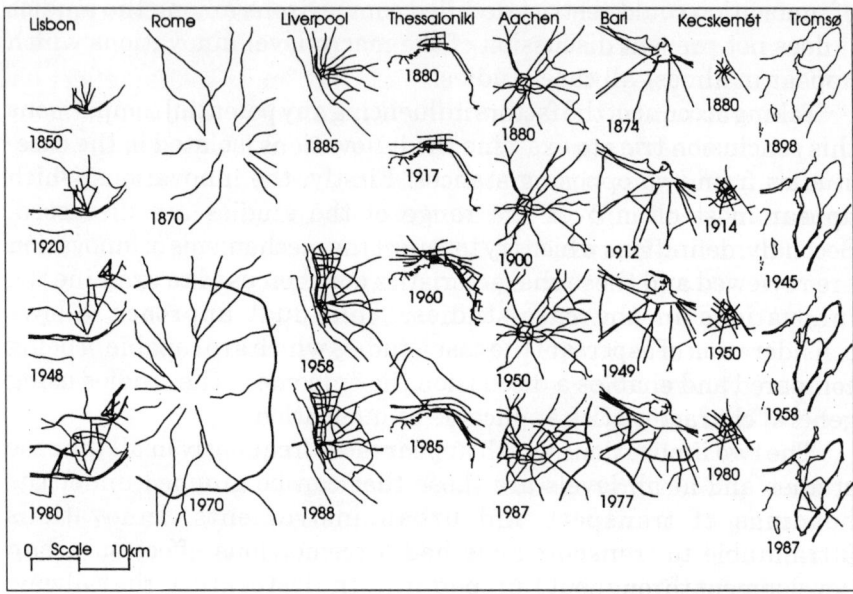

From the case-studies there is no obvious correlation between the production of land use and master plans, plans for extension or restructuring plans. However, what can be seen is groupings of plans in similar time periods, the mid-nineteenth century, the early twentieth and in the 1970s, which may be linked to waves of population growth. Urban renewal plans tend to appear when the fabric of an area has aged around one hundred years, although such initiatives can also be prompted by fundamental changes to the building structure and functions in a built-up area.

The other influential factor in this category of innovation concerns the ideas or theories which prompt legislation or produce a leading style which influences the form of the urban fabric. During the period of the study the most influential sets of ideas appeared to be those arising from the Garden City Movement and the Charter of Athens.

The Charter of Athens arguably represents the initial phase of the innovations which have had the greatest, perhaps most damaging, effect on the urban environment of the twentieth century. The segregation of uses and functions and the modern movement idea of the tower block standing in isolation reveal themselves in almost all of the studies, most commonly on the fringe of the city-centre. This single outcome illustrates, perhaps more than any other, how this type of innovation can have a fundamental effect rather than a partial influence on urban development.

There are many other innovations which have had an effect on urban development but within the range of the sample available in this project the two aforementioned are those which were recognised as having the most significant and widespread influence over the time-span, levels and geographic locations examined.

A definition of innovation which attempts to reflect the mechanisms that operate and may be used to examine the results of the case-studies could prove constructive in the process of preparing for inter-disciplinary cooperation. The three characteristics of innovation considered here concern the type of innovation, its origins and its geographic applicability or frequency.

The first is the distinction drawn at the beginning of the work between product and process innovations, the former dealing with a new thing to do whereas the latter implies a new way of doing something. This differentiation is very helpful in categorising the types of innovation suggested in the case studies. Within the context of this study an analysis of the innovations that have been isolated shows that for the most part they are process innovations arising out of the overwhelming importance of urban legislation and planning for the towns in the sample or out of the political, social and cultural ideals which were distinguished.

There is, however, a problem in the assumption that the majority of the innovations can be simply described as process innovations. Whether an innovation stems from a process or product there remains the issue of the life-course of the innovation and its effects. Examination of the duration of the effect of an innovation allows us to distinguish between innovations which originate from within the fields of built form, environment and land use and those which lie outside, with only the effects filtering through to become apparent in the urban development of a town. When examining the case studies in the sample the distinction between external and internal is important because, although some studies concentrate on internal innovations, most do not distinguish between the two. One of the major advantages of drawing this distinction is that the question of time begins to be a factor. By tracing the process of change in the morphology of a town the initial impetus may be diagnosed as a political or social innovation which occurred some years before. The time-period used here illustrates the life-span of an innovation from initiation through implementation to obsolescence. It is to clarify questions which arise concerning innovations extending over the boundaries between working groups that there is a need for interdisciplinary cooperation.

The third characteristic of an innovation to be discussed arises from the issue of the internationalisation of style and cultures and whether an innovation is specific to a local area, town or country. This category also allows us to consider the question of time in the innovatory process through the examination of how long a particular innovation took to become established in countries other than that of its source. However, the issue of time is not as important as the ascertainment of how internationally acceptable innovations are and whether there are examples of innovations which have not been accepted internationally yet have had a significant local impact. From the comparison of macro-level aspects of the case-studies the innovations which were picked out tended to be internationally accepted. Those which were significant locally tended to appear in response to a specific problem or threshold peculiar to that settlement.

Using these three characteristics as a basis for examining the selection of innovations discussed in the case-studies allows tentative suggestions to be made about the nature and type of innovation which has effected the built form, environment and land use of the towns in the sample over the last hundred years. Those which have had the greatest influence over the whole range of the sample appear to be process innovations. There appears to be a fairly even distribution of innovation originating from within and without the subject area and the most significant of these seem to have had an impact internationally. However, the value of these characteristics does not lie in a generalised appraisal from a limited range of studies but rather as an indicator which assists us in the development of further research questions. At this stage of the project the interest lies in the variety of methods and results emerging from the case-studies and in their independence as records of urban development and innovation in these towns.

3. LISBON

Vitor Matias Ferreira, Istituto Superior de Ciencias do Trabalho e da Empresa (ISCTE), Lisbon, Portugal*.

Introduction

This paper is the outcome of part of an extensive research project, directed by the author, financially supported by J.N.I.C.T. (National Institute of Technological and Scientific Research) and with the collaboration of Lisbon's City Council. The main purpose of this project is the prospective analysis of the process of social and urban change in the city of Lisbon.

Starting from this basis the paper has three principal parts connected with the main levels of the URBINNO Project and with the long period, 1890-1980, of urban change. These levels are: metropolitan area, core continuous built up area and functional units of the metropolitan area. For practical reasons the first two levels of analysis in this text appear in inverse order. Bearing in mind the historical and sociological approach the analysis of the core-continuous built-up area is examined in four periods:

---

* The author would like to thank R. Hestnes Ferreira, Teresa V. Heitor, Cristina Gomes, Teresa Requejo, J. Luis Casanova (ISCTE) for their contribution to the research. The chapter "Level 3C: Major axis, Ava. da Liberdade - Ava. Fontes Pereira de Melo - Ava. de República" has been written by Raul H. Ferreira.

1850-1920 Expansion of the city to the north and densification of the core.

1930-1948 Political control and urban planning.

1950-1967 The city's demographic intensification and the suburban expansion of the peripheries.

1969-1980 The structuring of Lisbon's metropolitan area.

Demographic growth and territorial expansion in Lisbon's metropolitan area, from 1890 to 1980, is also considered in four phases:

1890-1920 Analysing the dispersion and dependence of Lisbon's agglomeration.

1930-1940 Differentiation and density of Lisbon's agglomeration.

1950-1964 The suburbanization process of the agglomeration.

1965-1980 Structuring and expansion of the metropolis of Lisbon.

This is followed by an approach to urban changes in the spatial structure of the city, analysing the main expansion of Lisbon to the north, along a specific axis, which structures the capital. In each period of analysis the approach centres on the rhythm of urban change, both in relation to urban design and built environment.

An historical and sociological approach to the city of Lisbon

General aims of the research

This study develops a general survey and a description of the city of Lisbon as a whole over the last hundred years, taking into account the different stages of its evolution. Within the context of the physical and socio-economic evolution of the city, the main directions taken in the process of urban transformation are evalued through comparative analysis of documents and charts, including statistical and land-revenue data. The axes and the areas of the city subject to greater demands and pressures by the systems of activities which take place within them are also considered.

Considering the importance of Lisbon and its metropolitan area with about two and a half million inhabitants from a social, economic, administrative and political point of view, and also the dynamics of its growth and the evolution of the activities concerned, the first phase of this project has centred on the study and overall description of the city's development during this century. This has enabled us to further our knowledge of its systems and subsystems which will be the basis for surveying the evolution of urban morphologies in the areas chosen for further investigation.

In spite of the existence of several studies on this subject, it was considered essential to integrate and optimize them on the basis of an overall perspective which could provide a more profound interpretation of the evolution of the city of Lisbon in the period concerned.

Historical moments of urban change

In the last hundred years, the city of Lisbon has undergone three major changes in its urbanization process, following its reconstruction after the 1755 earthquake. Both socially and historically these three stages are related. First, the urban disruption to the north of the city during the 1880s; secondly, the rise and fall of an authoritative and modernist urban policy between 1930 and 1940; and finally, the birth and development of a metropolitan area, centralized and polarized by the city of Lisbon, from 1960 to 1980.

These stages are part of the birth and development process of modern Lisbon, which began with Pombal's intervention in what nowadays constitutes Lisbon's historical centre. On the other hand, each of these phases can be characterized according to three fundamental dimensions of socio-historical analysis:

1. Corresponding to the city's 'mode of urbanization'.
2. According to the processes of territorial structure and urban differentiation.
3. According to the city's 'urban image' at each of these specific stages.

1755-1850 The earthquake and Pombal's urban intervention

1. A 'centripetal' mode of urbanization, directed towards its own centre. The illuminist rationality in Lisbon's urban mesh, directing and centralizing Portugal's capital, from the kingdom's political centralization to Lisbon's urban centralization.
2. From Terreiro do Paço to Praça do Comércio, the 'royal square' as a structural element of the chequered urban mesh, but also as a symbol of the 'new mercantile winds', in contrast to Rossio,

35

the 'people's square'. Both these squares structuring the city but accentuating Lisbon's urban difference.

3.   The urban image of a riverside town, closed within itself, in which the square, dividing mesh of its centre highlights Lisbon's urban modernity, in parallel to the irregularity of its former built environment.

1890-1930 The disruption and change of the urban process

1.   The inversion of the urban process. The disruption to the north of the city, not only allowed Lisbon to expand but also determined a 'centrifugal', though limited, mode of urbanization. The territorial intervention of merchant capitalism and the regeneration of the capital, urban change in a socially and politically conservative context.

2.   Ava. da Liberdade, the umbilical cord of Pombal's urban mesh, and the Rotunda, urbanist platform, clearing the way for urban and suburban expansion of the city and establishing a new mode for urbanizing the capital. The 'short breath' of the urbanization process and Lisbon's socio-urbanistic difference, the paradigm of Avenidas Novas.

3.   The urban image of a 'naturally centralized city', although not exactly confined within its historical centre. The multiplicity of images as a result of the parcelling out of the urbanization process. The plan of expansion and urbanization of Lisbon in contradiction to the practice of private urbanization.

1930-1940 Lisbon of the empire, an authoritarian policy in a modern urban context

1.   The mode of urbanization following an increased control of rent levels for urban land, as a condition for establishing a specific urban process in Lisbon. The modernity of the urbanization and the authoritativeness of the plan, the first master plan of Lisbon. From the leadership of the urban process to a directional and centralized urbanisation.

2.   A fairly accentuated socio-urbanistic differentiation, the socially privileged expansion to the west and the industrial and working class 'vocation' for the east side of Lisbon's river zone. The centralisation of the capital and the nature of the suburban expansion within its own municipality. A radio-concentric

structure of the urban process.

3.   The image of the city as an empire's capital. The urban centralization was the city itself. Modernism and authoritativeness 'Portuguese style'. From Praça do Imperio to Areeiro, the rise and fall of an imperial model for the country's capital.

1960-1980 From urban concentration to the development of the metropolitan area

1   Urban expansion and suburban concentration. Industrialisation versus urbanisation and urbanisation versus tertiarisation. An intense changing process of land use. Concentration and 'desertification', the ethnological breakdown of the environment. The urban-metropolitan mesh and oscillatory displacements.
2   The centre and the periphery, urban and metropolitan exteriority. The socio-productive system and the territorial structure of Lisbon's metropolitan area. The axes of expansion and socio-urbanist differentiation, the 'five fingers' disparity. The making of the metropolitan area and increasing urban decay.
3   The urban image was (and still is?) an urban mirage, mediating space between the 'rural world' and the 'urbanized promised land'. The metropolitan image is above all the sum total of partial images in which the 'urban' dilutes itself. And hence the city of Lisbon's paradox, it is at the same time its own metropole and its own metropolitan centre.

Around this historical and sociological approach, of which the last two stages were amply developed in a published study (V. M. Ferreira, 1987), this research project has been developing a methodological perspective centred on the changing process of the urban form and the built environment of the city of Lisbon in the last hundred years.

Level 2: Core-continuous built-up area

The most important analytical dimensions in Lisbon's urbanisation were established as the following: urban design and land-use, built environment, infrastructure and transport, and urban instruments of planning. In view of the systematization already mentioned, the approach at this level is centred on Lisbon's urbanisation process during the last hundred years giving consideration to four fundamental stages. Figure 3.1 shows the stages of development of the physical structure over these time periods.

1850-1920 Expansion to the north and densification of the core.
1930-1948 Political control and urban planning.

Figure 3.1. *Core-continuous built-up area: stages of development*

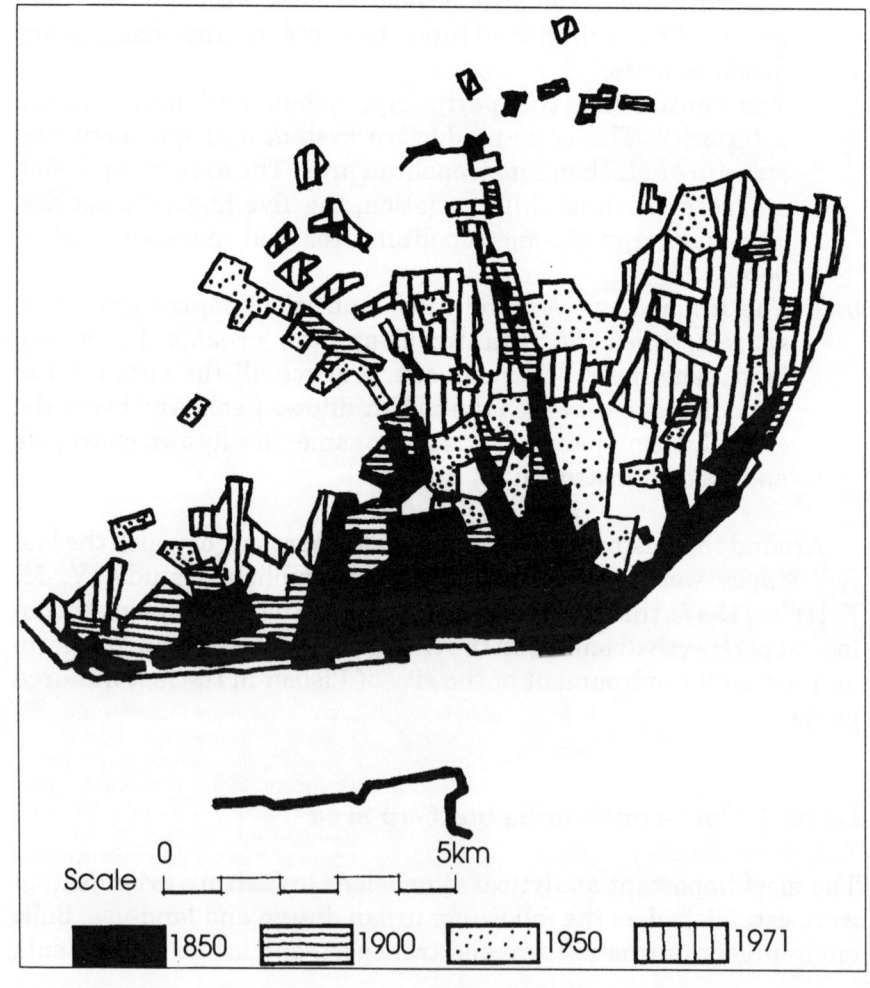

| 1850 | 1900 | 1950 | 1971 |

Scale 0 _____ 5km

Table 3.1. *Expansion to the north and densification of the core 1850-1920*

Element of Change	
Urban Design and Land Use	Destruction of the fundamental element of Pombal's urban model, Passeio Público (Public Garden). Directional change of the urbanisation process. Avenida da Liberdade, main axis of the city's structure (together with Avenida Almirante Reis), support zone for the new urbanisation (north). Densification of the historical centre and of the riverside zone. Expansion mainly to the west and along the river bank. The city's enlargement to the north, Avenida Novas, local circulation system and residential areas.
Built Environment	Dichotomic patterns of land-use. Planned areas, zones under the centre's influence and non-planned areas, peripheral nuclei. Eastern and western industrial zones. Discontinuous character of the urban fabric, empty spaces filled later as a result of new roadways. Different types of building, mansions, flats for rent and villas.
Infrastructure and Transport	The building of urban and suburban railways lines. The Eastern Line (1863), the Northern Line (1887), the West Line (electrified in 1907) and the Ring Line, with three stations, Sta Apolónia (Oriental), Rossio (Central), Alcantara (Ocidental). 'Americanos' taken out of service in 1917. Main public transport line, the centre and riverside zone. Tramways (1901) appear simultaneously with the city's development to the north.
Urban Instruments	The Avenida Nova's plan, a strict geometric design, the exteritority rule in relation to the centre, the hierarchy in the urban model. The urban network in articulation with the axes that begin in the squares with the centrality/dispersion function. Industrial growth without a large scale industrialization process, in the dock zones from Sta Apolónia to Grilo, Alcantara. Transfer of industrial units to the southern bank later more intensively. Industrial growth on the south bank, Barreiro, Moita, and Seixal, independent of the city. Legal instruments of expropriation and urban planning.

1950-1967 Demographic intensification and the suburban expansion of the peripheries.
1970-1980 Structuring of Lisbon's metropolitan area (LMA).

1850-1920 Expansion to the north and densification of the core

The inversion of Lisbon's urbanisation process, 'turning its back' on the River Tagus. The transport infrastructures, the country's economic development and the inevitable urban expansion of Lisbon. The first industrial units in certain areas of Lisbon's riverside zone. An urbanisation plan that repeated the geometric design of the city's historical centre. The Valmor Prize for modern architecture, Avenidas Novas. Growth of Lisbon's built environment by means of urbanisation companies, private entities which decided on the urban design of each zone.

Vicissitudes of Lisbon's urban planning, the city in search of an urban model (Table 3.1.)

1930-1948 Political control and urban planning

Two fundamental stages in this period, the first characterized by urban modernity, and the second by political authority. The first master plan for the city of Lisbon. A directional and controlling instrument of the urbanisation process, closely related to heavy intervention in the land structure. At the same time the plan was confined to highlight the main axis of the city's structuring. Zoning of the city with a clear social stratification. An imperial conception of the country's capital in which its urban monumentality veiled a stratified organization of space. An urban concept half-way between a German model and the Italian proposals. Some important projects in modern architecture (Table 3.2.).

1950-1967 Demographic intensification and the suburban expansion of the peripheries

Increase of the built environment in the city of Lisbon, as the result of the processes started over the last twenty years, 1930-40. Big increase in house building in the city and in Lisbon's peripheries. The first illegal allotments and buildings appear on the northern bank of LMA. Growth of the city outside its limits in its peripheral areas. Intensive growth of the tertiary section of the urban core and the beginning of this sector's expansion to the north of the city. The building of the bridge over the River Tagus simplified access to the south bank. Increase in density of residential building. Birth of new

Table 3.2. *Political control and urban planning 1930-1948*

Element of Change	
Urban Design and Land Use	A radial-concentric structure, wide radial streets from the centre to the periphery. West, Avenida Marginal; east, along the river bank with an industrial function; north, three axes which continue the new roadways of the former period and several ring roads.
Built Environment	Densification of the peripheral zones near the historic centre. Residential use of the connecting/articulating axis between the centre and the corresponding areas of Lisbon's agglomeration. Densely built up area, a fourth of Lisbon's total area. Intensive land-use along the riverside zone, especially in the west. Two important urban interventions, the expansion of Avenidas Novas and the incidence of land-use in Benfica. Building of social quarters in peripheral areas and decline of private urbanisation. Control of land-use. An exemplary urban plan, the urbanization of 'O Sítio de Alvalade'.
Infrastructure and Transport	The first bus lines along the existing tramway lines or extending these same routes. The first bus lines along the existing tramway lines or extending these same routes.
Urban Instruments	Special legislation applied in a rapid expropriation of land. A 'stock exchange' of municipalized land (nearly 1/3 of the administrative land of Lisbon's city council. Several instruments of political control and urban planning.

industrial units in the south bank. Increment of social housing. Changing patterns of land-use due to the tertiary sector and a different social use of space (Table 3.3.).

1969-1980 Structuring of Lisbon's metropolitan area (LMA)

Expansion and consolidation of a territory with metropolitan characteristics. Decrease of the city's resident population and demographic increase on both banks of the LMA. High rate

Table 3.3. *Demographic intensification and the suburban expansion of the peripheries 1950-1967*

Element of Change	
Change of Land Use	The tertiary sector takes over housing areas mainly in Av. Novas, which according to the plan were intended as an exclusive residential zone. Increment of private urbanisation in previously municipalized land.
Infrastructure and Transport	New routes to the west (Ava. Infante Santo, Ava. de Ceuta, Ava. de Roma). Ring road network permitting Benfica's urbanisation. Opening of the first undergroun line. Electrified railways lines (Sintra and Sta Apolónia). Extension of the bus network (Benfica/Camp. Grande, Alvalade/Areeiro, Aeroporto/Olivais). Building and use of the bridge over the River Tagus. Strong incidence in the south bank during the first phase of development, Lisnave, Siderugia. Small scale industries in Loures and Vila Franca de Xira during the second phase. Transfer of the city's industrial units to the northern periphery.
Population	Decrease of Lisbon's resident population. Biots of demographic increase in the city's peripheral zones.
Urban Forms and Processes	Use of land in northern and south-west zone of the city (Campo Grande, Lumiar, Benfica, Charneca, Ameixoeira). Beginning of parcelling and illegal urbanisation on the city's northern periphery (Benfica, Damaia, Loures, Sintra, Amadora). Chela's urban plan (social housing) and Olival's urban plan (middle class housing).

employment in the tertiary section of the metropolis' centre; relocation of some industries from the fringes of the city to the peripheral areas and establishment of new industrial units mainly in Vila Franca de Xira, Seixal and Barreiro. The population's attraction for the new industrial areas was already mentioned in the actual Master Plan of Lisbon (1967-76), 'to drive the town industrial units to the periphery, and to create new ones'. The periphery's increase of population was not followed by any building plans concerning housing and facilities and resulted in the progressive growth of 'illegal urbanisation'. In the late fifties several illegal developments were already visible. They began in some municipalities on the north bank of LMA and afterwards on the south bank. Land use on the south bank was intensified with the establishment of heavy industries, Siderugia, Lisnave and Sorefame, made easier with the building of the bridge over the River Tagus in 1966. Great changes in the morphology of Lisbon's peripheral territory, rural areas changing into metropolitan and areas of 'potential urbanisation' (Table 3.4.).

Table 3.4. *The structuring of Lisbon's metropolitan area 1969-1980*

Elements of change	City of Lisbon	LMA-North Bank	LMA-South Bank
Change in land use	Residential replaced by the tertiary sector (services). Progressive deterioration of the built environment. Intense oscillatory movements between the centre and periphery.	Good agricultural land taken for house building. New medium and small scale industries.	Intensification of illegal urbanisation over large areas. Scattered housing. Heavy industries and medium-sized industrial units.
Infrastructure and Transport	Reinforcement of the existing bus network. An underground line with only two axes.	Reinforcement of existing networks (railway and buses) especially those in connection with Lisbon. Significant assymmetries in the way of access into Lisbon	Intensive use of the bridge over the River Tagus (road network only). Road and river transport.
Industries/ Localization	Along Lisbon's riverside, a strong link with the Villa Franca de Xira's axis.	Sintra, metallurgical and mechanical sector. Villa Franca de Xira, several agglomerates of small and medium size industrial units.	Seixal, Barreiro, Montijo, ironworks, metalworks, cork industry, canned goods, ship building.
Population	Decrease and stagnation, in contrast to the periphery's increase.	Increase of 58% (80%+ of the resident population in the metropolitan area).	Increase of 118% (scattered population with a high percentage of young people).
Social Processes and Urban Forms	Linear expansion (axis) to the peripheral zones. Urban centralisation and metropolitan externalisation.	Suburbanisation without urban equipment. Rich west periphery, and poor northwest periphery.	Scattered use of the territory due to illegal urbanisation. Intensive exploitation of land use and house building.

43

Level 1: Demographic growth and territorial expansion in the Lisbon Metropolitan Area (LMA)

The Lisbon Metropolitan Area (LMA) has certain specific characteristics concerning its spatial structure and functional organization. It includes several counties which surround the city and which suffer from its direct influence, and is divided into two big zones, north and south of Lisbon, separated by a natural barrier, the River Tagus (Figure 3.2). This geographic characteristic has conditioned the diversified development of both banks, particularly with respect to land use.

The Bridge of Vila Franca de Xira and especially the '25th of April' Bridge, linking Alcântara to Almada, were the source of the south zone's increased development, as they joined both banks, in terms of distance and time. In this zone the non-existence of a railway link, envisaged but never executed, yielded a dispersed form of land-use.

In the north zone, on the contrary, as the railway connections with Lisbon have existed for some time, there is a more continuous form of land-use, oriented alongside the three main railway lines, the Cascais Line, the Sintra Line and the Northen Line, up to Vila Franca de Xira. Although the structuring impact of the railway is noticeable in some of the North Zone's counties, others, where this transport infrastructure is non-existent, show the same dispersed and tardy form of land-use. This form of built environment took on a specific character, in particular on the south bank, although it can also be found on the north bank in the counties of Loures, Amadora and Sintra, in the form of illegal allotments and building.

The intense demographic growth of the LMA, mainly from the fifties onwards, has its counterpoint in a decrease of the demographic growth rhythm in the county of Lisbon, which attained negative rates during the sixties. At the beginning of the century the increase in demographic growth on the south bank took place along the river zones, reaching interior areas only in a more recent period. While this tendency was developing the focus of growth was evident in Barreiro, Montijo and Almada. The first of these nuclei reached its highest rates of growth at the beginning of the century, then gradually lost its importance to the advantage of other localities. In the south bank, the development of Barreiro, closely connected in its initial phase to the south-southwest railway line, and later on to heavy industrial concentration, gradually yielded its place to other counties which directly 'benefited' from the building of the Bridge over the Tagus.

Figure 3.2. *Metropolitan area: county towns*

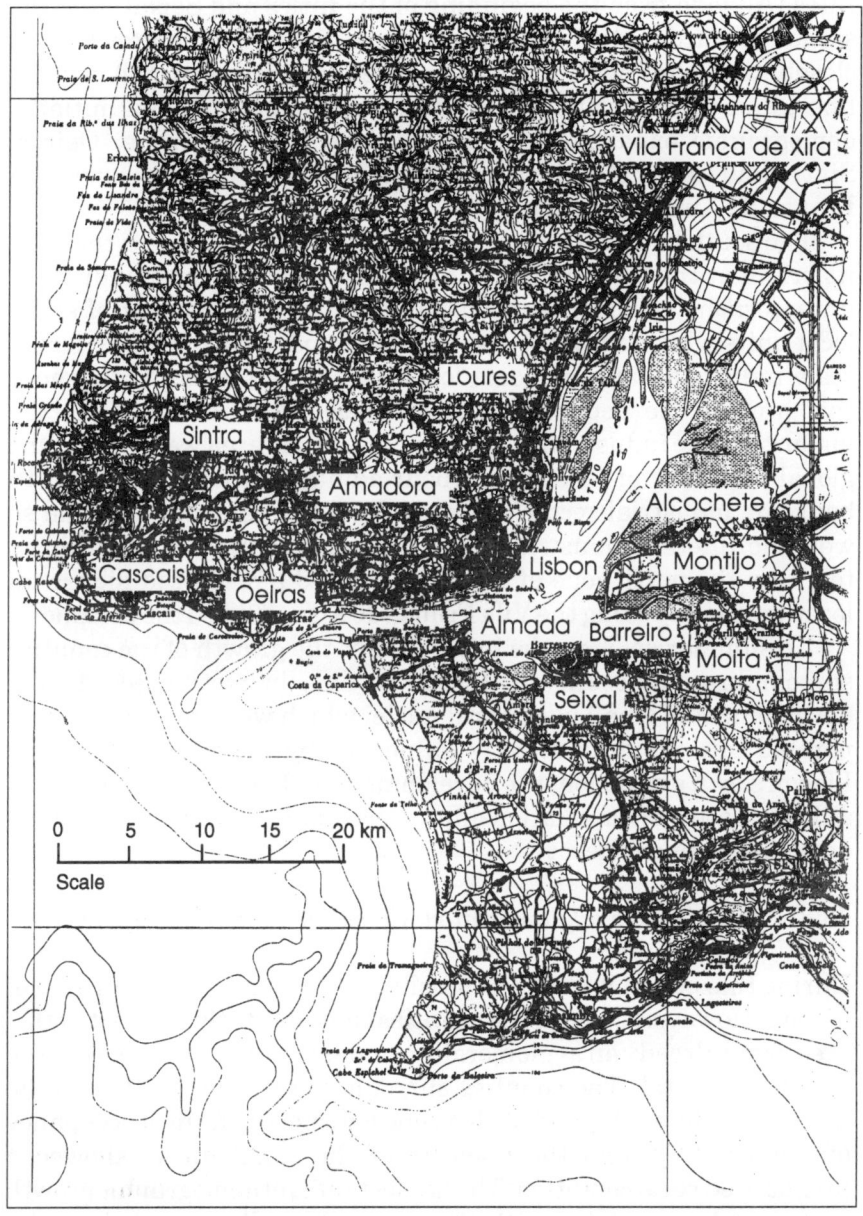

On account of their suburban residential function with the advantage of good road access linking both banks, Almada and later on Seixal were two of the counties with high demographic growth rates (Almada from the beginning of the fifties and Seixal from the sixties onwards). As to the north bank, demographic growth was closely connected with the appearance of nuclei along the railway lines, and the development of the rural agglomerations. Initially, demographic expansion proceeded along the west zone, Oeiras, Sintra and Cascais, and later on, advanced along the east zone, Vila Franca de Xira and Loures.

The south bank showed a specialized industrial process that had its origin at the beginning of the century: the transfer of industrial units formerly operating in Lisbon, and the development of small local units which, in the course of this period, developed into larger scale industrial units in some localities. Nevertheless, throughout this process the dependence upon Lisbon still remained, as the majority of the industrial infrastructures of the area were part of the global structure of Lisbon's harbour administration. On the north bank the industrial sector was considerably less important, and so were the different industrial zones. With no heavy industry within its territory the north zone of the LMA showed a dispersed industrial location, mainly along the Vila Franca de Xira axis and the periphery.

To summarise, it can be inferred that the growth of the counties surrounding the county of Lisbon is due, in most cases, to the development of small peripheral nuclei which were gradually incorporated into the area directly influenced by the city. Impelled by Lisbon's own expansion, the development of these counties was, and still is, based on a relationship of economic, social and political dependence upon the city.

1890-1920 Lisbon's agglomeration: dispersion and dependence

During this period it is not yet possible to find a metropolitan area of Lisbon. However, some of the counties that presently are part of this area, were already an agglomeration under the urban influence of the country's capital. The counties surrounding the city, either to the north or to the south, had at this time a dispersed form of occupancy of the territory, and the majority of the small nuclei showed a pronounced rural condition. The highest rates of demographic growth were closely associated with the new railway lines, and showed a tendency to be localized in the periphery nearest to the city of Lisbon.

The south bank, in spite of its difficult access to the city, showed remarkable development along the riverside and particularly in the county of Barreiro, which then began its industrialization process and at the same time became the terminal of an important railway line, the south-southwest line. The north bank, with the advantage of easy road and railway communication with the city of Lisbon, showed an increased rhythm of urban expansion. The most evident rise in population was registered in those counties which were recently served by the new railway lines, as was the case for Algés and Cascais.

During this period, the Lisbon agglomeration showed a dispersed form of land use and low demographic density, but was already structured by the railway and road networks planned in the late nineteenth century (Table 3.5.).

Table 3.5. *Lisbon's agglomeration: dispersion and dependence 1890-1920*

Elements of Change	North Bank	South Bank
Built Environment	Dispersed built environment, small nuclei of low demographic density surrounding Lisbon, alongside the railway axes.	Dispersed built environment, small nuclei alongside the river banks. Extensive forest area.
Demographic Structure and Dynamics	Oiras registers the highest rate of growth among the north bank's counties of the LMA, followed by Lisbon and Cascais. Loures, Villa Franca de Xira and Sintra register the lowest rates	Accentuated demographic growth in the counties of Barreiro, Almada and Seixal. Barreiro registers the highest rate (38%).
Infrastructure and Transport	Railway, Lisbon-Estoril/Cascais (1880), Lisbon-Vila Franca de Xira (1880), Lisbon-Sintra (1899).	South/southwest railway line with terminal in Barreiro. The connections with Lisbon were exclusively by water links.
Localization and Structure of Industrial Activities	Marble in Pero Pinheiro.	Transfer to Barreiro of a CUF factory formerly located in Lisbon. Development of a local industry connected to the agricultural sector slaughter industry.
Urban Instruments	--	--

1930-1940 Lisbon's agglomeration: differentiation and density

Throughout this decade the density of the existing nuclei conditioned the appearance of new areas of demographic growth that were to determine the organization of the territory.

In this way, the former tendency of demographic growth did not change. Almada and Barreiro together contained the biggest part of the south zone's population. The establishment of an industrial unit in the county of Barreiro explains in part its rates of growth, and contributed to its industrial 'specialization', which later on was to spread through the majority of this zone's counties.

The construction of the Lisbon-Cascais highway reinforced the transport structures of the north bank, contributing to the development of its west axis and, in this way, becoming one of the structuring elements of that territory. In the east zone, although the increase in population was not very high, some industrial units appeared in this period and expanded along the Vila Franca de Xira's axis.

The whole of this period was characterized by intense urban and political intervention in the city of Lisbon, the consequences of which were of lesser importance in the surrounding metropolitan area (Table 3.6.).

Table 3.6. Differentiation and density of Lisbon's agglomeration 1930-1940

Elements of Change	North Bank	South Bank
Land Use	Slight increase in density of former nuclei mainly along the riverside area, but with little expansion (highway and railway). Development of activities supporting Lisbon (agriculture and services)	Increment of land-use in areas of leisure (beaches) along the river and sea shore. Slight increase of land-use in former nuclei.
Demographi Structure and Dynamics	Decrease in demographic growth in the county of Oeiras. The parish of Amadora polarizes the demographic growth of the county of Oeiras. Break of growth rhythm in the counties of Cascais, Lisboa and Sintra. Increase in the county of Loures.	Within the demographic structure the counties of Almada and Barreiro still show the largest populations. Barreiro continues to show the previous tendency of demographic growth. Accentuated increase of demographic growth in the county of Moita (more than 30%).
Infrastructure and Transport	Lisbon-Cascais highway (1944).	A tendency for progressively substituting the rowing and sailing boats by motor boats. Building of an air-naval military base in Montijo.
Localization and Structure of Industrial Activities	Progressive industrial occupation along the Vila Franca de Xira's axis (mainly in Sacavém).	Concentration of the CUF industries in Barreiro and further expansion of this industrial unit. Development of the cork industry, more or less 13% of the country's cork factories were on the south bank.
Urban Instruments	Urbanisation Plan for 'Costa do Sol' (1948). Plan for Mafra (1946). Plan for Vila de Franca de Xira and Sintra (1949).	

48

1950-1964 Lisbon's agglomeration: suburbanisation

During the fifties and the sixties the demographic occupation of Lisbon's agglomeration intensified. Although the tendency was to expand along the main transport axes, at the same time, some of the dispersed nuclei started to fill the blank areas between the main polarizing nuclei.

Table 3.7. *Suburbanisation of Lisbon's agglomeration 1950-1964*

Elements of Change	North Bank	South Bank
Built Environment	Land-use along the railway axis. Illegal allotments in Loures, Sintra and Amadora, followed by illegal building. Dispersion of nuclei alongside the railway lines of 'Costa do Sol', Sintra and Villa Franca de Xira.	Intensive land use in several riverside nuclei (Costa da Caparica, Trafaria). The first, few, illegal allotments and buildings.
Demographic Structure and Dynamics	Demographic growth in the axis Amadora-Sintra. Beginning of intense demographic growth in the Loures Area. Lisbon registers its lowest rate of demographic growth.	Significant demographic growth in Almada (62%). The rhythm of demographic growth maintains its tendency in Moita (50%). A balanced demographic structure, where Almada is particulary in evidence (1690:36%).
Infrastructure and Transport	Electrification of the railway lines of Vila Franca de Xira and Sintra. Building of a bridge in Vila Franca de Xira linking the north bank of Lisbon with the south in the country.	–
Localization and Structure of Industrial Activities	Industries of non metalic minerals to the north of Sintra. Chemical industries and manufacture of rubber articles along the railway line of Vila Franca de Xira. Establishment of SOREFAME (metallurgy) in Amadora. Establishment of Tagus Cements in Vila Franca de Xira.	Industrial increment in several riverside nuclei (Barreiro and Montijo). Expansion of the cork industry, 28% of the country's cork factories were on the south bank.
Urban Instruments	First urban plans for the county towns.	–

The heavy increase in population was paralleled by the process of illegal allotment and building, which was to characterize this period and condition the later development of some areas of Lisbon's agglomeration. These rates of demographic growth were to be felt in an increasing process of suburbanization of the counties which surround Lisbon. In this way, the highest rates of demographic growth the north bank were found in the Amadora-Sintra axis, and also in the county of Loures, although the latter lacked transport facilities. On the south bank, the most flagrant example of demographic growth on the county of Almada, from which the future building of the bridge over the River Tagus cannot be disassociated.

On the whole, intense demographic growth and a progressive establishment of industrial units, led to the structuring of a metropolitan area which was urbanistically unbalanced and socially segregated (Table 3.7.).

1965-1980 Structuring and expansion of the metropolitan area

This period was characterized fundamentally by the general expansion of the counties which today are part of the administrative territory of the metropolitan area. In this context the various agglomerations on the south bank were particularly noticeable, in which demographic growth and land use reached the highest rates. The building of the bridge over the River Tagus in 1967 provoked intense changes in land use, therefore the high rates of potentially developable land in some of the zones was understandable, as was the accelerated rhythm of building. On the north bank there were also important changes. Enormous portions of rural areas became urban land.

The population explosion of this period, especially on the south bank, brought about an unbalanced demographic structure, where together with the big urban centres, Almada and Seixal, there also existed a dispersed form of land-use. The demographic increase on the north bank was slightly more homogeneous in Amadora, Oeiras, Sintra and Loures, while in the same period there was a decrease in the county of Lisbon.

The establishment of big industrial units on the south bank stressed its industrial vocation. With different dimensions and characteristics the establishment of new industrial units on the axis of Almada-Queluz and Vila Franca de Xira also contributed emphasizing the industrial tendency of these axes of Lisbon's north bank. In this way, both the north and south banks of Lisbon's metropolitan area show quite different rates and forms of built environment and land-use, though, as a whole, they form an area with urban-metropolitan characteristics. However, this social-urbanistic condition has not yet found a model which conforms to a metropolitan nature (Table 3.8.).

Table 3.8. *The structuring and expansion of the metropole of Lisbon 1965-1980*

Elements of Change	North Bank	South Bank
Built Environment	Progressive use of rural land. For the whole of the LMA the 'Sintra axis' registers the largest areas of development land (between 20% and 30%) and together with the 'Cascais axis' the highest rates of house building (26%). The growth rates of development land are higher in the Sintra and Loures axes during the sixties, and in Vila Franca de Xira and Loures during the seventies.	In the whole of the LMA the south bank registers nearly 40% of development land, and 20% to 30% of built environment. The growth rates during the sixties go up to 83% for the built environment, and 89% for development land.
Demographic Structure and Dynamics	Fast demographic growth in the administrative area of Amadora, which becomes a county in 1979. Increase of population in the counties of Sintra, Oeiras, Vila Franca de Xira and Loures. The demographic rate in the Lisbon administrative area continues to decrease.	Unequal demographic structure due to the counties of Barreiro and Moita, which register rates of 30% and 20%. Significant growth in the county of Seixal during the sixties (86%) and the seventies (134%). Almada and Barreiro also register high rates of growth.
Infrastructure and Transport	Reinforcements of the bus network leading into the city and the interior of the counties.	Bridge over the River Tagus, Alcantara-Almada link. South highway.
Localization and Structure of Industrial Activities	Major expansion of Tagus Cements in Vila Franca de Xira. New heavy industry unit (MAGUE) and Brewery (Central de Cervejas) in Vila Franca de Xira. Standard Electica. New industrial units scattered in the areas of Sintra and Vila Franca de Xira	New ship building docks (LISNAVE) in Marguira (Almada). Initial phase of great expansion. New iron work unit (SIDERUGIA NACIONAL) in Paio Pires, Seixal. The South Bank registers 31% of the country's cork factories.
Urban Instruments	Master Plan of Lisbon (P.D.R.L., 1964). Correction of the P.D.R.L. in 1972/74 (incomplete).	Master Plan of Lisbon (1964). Correction of the P.D.R.L. in 1972/74 (incomplete).

51

Level 3C: Major axis, Ava. da Liberdade-Ava. Fontes Pereira de Melo-Ava. da República

The main purpose of this section of the study is an understanding of urban form and its evolution, bearing in mind the function of the city as a whole and its support networks. The main changes in the built environment and infrastructure are documented and the adjustments to the morphological structure evaluated. The approach is centred on the rhythm of urban change, both in relation to urban design and built environment, in each period.

An analysis of the main expansion of the city of Lisbon to the north, along the axis Ava. da Liberdade, Ava. Fontes Pereira de Melo and Ava. da República - which, when the Passeio Público was sacrified, expanded towards another of the city's traditional green areas, Campo Grande to Alvalade - shows throughout a whole century the evolution of the occupation of the city's principal axis by the upper-middle class. At the same time consideration is given to the expansion of another main axis, occupied by the lower-middle class, Rua da Palma, Rainha D. Amélia, Almirante Reis. Together, these axes define a complex system of physical interrelationships and urban and sociological morphology. This urban system was characterized by a radial expansion and its links to the traditional city centre, by way of new convergencies and perpendicular links to the two axes already mentioned. It became a new city with new quarters totally different from the old ones, which provided a new way of living and dispersed the main activities, creating new centres. This stage, regarding Lisbon's development in qualitative and quantitative terms, is pre-metropolitan.

The situation in 1890

At the end of the nineteenth century, Lisbon had a population of about 300,000 people living in close association with the river, and its expansion to the north was limited by physical, technical, economic and legal factors. The level of occupation within the old boundaries had reached its limits, given the existing social structure, physical limitations and technical resources. Meanwhile, the rural areas to the north of the city were in a 'stand by' situation, half cultivated, punctuated by manor houses, palaces of the royal family and palaces of noblemen, surrounded by small communities of peasants and servants.

The inevitability of expansion to the north was reinforced by the location of new city facilities along these roads, including the Cattle Market at Entre-Campos, the Penitentiary at Campolide and the Bullfight Ring at Campo Pequeno. Mobility had increased with new mass transportation, the 'American' street car and elevators, in the hilly sites of the centre, while the railroad system that reached the centre and surrounded Lisbon on the north side in 1890 could not be adapted to the urban topography. In contrast to what was happening in other European capitals it was supplanted by an electric streetcar system, better adapted to the scale of the city.

Before the expansion of the city to the north the area of the Passeio Público (Public Alley) of the wealthy people at the limits of the Lisbon of Pombal was structured by old narrow radial roads converging on the centre of the city, Portas de Sto Antao, Andaluz, S. Sebastiao, Palhava, Benfica. The major one, Ava. da Liberdade, was traditionally related to agriculture and linked the Valverde Valley to Benfica. But the axis Rua da Palma, Anjos, Arroios, Arco do Cego, Campo Grande, Linhas de Torres, along the valley of future Almirante Reis, and between them the axis Campo de Santana, Picoas, Rego were also important links to the outside, leading not only to the residential houses of the aristocracy, but to the north of Portugal.

Curiously the destruction of the Passeio Público, the best solution for the extension of Lisbon towards the north, was associated with the rising importance of Campo Grande. This was the traditional grazing area of Avalade, market and place of leisure since the sixteenth century, and more recently the recreation of the underprivileged. It became the main target of new development and defined the scale of the planned extension with complete disregard for the old quarters. The importance of this space to the new urban system was confirmed by the first name given to the Av Pereira de Melo, Av do Campo Grande.

1890-1910

During this period the city developed outside its traditional limits for the first time. This expansion cannot be separated from the growing importance of the middle class in Portugese society throughout the nineteenth century, culminating in the Republican regime.

The importance of the urban changes between 1890 and 1910 was enormous. Ava. da Liberdade was extended to the Rotunda do Marquês de Pombal and from there to the Rotunda das Picoas/Saldanha and through Picoas/Ava. da República to the Campo Grande, securing the link between Baixa and this green area. This development was structured by new residential areas including the Bairro da Estefania in 1880, Bairro Camoes in 1890 and the Bairro between António Augusto de Aguiar, Fontes Pereira de Melo and Picoas and the development along Ava. da República to Campo Pequeno between 1900 and 1910.

The new street system followed the checkboard pattern that had marked nearly all urban developments in Lisbon since the sixteenth century. On a larger scale it adjusted harmoniously to ancient morphological units: the axis Portas de Sto Antao, Andaluz, S. Sebastiao, which was entirely preserved; the axis Campo di Santana, Gomes Freire, Picoas, partially dissolved near Saldanha; and the axis Rua da Palma, Anjos, Arco do Cego, Campo Pequeno, Entre Campos, also partially preserved but, little by little being dissolved between Arco do Cego and Campo Pequeno. Most of the buildings of this period were residential, but some public facilities were built, mainly schools, Liceu Camoes Veterinary School and a hospital, Rego.

Technical evolution and the improvement in public transport laid the foundations for city growth. Improvements in engineering made possible the building of viaducts, discreetly inserted in the most vulnerable points where the new road systems overlapped the old, and the erection of elevators in the old city. Electricity not only helped slowly to improve city and domestic illumination but also was the basis of the new transport system that linked the new to the old quarters of Lisbon.

1910-1926

This period corresponds to the years of the republican democratic regime after the fall of the monarchy and was a period of reform in every branch of Portuguese society, particularly social affairs and education, and also of consolidation of the new urban systems in Lisbon. These were unstable years both externally, with World War I, and internally, due to the opposition of the forces of the Ancient Regime and to political upheaval caused by several revolutions and right wing dictatorships.

The population of the city rose by about 100,000 and there was a big change in the administrative structures but not so much in the physical environment. The urban system on both sides of Ava. da Republica was completed and the main buildings constructed in the area of the study were hospitals, the Psychiatric Hospital Júlio de Matos, starting a new occupation and street system near the Campo Grande, the Maternity Hospital of Lisbon, next to Ava. Fontes Pereira de Melo and the small Ceramic Museum of Bordallo Pinheiro at Campo Grande. While the construction of private detached houses and residences proceeded in the area, the first social housing quarter, the Bairro do Arco do Cego, was started in 1919, marking a new policy in housing development. The public transport system was also improved with the completion of the electric streetcar network.

1926-1934

The first years of this period were marked by military dictatorship that led to a complete collapse of the economy and the subsequent rise of the Estado Novo of Salazar, in which the action of the powerful Minister of Public Works, Mayor of Lisbon and Director of the Superior Technical Institute (IST), Duarte Pacheco, was heavily felt in Lisbon and especially this area. Following a policy of public works, contrasting with the former period, the occupation of the Avenidas Novas by upper class houses went on and new urbanisation began to the east of Arco do Cego, between this street and the Estrada das Amoreiras, the future Ava. de Roma axis.

This new area, closely linked to Duarte Pacheco, included the complex of the IST, 1927, the Statistics Institute, 1931, and the new urban quarter, east of Campo Pequeno. While to the west, Ava. de Berna was extended further with the building of the Igreja de Fátima and surrounding residences in 1934. The new eastern development organised the second main valley of Lisbon, Ava. Almirante Reis, from Praça do Chile to Areeiro creating a new street system with the main axis of Ava. da Liberdade, Fontes Pereira de Melo, Ava. da República.

It is interesting to note at this early stage that the regime chose modernism as its trademark, upgrading it in relation to the Art Deco of former periods. Of special note are the institutional buildings of Pardal Monteiro (IST, INE) and the residential buildings of Cassiano Branco.

1934-1953

This period of 20 years covers several phases of Salazar's rule, starting with pure fascism before World War II, when Duarte

Pacheco continued his ruthless work in the urbanisation of Lisbon until his death in 1943, and the post-war stage with successive adjustments of strategy by the dictator in political as well as in the cultural and artistic fields. In a shift from the period in which modernism was favoured, the so called 'Portuguese style' was imposed on modern architects from 1940 until 1948, when it was strongly questioned by the first Congress of Portuguese Architects. This proved to be a turning point that slowly gave way to a new, more technocratic form of intervention in the city.

This was the period of bold action to expropriate large rural areas for the city, and the construction of new residential quarters along the new axis, the Ava. de Roma, that overlapped the railway loop in the direction of Alvalade. These quarters situated between Areeiro and the Psychiatric Hospital were the Bairro das Estacas, S. Miguel and Alvalade, first phase, framed by a new street system defined by Ava. da Igreja, Ava. dos Estados Unidos da América, Praça de Londres, Praça do Areeiro and Praça do Alvalade.

In this period the quarters to the south east of Av António Augusto de Aguiar became a model for new forms of direct intervention by the municipality and for implementing the newly conceived 'national' style together with the Praça do Areeiro. The system of Ava. da República and its parallels, 5 de Outubro and Defensores de Chaves, was completed with the opening of Ava. Praia da Vitória (west) linking the Praça do Saldanha to Ava. 5 de Outubro and altering the old axis of Picoas to define the site of the Monumental movie house, now destroyed. In addition, to conclude the system around IST, to the west, the Rua D'Estefania on the old axis of Rua do Arco do Cego, and to the east the big Alameda Afonso Henriques across the valley of Almirante Reis was created.

The occupation of this area by a new wealthy population, and its increased importance for city development owing to the ease of transportation among other factors, favoured a diversity of building typology concentrating activities of general interest to the city.

Little by little this area was becoming a second centre competing with the Baixa of Pombal and dislocating the centre of gravity of Lisbon along Ava. da Liberdade to the north.

In morphological terms, the duplication of Ava. da República with its parallel system of Ava. de Roma increased the strength of this area and merged the two main boulevards of Lisbon to the north, the Ava. da Liberdade, Ava. da República upper middle class system, and the Ava. Almirante Reis, Praça do Chile lower middle class system. In infrastructural terms the main factors were the crossing of the railway loop by Ava. de Roma and Ava. da República and the creation of a new public transport system, the bus, shortening journey times along the main radial avenues.

This area, following the tradition of preceding periods, concentrated a wider range of urban facilities, the Mint, a Monumental Luminous Fountain, markets, churches, the Monumental Theatre-Cinema, the Imperio Cinema and the National Civil Engineering Laboratory (LNEC)

1953-1963

These years showed the stability of Salazar's regime in spite of several political crises and strong opposition. The development of the study-area was almost concluded and the construction of the University City had begun in spite of the opposition of the dictator himself. The Gulbenkian Museum was a landmark of the period as a public building diverging from the Estado Novo model. Among the most important buildings of this period, were the ones in the University City, including the big University Hospital, in the German tradition, the National Library, several secondary schools, new sports grounds, churches, markets and movie houses that completed the facilities of the zone. The second phase of Bairro de Alvalade was completed and the Praça de Londres and Praceta and Ava. dos Estados Unitos da America were almost completed in a continuous process of tertiarization that slowly was changing the character of the area.

In public transport the opening of the subway system, which followed the traditional radial line of Lisbon's valleys, was a great improvement, occurring about a hundred years after the systems of other large European and American cities.

1963-1980

With the wars in Africa, the illness of Salazar, and the rule of Marcelo Caetano unable to change and give credibility to the regime, a democratic regime finally arrived on the 25th April 1974. This period covers all these events and while in the area of study development was almost concluded, the trend was for the reinforcement of the tertiary sector and the physical change that followed.

In transportation there was a shift to individual car transport in the sixties with the building of several viaducts to favour traffic flow:

57

Campo Grande to improve the second ring road, Ava. dos Estados Unidos under Ava. de Roma, and Ava. da República under Praça da Guerra Peninsular.

The tertiarization of this area increased to a point that in a short time the Ava. da República and the other main avenues might lose their residential function. Before the urban texture was completed, a process of indiscriminate speculation began. Redevelopment was not limited to buildings of poor construction or lacking basic amenities, but also included the demolition of single dating houses from the beginning of this century. This process ignored qualitative appraisal, civic conscience and reflection about urban development in favour of profit and led to the destruction of buildings of good quality construction, with high civic and cultural value, such as the Theatre Monumental.

Conclusions

The conclusions are drawn in accordance with the two definitions which formed the basis of approach: (i) The social and territorial achievements which have taken place in Lisbon, and which the research tried to emphasise. (ii) The conditions and specifics of social innovation which are highly significant within the process of urban change.

The long cycle involved in this approach obscured any punctual conjunctive alterations to the urban structure within the period examined and, in contrast, the study centred itself on the fundamental points of social and urban metamorphosis in the city. Through this approach, four phases of effective social and territorial change were ascertained during the period of Lisbon's 'historical modernity'. These stages relate to the reconstruction following the earthquake in 1755, the disruption at the end of the nineteenth century, the upsurge and downfall of a modernist yet dogmatic trend of urbanisation which took place in the 1930s and 1940s, and finally the metamorphosis of the last few decades in which a metropolitan area has been formed without the corresponding political or institutional legislation.

The historical picture of Lisbon's urban change formed the basis for the three levels of analysis: (i) Continuous core, with respect to Lisbon's own urban space in terms of historical settlement and urban consolidation. (ii) Metropolitan area, the expansion and formation of a characteristically urban metropolitan territory, which may be designated as Lisbon's metropolitan area. (iii) A major axis, the New Avenues, which represents the city's urbanisation process in terms of function and space.

During the various periods of change it was possible to identify the empirically representative constituents of variation on each level. This identification was based on the analysis of problems concerning social metamorphosis and urban innovation which guided the approach. A horizontal interpretation of the thematic lines of approach was used to identify the roots of the constituents on each level. This not only confirmed the basis, but also acted as a guide for understanding the empirical information. A horizontal interpretation made it possible to clearly understand the synchronous nature of the constituents and, at the same time, helped to codify that information from the point of view of a diachronic perspective on urban change and social innovation. This perspective reflects the thematic vectors favoured in the empirical analysis. The following three sections explore each of these themes.

Urban form and space structure

Throughout all the stages of analysis there has been a steady structural form which prevented the upsurge of urban innovation. The disruption north of the nineteenth century city and the so-called 'Pombal Style' of reconstruction at the end of the eighteenth century, although not part of the time period considered, were largely responsible for conditioning the urbanisation process at the start of the twentieth century. The structural characteristics of a dogmatic city government's 'modernist model' should not be minimised since this trend profoundly influenced Lisbon's urban structure from the 1930s. The idea behind this conclusion is to illustrate that this model has not contravened the former logic of the urban pattern. In fact it served as a revival of one of the main elements of Pombal's reconstruction, the paved public promenade or walk. During the 50s and from the end of the 60s the territorial expansion process and the process of urban occupation followed the same radial-concentric logic, although more radial than circular, clearly emphasising the shaping or forming nature of earlier urbanistic approaches. Also, it is worth noting the relationship between the urbanisation of the city and the organisation of the metropolitan area. Prior to the precise formation of the metropolitan territory occupancy was very sparse, showing no precise urban shape but rather small rural gatherings dependent upon the primary formation of the city of Lisbon.

From a diachronic point of view and in the ambient conditions of the long cycle of urbanisation it is possible to conclude that the organisation of the urban space and the metropolitan area, within the previously referred to inference, has resulted in equally enduring and extensive changes to land use, for example, rural areas becoming densely occupied, potential development sites taken over by illegal occupancies, residential areas turned into office and service areas. In short, functional zoning acted primarily as a social discrimination device for the occupancy such areas.

Built environment and the transformation of land uses

Although this theme is clearly incorporated in the conclusion to the previous section, the example of the New Avenues aimed to unify the characteristics of a space, an attribute which figures frequently in Lisbon's urbanisation process. When comparing planned and non-planned areas the clarity of this example is evident up until the end of the nineteenth century. The plan for the New Avenues was the turning point of the century and, in contrast to the preceding urban expansion, introduced a reticulated mesh similar to Pombal's into a prospective development area. On the other hand the mesh of the New Avenues was built up around a central axis, the Avenue of the Republic. Radially, this avenue formed one of the dominant axes of the city from the end of the nineteenth century and its basic influence has continued up to the present.

In the space built up within the city and in the metropolitan area very different types of development and construction were adopted. These were reached by means of progressive technical innovation within each scheme influenced by the characteristic forms of the sites. While it is important to stress the distinct characteristics of the built up space, it is not possible to consider them without relating them to the transformation of uses discussed earlier in this analysis. In the case of the New Avenues there was a drastic change in land use that should be recognised as a definite ranking process that changed a preferential and historically occupied space into a area dominated by residential use. The social costs of this economic innovation and urban change have yet to be estimated, particularly in terms of urban quality. The territory grew up severely restricted in its central area, thus losing urbanism, and expanded speedily in the outskirts, breaking down the precarious balance of the existing rural zones.

Productive space and transport sub-structures

Current practice associates long-term urban change with advanced technologies being introduced into the systems and means of transportation. In addition, there is a background of urban history, almost entirely based on these modern technological and economic means, which is related to a vital need of all social groups, free circulation. This is another theme which points to the previously described technical, social and urban changes. It is true for the railway at the end of the nineteenth century and its electrification during the first decade of the twentieth. Later there was the upsurge of the motor car, becoming competitive with, then overtaking public transport and the similar development of goods transportation. Finally, there is the underground network which is still inadequate.

It is obvious that the established sub-structure of the territory has effected the movement and circulation system in the urban-metropolitan area. On the other hand, that structural process is differentiated in accordance with the city's own functional zoning. The urban and social geography of the territory, with the east-west and north-south divisions of the metropolitan area, clearly illustrate this social and functional discrimination process in combination with an unequal and asymmetric sub-structure of that same territory. It should also be noted that this lack of symmetry and proportion in the substructural processes strongly effects or even shapes the previously radial-concentric mesh. The five fingers which form the main arteries of Lisbon's urban-metropolitan expansion are also the main substructural angles of that territory, three of them with railways almost from the beginning of the century. It was the lack of this means of transportation in the remaining fingers, Loures and the south bank of the Tagus, which was responsible for the sparse and untidy occupation of these areas. Spaces for production, especially those along the river in the early stages and later in the outskirts, were not only part of the metropolitan process but also grouped themselves according to functional zoning and thus added to the discrimination within the social and urban segregation process of the city and the metropolis.

The formation of the industrially productive space of this urban-metropolitan territory was historically based on a complementary and interdependent principle. This relates to a sector which was strongly linked to the production of consumption means and mainly located, in the first place, in central areas and on the west side of the river bank. Secondly, it relates to sectors essentially linked to the production of means of production which were concentrated on the south bank as well as scattered here and there in the north and outskirts of Lisbon.

In the last few decades there has been a strong tendency to favour road transportation inside the metropolitan territory, to the detriment of a steady progress and expansion of the railway system. The most significant example of this is the 25th of April Bridge which has no railway. This has reinforced the metropolitan process instead of the regional zoning of the south bank and is undoubtedly the most decisive aspect of the territory's own sub-structure.

Urbanistic instruments and forms of intervention

There is no doubt that from the end of the nineteenth century the various interventions in Lisbon's urban and metropolitan areas are a paradox in comparison to the highly specific urbanistic instruments which the expert literature made known as urban planning. In fact, the continuously recurring rhetoric has accompanied an actual deterioration, if not an absolute lack of methodology, in the assumptions and the effective mechanisms within the regulating instruments for the area.

Recently there have been a few exceptions to this rule, the plan for the New Avenues and the Plan for Avalade in the 1940s. But these are in contrast to the general overall situation.

More recently, some of the quarters in Lisbon's metropolitan area have attempted to organise their own municipal plans and the political attitude aims to fight against this urbanistic 'curse'. The exceptional plans which have been carried through, point to or were the result of the city's own urban conception at each of these moments. They had a similar impact on Lisbon's own urban structure and have resulted in effective urban change within the country's capital city.

The case of the two models known as urbanisation plans is different:

The Master Plan for the Region of Lisbon, concluded in 1964 and

still to be officially approved.
The Master Plan for the City of Lisbon, completed in 1967 and officially approved in 1977.

It was not considered appropriate to analyse the contents of the official documents but rather to put an emphasis on the strong double meaning of the 'plan' model. Accepting the propriety of these documents, not only diagnostically but also in the light of the philosophy inherent to the proposals, the official non-approval of the original master plan for the region can be seen to emphasise the antagonism between what has been said and what has been done, intention as opposed to reality. The revision carried out between 1972 and 1974 did not make any further progress, although the events of 25 April 1974 can be considered as the cause of the disruption to the proposals.

In Lisbon's master plan, despite official approval, the same ambiguity exists since throughout its contradictory and unsteady implementation period it was obvious that the proposals were based on the empirical knowledge of the city in the 1960s.

This analysis also considered other urbanistic instruments such as partial plans, regulations governing expropriation and control of landed property, but they bore no close relationship to the specific area of urban planning. The plans which have been discussed were regarded as a means to solve the territory's problems, but in effect they have been merely a constant background to the sequence of events. It is certainly true that when speaking of the assumptions of Lisbon's urban planning there was at least at the level of speech some urbanistic innovation, but not so much in terms of new forms of intervention as an effort in attempting to adapt to other more comprehensive forms of that type of approach.

References

Cruz M.A. (1973), *A Margem Sul do estuario do Tejo. Factores e formas de organizacao do espaco (The South Bank of the Tagus' Estuary. Elements and Forms of Spatial Organisation)*, unpublished. Departamento de Geografia e Planeamento Regional (1978), *FCSH - Universidade Nova de Lisboa, INIC / JNICT - Seminario Internacional, Area Metropolitana de Lisboa; que futuro? (International Seminar. Lisbon's Metropolitan Area; what future?)*, Fundacao Calouste Gulbenkian, 13-16 Outubro, Lisbon.

Ferreira A.F., Guerra I.P., Ferreira V.M. (1991), 'An evaluation of the economic and social impact of the April 25th Bridge over the River Tagus in Lisbon', in Montanari A. (ed.) *Under and Over the Water. The economic and social effects of building bridges and tunnels*, Edizioni Scientifiche Italiane/C.N.R., Naples.

Ferreira A.F., Guerra I.P., Matias N., Stussi R. (1985), *Perfil sociologico e estratégias do 'clandestino'. Estudo sociologico da habitacao clandestina na AML (Sociological Outline and Strategies of Illegal Building. Sociological study of illegal housing in the LMA.)*, Centro de Estudos de Sociologia/ISCTE, Lisbon.

Ferreira V.M. (1987), *A cidade de Lisboa: de capital do Imperio a centro da Metropole (The City of Lisbon: from Capital of the Empire to Centre of the Metropolis)*, Dom Quixote, Lisbon.

Franca J.A. (1977), *Lisboa Pombalina e o Illuminismo (The Lisbon of Pombal and the Enlightenment)*, Bertrand, Lisbon.

Franca J.A. (1980), *Lisboa: urbanismo e arquitectura (Lisbon: Urbanism and Architecture)*, Bertrand, Lisbon.

Henriques J.M. (1980), 'Crescimento economico e desenvolvimento urbano-metropolitano de Lisboa' (Economic Growth and Urban Metropolitan Development in Lisbon), in Ferreira V.M. (ed.), *Cidade-Metropole de Lisboa. Materiais para uma abordagem interdisciplinar (Metropolis-city of Lisbon. Materials for an Interdisciplinary Approach)*, Nucleo de Estudos Urbanos e Territoriais/ CIES do ISCTE vol. 3, ISCTE, Lisbon.

Ministerio da Habitacao, Obras Publicas e Transportes (1982), Secretaria de Estado da Habitacao e Urbanismo. *Ordenamento da Peninsula de Setubal. Dinamica de ocupacao do solo (Land Structuring of the Setubal Peninsula. Dynamics of Land Use)*, Dir. Geral do Planeamento urbanistico, Lisbon.

Ministerio das Obras Públicas (1964), Plano Director da Regiao de Lisboa, Anteplano. *A: - Resumo do Inquerito e analise Regional*

(Master Plan of Lisbon, A: - A Summary of the Inquiry and Local Analysis), Dir. Geral dos Servicos de Urbanizacao, Gab. do Plano Director da Regiao de Lisboa, vol. 1, Memoria discritiva e justificativa, Lisbon.

Nucleo de Estudos Urbanos e Regionais (NEUR/IACEP) (1983), *Tendencias recentes na urbanizacao da Area Metropolitana de Lisboa - Identificacao de areas densamente urbanizadas (Recent Tendencies of the Urbanization of Lisbon's Metropolitan Area - Identification of Densely Urbanized Areas)*, IACEP, Lisbon.

Rodrigues M.J.M. (1979), *Tradicao, transicao e mudanca. A producao do espaco urbano na Lisboa oitocentista (Tradition, Transition and Change. The Building of the Urban Space in the Lisbon of the Eighties)*, Boletim Cultural/Junta Distrital de Lisboa, (84) Serie III, Junta Distrital de Lisboa, Lisbon.

Monografia de Lisboa. Ex-Administração da Indústria e Obras
Analíticas. Dir. Geral dos Serviços da Urbanização Obr. de Plano
Director de Região de Lisboa (vol.). Memória Descritiva
(...)(1968). Lisboa.

Núcleo de Estudos Urbanos e Regionais (N.E.U.R.A.C.R.) (1984).
Tendências recentes na urbanização da Área Metropolitana de
Lisboa - in: Prácticas de ordenamento e ordenamento. Recent
Tendencies of the Urbanization of Lisbon's Metropolitan Area.
(depositado no Dias et.) (Polúmen d'essai) I.A.C.E.P.L. Lisboa.

Rodrigues, M. M. (1976). Trânsito, transferentes e transporte. A produção
do espaço urbano nas Lisboa et cetera (Vasc. don, Tradução and
Grana). The Parking of the Urban Space in the Lisbon of the
periferia. Relatório. Colecção Política territorial de Lisboa. Relatório
III. Junta Distrital de Lisboa. Lisbon.

4. ROME

Armando Montanari, National Research Council (CNR), Naples, Italy.

Introduction

The centre of Rome has had its own particular development over the last century, conditioned above all by the wealth and fame of the city's history. First the capital of the Roman Empire and then the centre of Christianity, the city has always attracted the interest and attention of the entire world. Even when it was still a semi-rural city in 1870, it was at the centre of a series of international relations. Therefore, it was considered worthwhile to examine the innovative transformations and processes which affected the historic centre, the only part of the city which has had continuity with its past.

Methods

This analysis focused on three levels, the city's historic centre, a quarter in the historic zone, and a building in that quarter. These correspond to levels 5, 6 and 7 of the initial programme. Within each level the periods examined vary according to selected stages of development. The spatial and temporal limits common to the entire project were examined before the most important types of innovation in the process of morphological change were identified. The morphological transformations were complicated by the fact that evidence of growth or change were flanked by innovations linked to transformations that were less easily recorded because they did not involve changes in form, but in intended use or residential social structure.

Level 5: Specific area, the historic centre

Geographical dimension

No one, unequivocal definition of the historic centre exists, since every study on this part of the city has produced its own qualitative and quantitative interpretation of historic centre. It may be useful to refer to the meaning given to the term historic centre in Italian town planning legislation, although, here again there is no one, unequivocal interpretation. Law 765 of 1967 gives a first very general definition in which historic is any urban agglomerate having a, '...historic, artistic character or of particular environmental worth.'. The Ministry of Public Works, with a view to ensuring, as far as possible, that regulations be uniformly applied, indicated the criteria for defining urban agglomerates of, '..an historic, artistic character or of particular environmental worth.':

urban structures in which most of the blocks contain buildings built prior to 1860, even in the absence of monuments or buildings of particular artistic worth;

urban structures within the ancient walls, entirely or in part conserved, including possible external extensions which come under the previous definition.

urban structures, built after 1860, which as a whole are evidence of quality building.

In the Town Plan of most Italian municipalities the historic centre coincides with the area limited by the surrounding walls, whether still existing or demolished. The decision in this case is therefore made in relation to one of the fundamental elements characterizing the urban structure. Included within the historic centre are areas previously used as orchards and gardens, or built up only in recent decades, or again, ancient areas, that have been demolished and then reconstructed.

Other municipalities have used the period of construction as the point of reference for defining the historic centre, the period in which the settlements were established or features of the town planning layout. This is the path followed by Rome City Council which in the general Town Plan, initially drawn up in 1962, defined the historic centre as Zone A, the part to be conserved and restored, including the area built prior to 1870 and those areas which, though built after that date, are of interest from the point of view of town planning or of building structure. Thus defined, the historic centre of Rome has a surface area of 650 ha., which reaches approximately 1000 ha. including the Vatican City and the archaeological zones.

In this study the historic centre is defined as that part of the city of Rome which is enclosed within the walls built by the Emperor Aureliano around the year 275 A.D.. It is an area which includes the built up historic part, the archeological zones and the zones built up in the last century upon the areas previously occupied by villas, orchards and gardens, a total of 1400 ha. (Figure 4.1.).

Time dimension

The starting point for the investigation was 1870, the year in which the Italian Army invaded the city of Rome, after breaking its way through the Aurelian Walls near Port Pia, and annexed the city to the new Kingdom of Italy. On 1 July 1871 Rome became the capital of the Kingdom.

During the Roman Empire, under Emperor Augustus, (27 B.C.-14 A.D.), Rome, the capital city, had just under one million inhabitants, this figure increased to just over one million at the time of the Emperor Domitianus, 81-96 A.D.. The population gradually decreased following the construction of the walls around Rome in the third century, and eventually came down to a few thousand. When the popes were in Avignon between 1305 and 1377, the city had approximately 17,000 inhabitants. In the middle of the sixteenth century, the population started to increase once more and reached 80,000. By the seventeenth century there were already 150,000 inhabitants and in 1870 Rome had a population of 244,484.

In the last century, Rome and Italy experienced a number of very different situations from the point of view of their socio-political history. The Savoy monarchy, the Fascist Regime, the Resistance, the fall of Fascism and the parliamentary Republic have all left their mark on the development of Rome and been elements of great innovation. Another important point to bear in mind is the fact that every twenty-five years there has been a Holy Year in Rome, and that the Olympic Games were held in the city in 1960. Significant funds were channelled to the public works sector in relation to these events which also required the reorganization of accommodation facilities in order to house the many pilgrims and tourists who visited Rome on these occasions.

There have also been many different pieces of urban legislation with the Town Plans of 1873, 1883, 1909, 1931, 1962, and the Variation to the 1962 plan, of 1974. Because of the many changes and events that have occurred in the course of this last century, it seemed opportune to divide the last one hundred years of Rome's urban history into five periods, 1870-1924, 1924-1950, 1950-1962, 1962-1978, 1978-1988.

1870-1924

This period was dominated by the need to transform the city into the capital of the Kingdom of Italy. It may be useful at this stage to say a few words on what led the Savoy monarchy to move the capital of the Kingdom to Rome. History books generally refer to reasons of domestic policy, of foreign policy, and also to the ideological significance of the mythical idea of 'Roma caput Mundi'. The more profound reason for the choice of Rome as capital city can probably be linked to its weakness. It was a relatively small, isolated city - Naples was four times bigger than Rome, Milan and Turin twice as big - surrounded by a relatively uninhabited territory with an economy that was still essentially agricultural. Nineteenth century writers describe it as a big country village. Most importantly, it was considered to be a relatively peaceful city, far removed from the turbulent climate of the urban areas of the north and of Naples.

There was, however, still the problem of disagreements with the Church State. Although the Italian troops had not gone beyond the walls surrounding the Vatican City, the occupation of Rome was per force considered an act of violence which led to a series of problems and difficulties in the relations between Italy and the Vatican. Pope Pius IX considered himself a prisoner and refused to leave the Vatican. This situation of conflict was solved many years later, after

70

Figure 4.1. *Map of the metropolitan area of Rome - Photo Analysis of SPOT 1 (19 May 1986) satellite Image[1]*

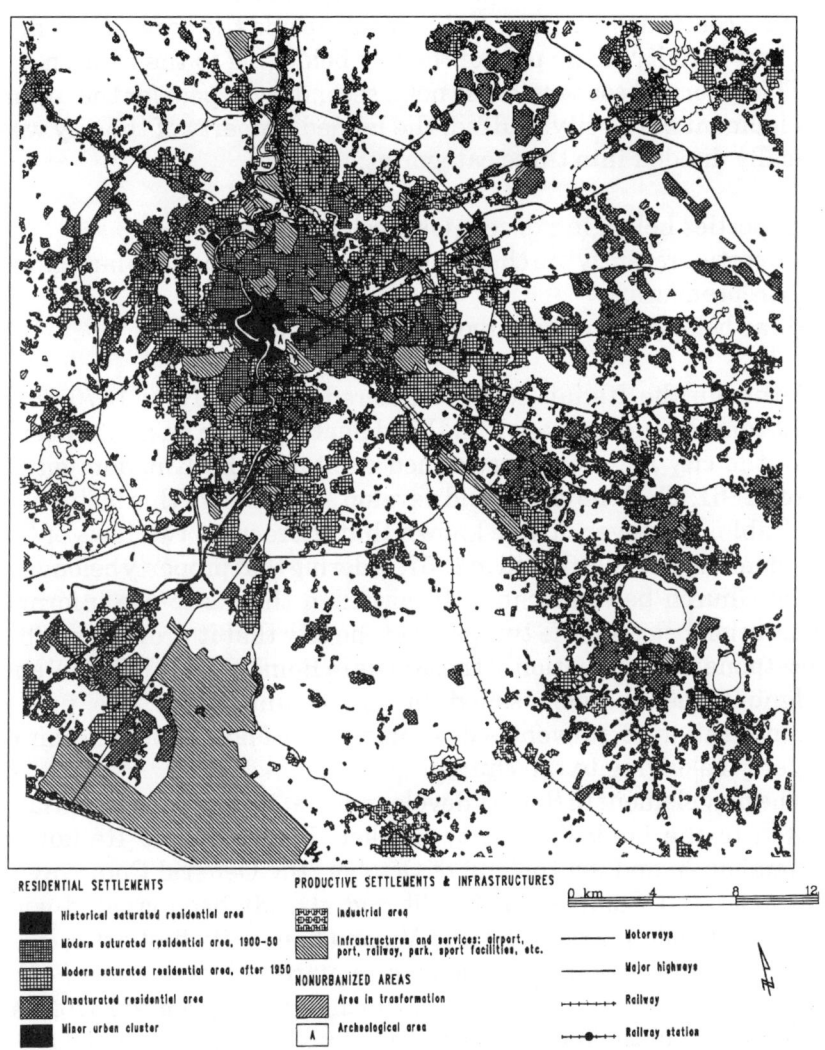

RESIDENTIAL SETTLEMENTS

- ▓ Historical saturated residential area
- ▓ Modern saturated residential area, 1900-50
- ▓ Modern saturated residential area, after 1950
- ▓ Unsaturated residential area
- ▓ Minor urban cluster

PRODUCTIVE SETTLEMENTS & INFRASTRUCTURES

- ▓ Industrial area
- ▓ Infrastructures and services: airport, port, railway, park, sport facilities, etc.

NONURBANIZED AREAS

- ▓ Area in trasformation
- ▓ Archeological area

0 km   4   8   12

———— Motorways

———— Major highways

+—+—+—+ Railway

+—●—+ Railway station

[1] The work was prepared by A. Montanari with the informatic assistance of C. Magnarapa as part of the CNR Strategic Project on Italian Metropolitan areas. The equipment was provided by CNUCE/ CNR.

71

what is known as the 'Conciliation', with the signing of the Lateran Treaty, February 11 1929. Despite the political contrasts in this period the two cities, the lay and the religious, continued to live side by side without friction. The parliamentary activity of the Italian part of Rome continued without problems alongside the ecclesiastic and religious activity of the part of the city belonging to the Church.

This informal cooperation is not a secondary aspect for the urban development of the city. Indeed, the landed property of the city was basically divided into three categories:

properties belonging to religious orders;
properties recently purchased for urban development by farsighted entrepreneurs;
villas belonging to the important Roman families.

Although the Italian soldiers entered Rome only in 1870, as a direct consequence of Napoleon III's defeat in Sedan on September 1 1870, the city had been proclaimed capital of the Kingdom on 27 March 1861. That the city would develop as a result of this factor was inevitable. Moreover, it was known that there were two laws of the Kingdom, issued in 1866 and in 1867, ordering that property belonging to the Church be confiscated. It was then thanks to the informal cooperation between the two parts of the city that it proved possible to postpone the application of these laws in Rome to 19 June 1873, and to ignore the decrees issued in the meantime by the public administration to prevent sales, mortgages and the alienation of property belonging to the church.

The new element in Rome's development at this point is essentially related to the landed property market. Rome became Italian on September 20 and, by the end of October, the General Company of Provincial and Commercial Credit had already been created with capital in London, Frankfurt and Vienna banks. Its first action was the purchase of one hundred thousand square metres near the Baths of Diocletian. The landed property, which, already prior to 1870, was concentrated in the hands of few, ended up being the monopoly of a few Italian and foreign financial groups and large banks. It is estimated that in 1875 one company, the *Compagnia Fondiaria Italiana* possessed just under 500 ha., all of which was destined for real estate development. In these conditions, there was no market law for determining the price of land. The property of the important Roman families and religious orders was sold towards the end of 1870 at 5 to 6 lire per square metre, resold in 1871 at 15 lire a square metre, at 30 lire in 1876 and at 100 lire after 1880.

One of the entrepreneurs was Monsigneur De Merode, the former Protominister of Arms of His Holiness who provided both the land and the covering of the church for those who wanted to build, which clearly was an advantage in helping overcome the reluctance of the Roman Catholics to deal with the Piedmontese laymen. By February 2 1871, a first agreement had already been signed between Mons. De Merode and the City Council for building on the area near the Baths of Diocletian.

The Papal city consisted of houses, palazzi, churches, villas and wondrous ruins. It had to be rapidly transformed into the capital city of a modern kingdom. New roads had to be built, government departments established, there had to be housing for civil servants, barracks for the military garrisons, and a decorous seat had to be found for Parliament. The old and the new buildings had to be protected from the Tiber which flooded the city at regular intervals. But despite these many problems, the capital of Italy was moved to Rome in 1871, although the public administration had still not had enough time to prepare for such an important event. The Town Plan of 1873, rather than being an instrument for planning the urban development of the city, became an instrument for ratifying the urban development decided on the basis of the demands of the landed property owners. Demographic growth, the population almost doubling in 30 years, a law of 1881, law no. 209, which set a state contribution for building in the city, and property speculation were the causes of the building boom called 'building fever' which began in 1880 and continued until 1885. Two types of houses were built in this period: small villas surrounded by a small garden for the well to do, and four or five storey blocks, rented to civil servants.

This period came to an abrupt end with the 'great crisis', between 1887 and 1892, caused by the sudden drop in the shares of Roman real estate enterprises which up until then had been very well listed in Italy and abroad. The crisis was caused essentially by the enormous increase - 100 to 200% between 1883 and 1887 - in the prices of areas to be developed for real estate, the result of speculation which had completely distorted the economy of that entire sector. Another reason was the international financial crisis which led to the sudden collapse of an activity based on the transfer of bills, renewed each month, until they could be discounted abroad. These entrepreneurs, most of whom had no previous experience in this sector, went bankrupt as soon as they were asked to pay the bills in cash. One entrepreneur after another went bankrupt, as well as a number of large banks, and many families of the upper and middle bourgeoisie suddenly found themselves broke. As a consequence there was unemployment among labourers in the building sector and this in itself was a serious social problem which led to street demonstrations. 10,000 unemployed were sent back to their place of residence by a compulsory expulsion order. On 8 February 1889, Rome witnessed its first protest of jobless people. The shares of the major building companies, the *Societá Generale Immobiliare*, dropped from lire 1260 in 1887 to lire 114 in 1892. Whereas the number of rooms built had reached approximately 13,000 units a month in the two year period 1886-1887, this decreased to around 800 in 1888-1889.

Meanwhile the fundamental infrastructure for the further development of the areas to be built on were being created. Between 1887 and 1891 three new bridges were built over the Tiber, and another three were built at the beginning of the twentieth century. Most of the bridges built during the Roman Empire were out of use and only one bridge had been built during the time of the popes. With the new bridges, a new road network had also to be developed and many of the old buildings had to be demolished. At the beginning of the century, Rome recovered from the great crisis and many new initiatives were begun. This period, which lasted until the first World War, was marked by a housing supply which, at least in part, started to escape the control of private enterprise. In 1903 the Institute for the Construction of Low Cost Council Housing was created and the activity of cooperatives for the construction of housing increased, starting with the cooperatives of the tram workers, the railway workers and civil servants. This trend was supported by two national laws, one passed in 1904 and the other in 1907. The purpose of the former law was to bring some order into the complicated financial situation of Rome City Council and, among other things, introduced a tax on housing areas, which was equal to 1% of their value. The second law gave a more precise and binding definition of a housing area and increased the tax to 3%.

The second innovation in this period was the victory in the elections of a coalition of parties, consisting of radicals, republicans and socialists, led by Ernest Nathan. Nathan, who was mayor of Rome from 1907 to 1913, was of British origin, a fervent 'Mazzinian' and mason, known to be a respectable professional and a man of great honesty and moral virtue. Nathan's programme was based on the increase of primary education, the protection of public health, and a town planning policy to limit speculation and, finally, the participation of the people in the decisions of the local administration. Thus 16 new primary schools were built in this period, centres for educating people to combat malaria in the rural areas around the city were created, steps were taken to establish that certain areas be state property and belong to the City Council, and the activities of the Institute of Popular Housing were supported. On 20 September 1909, the citizens of Rome were called to express their opinion, in a referendum, on a proposal to municipalize public transport by rail. This was Rome's first and only referendum. The results of the referendum were in favour of the project and, in 1911, ATAC, the Enterprise of Communal Trams and Buses, was created. It was followed in 1912 by the creation of ACEA, the Communal Enterprise of Electricity and Water. Electricity had been introduced into Rome in 1890.

In drawing up the general Town Plan, Nathan also tried to break with the past and above all with the technical office of the Council which had certainly been involved in all the property speculation and deals. An outsider was called in to draw up the plan, a certain Edmondo Sanjust di Teulada, at the time chief engineer of the Milan Civil Engineering Office. The plan was approved in 1909 and will remain in the history of Rome town planning as one of the best plans ever to be approved both from the point of view of its technical and town planning quality and clarity, and for its applicability. The plan aimed essentially at regulating the areas immediately alongside the ancient walls. For the historic centre, the plan simply eliminated most of the demolition which had been planned in the previous decades, preserving only four sites. In a period that was of particular importance for the life of the nation and favourable for the economy, the Rome City Council made great progress in its efforts to bring the living standards of the city up to those existing in the other urban centres of the nation.

Despite his many successes, Mayor Nathan, who used, among others, the following slogan, 'We have the weapon of taxes on areas firmly in our hands and will not hesitate to use it.', was clearly not popular with the land owners, who created an association and opposed him fiercely. A coalition was formed for the elections of 14 June 1914, and successfully overcame Nathan's popular front. A member of the aristocracy thus became mayor of Rome, the tax on the areas to be built on was abolished, a stop was put to municipalization and the most important innovations of the Town Plan were thwarted.

With the First World War business in Rome came to a standstill, very little manpower was available and building activity slowed down. The post-war crisis was not as serious as that of 1887-1892 and every effort was made to overcome it by building as much as possible wherever possible. The building density of plots was increased, the tax on areas to be built on abolished, and the Town Plan of 1909 rendered ineffective. In 1919 a package of urgent provisions for the city of Rome was passed. Funds were made available for the construction of municipal housing, roads and public utilities. The Depots and Loans Fund was authorised to finance the construction of low cost municipal housing by the National Building Union. The building of municipal housing and of other public services was declared to be of public utility so as to benefit from the mechanism of expropriation and compensation established by law.

In 1920 some 24,000 rooms were built, approximately double the average number of rooms built in the pre-war years. A new element to appear in this period was the distinct separation between the part of the city inhabited by the middle classes and the part by the lower classes. Under the popes, Rome was a small city and the integration of the various social classes was possible because they knew each other and lived side by side in a paternalistic type relationship. By 1921, Rome had almost half a million inhabitants, but had not become an industrial city and never would become one. It thus had none of the modern instruments for developing the integration of the various social classes. The middle class areas are in the central part of the city, on the outskirts are the lower class areas consisting of outer suburbs and shanty towns.

1924-1950

1922 was the year of the fascist coup. By 21 April 1923, only six months after the march on Rome, Mussolini, in a speech from the Capitol, was already voicing his ideas for transforming Rome into the capital city of fascist Italy. In addition to being head of government, Mussolini believed himself called upon to perform, at least *de facto*, the function of Mayor and Town Planning Councillor of the City of Rome. By 1924 his ideas were already starting to be put into practice. The houses around the Trajan Markets, the Trajan Forum and the Forum of Ceasar and Augustus were destroyed. No indication of these works was given in the 1909 Town Plan; Mussolini had simply developed his own personal plan for the city of Rome which he started to describe in a number of his speeches in 1925. 'Rome must appear wonderful to all the peoples of the world: huge, orderly, powerful as in the time of the first empire of Augustus.'. To reach these objectives, Mussolini singled out two main types of problems, '...those linked to need and those linked to grandeur...'.

Many buildings were demolished and many areas of the historic centre underwent great changes in the period between 1924-1950. Between 1924 and 1940 not a year went by without an old area of the city being demolished. Thousands of people were forced to move into the suburbs on the extreme outskirts of the city. This process came to a halt during the war, but continued immediately afterwards, despite the change in the political, economic and social situation of the country, as the city readied itself for the 1950 Holy Year.

Mussolini's town planning policy for the historic centre consisted of slogans which were then quickly taken up by the various individuals constantly hovering around him. These in turn, used the slogans as a cover and saw to it that the areas were demolished to the best of their ability and to their best advantage.

Referring to the demolition of the Borghi, an area just near St. Peter's, a newspaper of the time wrote, '... April 21 1934: an historic date which marks the beginning of a grandiose task. As he came out of Hadrian's Park, which he had just donated to the people of Rome, the 'Duce' stopped to gaze at the entrance to the Borghi and at St. Peter's. He stood without moving, deep in thought. Undoubtedly it was then that he decided to address and solve the problem which had remained unsolved for centuries ...'. The design was completed in 1936 and demolition started in 1937. The whole job was finished for the 1950 Holy Year.

What today is known as Via dei Fori Imperiali was built by demolishing most of the buildings, in some case disassembling them bit by bit, and building them elsewhere, and flattening an entire hill (Figure 4.2). The hill was called Velia and stretched along the Basilica of Massenzio from Piazza Venezia to the Colosseum. Like other orographic elements, the hill had conditioned the urban development of the city of Rome and therefore was an integral part of its history. Anything which according to the canons of the period was not considered monumental was destroyed. No document has survived on the many houses, palazzi and churches that were demolished, very few photographs, no surveys. There are only paintings reproducing them.

Figure 4.2. *Via della Conciliazione*

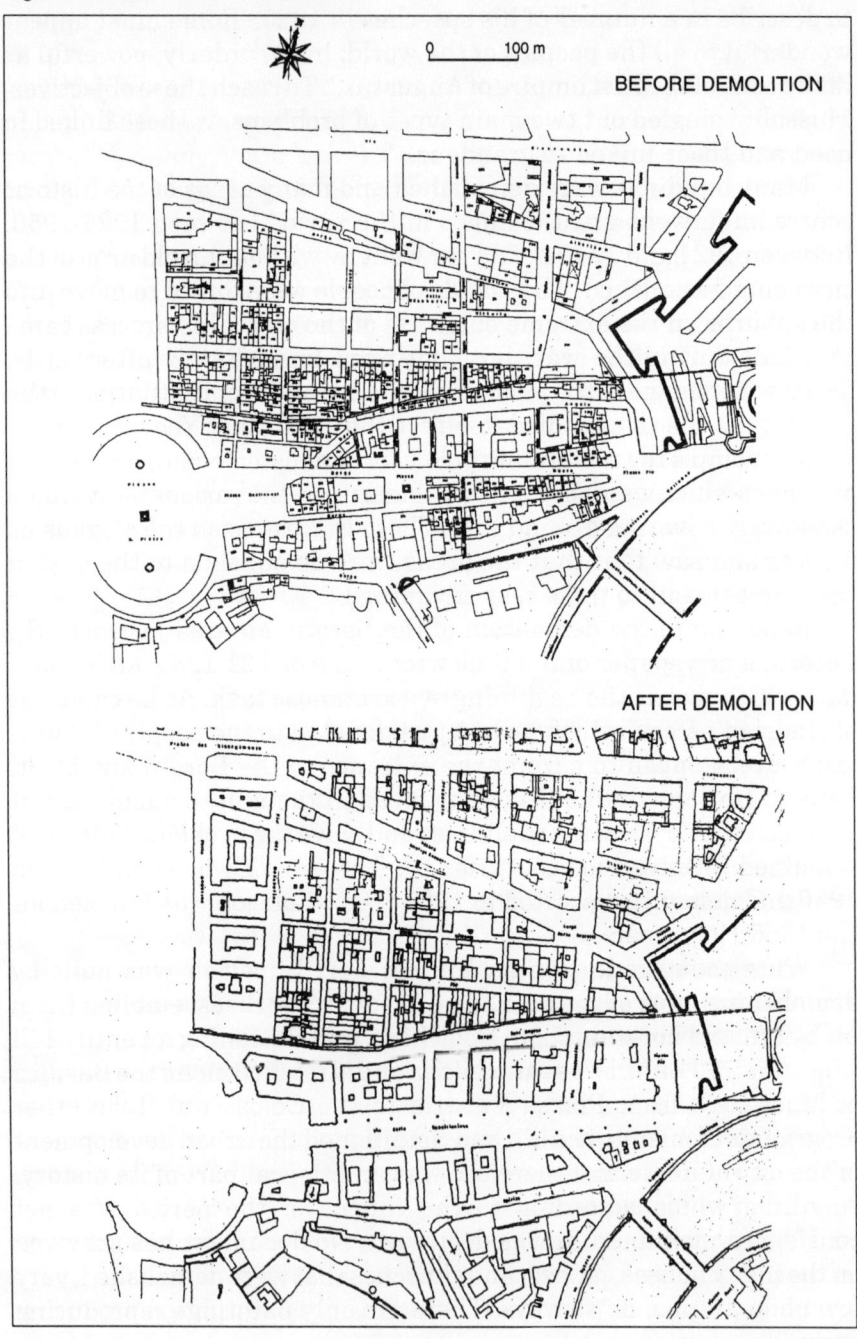

BEFORE DEMOLITION

AFTER DEMOLITION

After 1950, the grandiose demolition works planned in the previous period were, at last, halted and any attempt at starting them again met with the opposition of public opinion, which was now more aware and at last free to express itself. In this context it should not be forgotten that even in 1951 the Technical Office of the City Council had drawn up a detailed plan for the area of Via Vittoria, in which more buildings were to be demolished. Initially very few opposed this plan, but those who did were sufficiently vocal and incisive to force the Ministry of Public Works to issue a communique on 3 August 1952 expressing the need to put off the project and the plan was stopped.

But in the opinion of many, this response from the Ministry of Public Works was not an indication of a change of policy but simply an incident. Councillors and the technical offices of the Council continued to believe that an old building was synonymous to an unsafe building and that for moral, hygienic and aesthetic requirements it would be better to move all the inhabitants of the old buildings in the historic centre to the outskirts and then undertake the necessary works of demolition and reconstruction. Theories such as these still had a great hold on public opinion in the fifties, for reasons of ignorance and also of dishonesty of the few who were to draw great advantage from works to improve and re-adapt the old areas. The demolition of buildings in the fifties was always linked to the need to free the historic centre from the obstacles which prevented the roads from being widened and therefore the passage of more traffic. The fact that it was not possible to travel by car in the old parts of the city where the roads had been made for pedestrians, or at most a carriage, was seen as an obstacle preventing the development of the historic centre as a business centre and therefore from increasing in value. Discussions on the pretext of adapting the ancient buildings to the requirements of modern life continued in the fifties, also in relation to works for the demolition of two other areas. In 1955 Rome City Council had drawn up two detailed plans for demolishing two areas in the historic centre, but fortunately neither of the projects were ever implemented.

If it proved impossible to enact the detailed plans for radical works of demolition and reconstruction because of the firm opposition of a part of public opinion, mainly intellectuals, the situation as regards individual initiatives of demolition and reconstruction was different.

In this case the relationship between those asking for a building permit and Rome City Council, which granted such a permit, was a private relationship in which public opinion could only intervene once the work had been done. In fact, in the area of individual initiatives of demolition and reconstruction, the freedom of interpretation of the few rules and regulations which should have safeguarded the historic centre was directly proportional to the importance of the person requesting the permit.

In these twelve years many permits were granted for the demolition and subsequent reconstruction of individual building units and for the demolition and subsequent reconstruction of internal structures leaving the facades intact. In the same period many buildings underwent more or less extensive works to raise their height or restructure the roofs, terraces and lofts. Certain categories of property can be identified by the type of intervention carried out in those years.

Intervention by public bodies This category involved the restructuring of several hospitals. They were old buildings built as hospitals towards the end of the last century, or even earlier, and therefore with all the likely shortcomings from the point of view of hygiene, function and size. It would have been more advisable at the time to make the necessary modifications and use these structures at the most for local health services. Instead important restructuring works were undertaken, and the structures of the old buildings were destroyed without solving the problem of the city as regards health services. Following these works of extension, the health services for the entire city came to be concentrated in the historic centre. A huge 120,000 cubic metre structure was put up alongside the old hospital of St. Giovanni. The public bodies started up a more intensive programme to demolish and restructure their office buildings thus starting a process to turn the historic centre into a service centre.

Intervention by religious bodies Two old convents in the historic centre were turned into hotels.

Intervention of private companies Here too there were clear signs of the start of the process of turning the centre of Rome into a centre for service industries. A large company, IBM, successfully had an ancient building demolished and reconstructed to house its offices. Other companies did likewise.

1962-1978

The most advanced cultural battles of the nation were fought in the fifties over the Town Plan of the city of Rome. A general town plan was drawn up in 1962 which, despite the many hopes it raised and the subsequent inevitable disappointments, was essentially the result of these battles.

For the part included within the ancient walls, the plan basically distinguished three different categories.

Zone A: Conservation and recovery, the area defined as a historic centre, intending by that the urban area built before 1870.

Zone B: Transformation of buildings with special volumetric limits. This included the whole zone built after 1870, and works of demolition and reconstruction could be effected, even on single buildings.

Zone C: Reorganization of road network and buildings. Certain areas built after 1870 are expected to be reorganized on the basis of a detailed town plan.

For zone A, the regulations established, among other things, that, '... any intervention to be effected in the historic centre must be incorporated in a detailed plan; in the absence of such a plan, only isolated interventions aiming solely at conserving, restoring or improving the sanitary conditions of property units is allowed, in exceptional circumstances and in cases of proven need...'.

In actual fact, as Rome City Council had not drawn up detailed plans for the historic centre, this ruling opened the way to fragmentary, isolated initiatives, with the equivocation of maintaining the external appearance and local colour. On 5 April 1973, the Mayor of Rome wrote in *Il Corriere della Sera*, '... the historic centre of Rome is the largest and best conserved historic centre of Europe...'. It was possible for him to make statements such as this, despite the criticism of some small minorities, because most of the press and other media, and hence public opinion in general, was convinced that the situation of the historic centre was, if not perfect, at least acceptable.

An inspection in early 1974 by an association of citizens, *Italia Nostra*, of a sample of approximately 100 buildings in which building activities were under way, revealed that only a very small percentage of the works, about 18%, were actually to be considered restoration works. 55% were violations completely altering the typological structure and construction features of the buildings, while in 27% of cases the buildings were actually demolished and only the external facades remained intact. The survey conducted by *Italia Nostra* revealed other surprises: 44 buildings in the historic centre almost all belonging to the state and covering an area of more than one million cubic metres, were discovered to be in a state of abandonment.

When in the spring of 1976 the Magistratura confiscated, for the first time, a construction site in the historic centre for serious building irregularities, it did not come as a surprise to public opinion. For months cultural associations and the media had been publicising the irregularities and abuses committed on buildings in the historic centre. In only two years, the situation had undergone a radical change. What was surprising was the continued inertia of the administration of Rome City. Despite the polemics, it continued to take no interest in the historic centre, or rather to let others take an interest in it for their own personal profit.

In the period 1976-1987, the Magistratura confiscated more than 60 buildings undergoing restoration works in the historic centre. Those actually guilty of performing the illegal works were certainly directly responsible for this unfortunate situation but, undoubtedly, the administrations of Rome City Council up until 1976 had their share of responsibility. The abuses committed in restoring buildings in the area is but one aspect of a situation in which the historic centre was in effect abandoned to its destiny by Rome City Council. With no surveys, no studies, no plans, the historic centre was unknown territory. There was only one technical office for issuing permits and nothing else, no checks, no controls, no serious on-site inspections. The determination with which the Magistratura addressed the issue and the victory in the local elections of a coalition of left wing parties, for the first time since the end of World War II, led many to hope that the situation had finally changed and that there would be a new policy for the historic centre of Rome. Unfortunately the new administration of the City Council was slow to act and excessively cautious, and therefore shares the same responsibilities as the previous administrations. There was no real policy, no people to implement it and the voters did not confirm their confidence in the left wing coalition in the 1984 elections.

During the period 1976-1987 the destruction of the historic centre was greatly curbed, both thanks to the action of the Magistratura in 1976 and also to a new awareness of public opinion for Rome's cultural heritage and therefore its historic centre. The price of apartments in the area has become so high that speculation is no longer possible.

The myriad of public bodies, Parliament, the City Council and the government ministries continued to act on their own behalf, selling, buying, occupying and abandoning buildings without any reference to an overall policy and without paying too much attention to the problems of the buildings in which they operated.

1978-1988

In November 1978 a number of fragments decorating the Column of Marcus Aureliius fell to the ground. This incident started off a series of careful surveys and controls on the monuments in the centre of Rome. The results of these tests were publicized by the Sovrintendenza ai Beni Archeologici of Roma a few days before Christmas 1978. The situation was extremely serious, air pollution was corroding the ancient buildings and the experts forecast that if pollution did not decrease within a few decades everything documenting the history of art of Rome would be lost.

Undoubtedly pollution was not a new phenomenon, its effects on buildings and, above all, on old buildings were certainly not unknown. Unfortunately the phenomenon had not been taken into due consideration by the public administrations of Rome; pollution continued and the cries of alarm sounded by the environmental associations were considered to be annoying alarmism. The problem of air pollution was wilfully ignored for many years, despite the unmistakeable signs of deterioration which had for some time marked the city's main archeological monuments. As often happens in this city, for a problem to be considered in all its gravity an incident, an exceptional event, had first to take place.

Things started moving immediately, special funds were allocated for restoring monuments found to be in the worst condition, the first steps were taken to reduce pollution. Above all, private traffic was forbidden in the historic centre, first in one or two streets, then a whole area and finally, in 1988, in almost all the historic centre. Ten years after the original incident, too little has been done and pollution continues to increase above all in the historic centre, as a consequence of a mistaken urban development policy which had concentrated the main business activities of the metropolitan area in this sector, without taking measures to improve public transport. It has become an issue not so much of safeguarding the integrity of ancient monuments but also, and above all, the health of citizens.

At this point, the problem is such as to be certainly beyond the sphere of the City Council of Rome. Environmental pollution cannot be considered, addressed or solved at the local level. In this field much should and can be done through the exchange of information and experiences both at the national and international levels, in an effort not only to acquire new analytical instruments or reach new technological levels, but also to ensure the establishment of new cultural concepts and styles of life in the urban areas, which are more suited to the protection of the environment.

Level 6: Block, Tor di Nona Quarter

This is an ancient quarter located in the historic centre near the River Tiber and therefore it has been exposed to the recurrent risk of flooding for centuries (Figure 4.3). The consolidation of the course of the river, the construction of protective embankments and a broad tree-lined avenue at the end of the nineteenth century involved the demolition of many buildings and put an end to all physical relationships between the quarter and the river. It is a particularly old district with the centre lines of some streets dating back to Roman times. The entire area underwent substantial construction activity after the papal court was transferred from the Lateran to the Vatican. A consequence of that growth was that the market was moved to the vicinity of Piazza Navona in 1477. The area was also an important centre for receiving pilgrims visiting the Vatican and had many inns and hotels. In the fifteenth and sixteenth centuries there was further construction activity, but the street structure remained essentially unchanged.

This quarter was selected for study because part of it was influenced greatly by city planning developments over the last century. Beginning with the Town Planning Scheme of 1873, the district was subject to a series of clearance and restructuring projects resulting from a reorganisation of the historic centre's street network. The debate reached its conclusion with the Detailed Plan of 1933, which provided for almost the total demolition and reconstruction of ten buildings in the district. However, the demolition work was not carried out and in the 1950s the debate was reopened in an effort to protect Tor di Nona. This protection was sanctioned in the 1957 variation to the Detailed Plan. A new project was launched in the middle of the 1970s, and the first restorations were completed ten years later. Four periods for examination were identified according to

four phases of innovation, 1870-1933, 1933-1957, 1957- 1976 and
1976-1988.

Figure 4.3. *Tor di Nona*

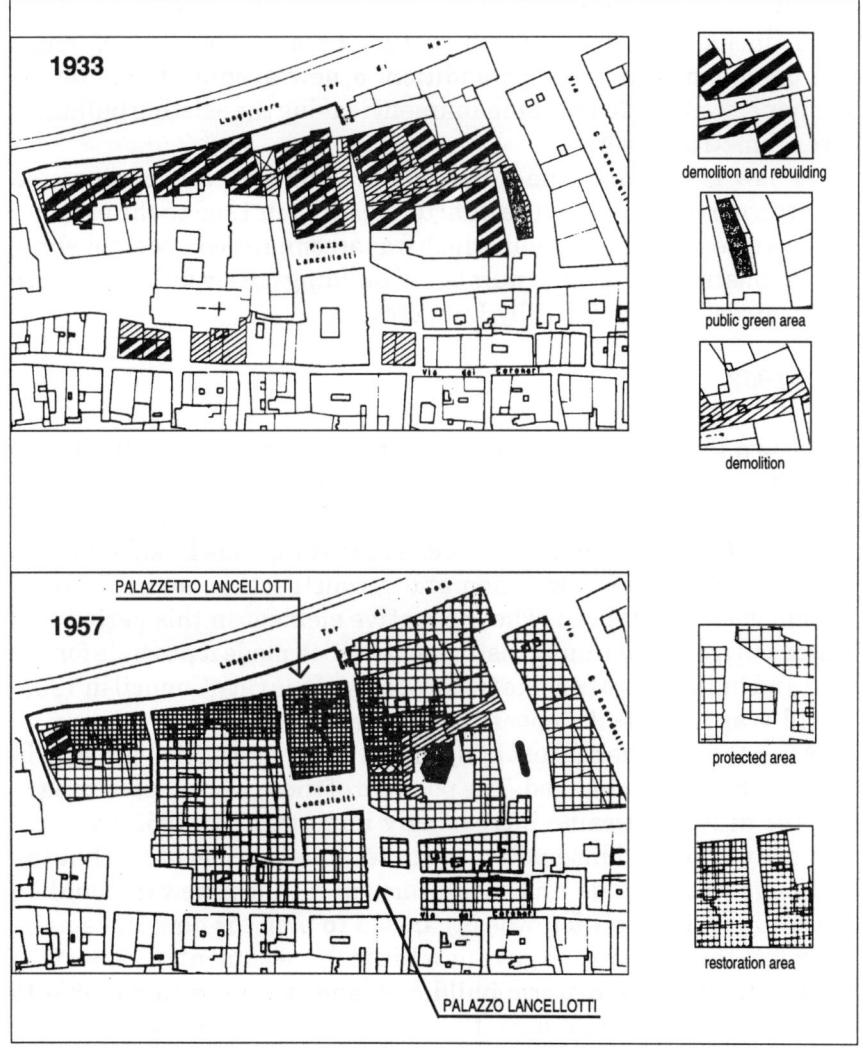

1870-1933

The innovation of this period was the construction of embankments to contain the waters of the River Tiber, work which began immediately after the city became the capital of the Kingdom of Italy. The river overflowed its banks regularly and the walls of buildings in the area still bear plaques indicating the levels reached by the water on occasions of exceptional flooding. This construction required the demolition of all buildings between the Via di Tor di Nona and the Tiber, including a seventeenth-century theatre, the Apollo, which was demolished in 1889. In addition, a new avenue, Lungotevere Torinona, was built on the embankment leaving the existing buildings several metres lower.

A new bridge over the Tiber, Ponte Umberto, was constructed near the northern part of the quarter, leading in 1906 to the opening of a new street, Via Zanardelli, which was accomplished by demolishing and reconstructing a large number of buildings. New buildings were at the street level defined by Lungotevere.

1933-1957

The Town Planning Scheme of 1931 provided for the demolition and reconstruction of all ten buildings. In 1933 a Detailed Executive Plan for the redevelopment of Tor di Nona was produced. On the basis of this plan Rome City Council proceeded to expropriate land, buildings were designated for demolition and reconstruction, and the street system was rationalised. The innovative element in this period was the executive city planning instrument, which made it possible for the land and buildings to become the property of the City Council in 1939. World War II made it impossible to carry out the plan.

After the war the context of the debate over demolition projects in the historic centre changed. The position of those who wanted to save the Tor di Nona became increasingly important. In 1955 the local authority appeared insensitive to these signals by deciding that the buildings were unsafe and insanitary, and that renewal would be necessary. A competition was organised to plan for new office uses. However, the competition gave new life to the arguments of those who wanted to save the historic buildings, and it proved impossible to award the first prize because the submissions were so mediocre.

1957-1976

The proponents of demolition did not give in easily and in a surprise move in 1957 the buildings of Tor di Nona were emptied with the aid of the police and the inhabitants transferred to a new district about twenty kilometres from the centre of the city. Consequently, the quarter remained uninhabited for about twenty years with the windows of its buildings walled up.

Innovation in this period was in the form of a careful study and subsequent restoration plan produced by a commission headed by the superintendent of the monuments of Rome. Surprisingly, the commission found that the buildings were not in a state of collapse and that once restored they would be quite safe to live in. The Variation of 1957 to the Detailed Plan of 1933 provided for the protection and restoration of the entire area. However, this only served to prevent the demolition of the quarter, since the local authority lacked both the will and funds to carry out the restoration.

1976-1988

The City Council which took office after the elections of 1976 resulted from a broad debate which had involved the city's entire community, and had included the protection and restoration of the historic centre as one of its most important elements. One of the new council's first initiatives was to start work on an area plan for Tor di Nona. Historic analysis and archive material were used to identify typological categories, then differentiated maintenance, restoration and restructuring interventions were developed in relation to these categories. Finance was assured by using funds and standards for low cost and municipal housing. This allowed for the inclusion of small workshops and apartments for families with limited incomes who already lived in the historic centre. The first building contracts were awarded in 1978, for the restoration of 111 residences and 42 workshops, with a total investment value of four billion lire, at 1978 prices. The work took longer than expected, partly due to administrative difficulties, partly because the municipal council which took office in 1981 did not assign the same importance to this quarter despite having the same political constitution as its predecessor. This meant that the first apartments were not handed over for occupation until the second half of the 1980s.

Level 7: Plots and buildings, Palazzetto Lancellotti

Palazzetto Lancellotti is a block of twelve buildings, each served by its own stairwell, with a seventeenth-century facade (Figure 4.4). From the outside it appears like a single building, but in fact the block is composed of twelve 'palazzetti' or small buildings constructed and expanded in different periods beginning in the fifteenth century. The seventeenth century facade was constructed to satisfy aesthetic and ornamental needs when the Lancellotti family decided to build a piazza in front of its own palace, which was opposite, and so demolished part of the palazzetto to create the necessary space. There is no other precise information on the morphological transformations of the building. It was used for residential purposes until 1957 when it was emptied because it was considered unsafe. The most significant innovative elements in this century occurred in 1957 and 1978.

1870-1957

The organisation of space showed the presence of twelve building units which were presumably constant for the whole period except for minor, superfluous additions.

1957-1976

The owner, Rome City Council, emptied the buildings in 1957. The administration wanted the buildings empty since it would be easier to start demolition work. In the same year a restoration plan was produced in an effort to demonstrate that the building was still structurally sound and that the spatial organisation remained valid. This plan constituted an innovative element because it obliged the people in the government to contend with a concrete solution for which the costs and benefits could be calculated. The plan proposed a total reorganisation of the interior space while saving the exterior. Some load bearing structures were preserved but the way the building functioned was changed. Only eight of the twelve stairwells were preserved, resulting in twenty one largo apartments of between 200 and 300 square metres. The plan was never put into effect but it served in an excellent manner to protect the building.

1976-1988

The preparation of a restoration plan ordered by Rome City Council in 1976 used the methodological results from Europe's Year of Architectural Heritage and the results of the national and international

Figure 4.4. *Palazzetto Lancellotti*

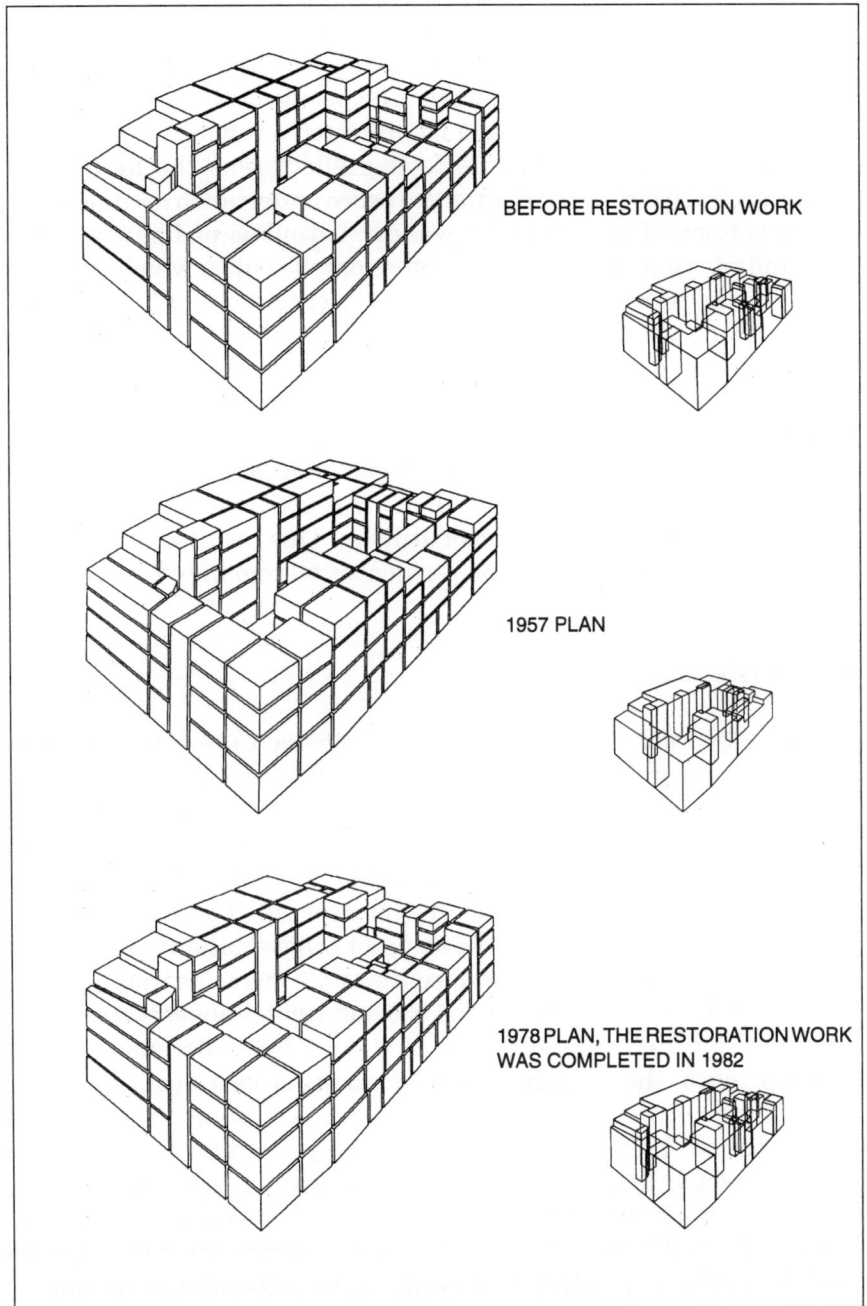

BEFORE RESTORATION WORK

1957 PLAN

1978 PLAN, THE RESTORATION WORK
WAS COMPLETED IN 1982

debate over Bologna's experience. In presenting the first plan for the historic centre in July 1977, the Mayor of Rome, art historian C.G. Argan, declared that, '...culture is not saving the historic centre, but saving it in its entirety: that is, respecting not only the houses and the streets, but also those who inhabit them. It is necessary to have an anthropological concept of culture. The people of the historic centre are a cultural asset and should be protected as such...'.

The restoration of Palazzetto Lancellotti was the first public restoration project for residential purposes undertaken in the historic centre of Rome. The restoration was carried out using finance for low cost and municipal housing, therefore only families which satisfied the criteria established for this type of intervention could live there. The contracts were awarded in 1978 totalling 1.1 billion lire. The plan respected the building's typological characteristics permitting the restoration of forty six apartments, fourteen workshops, a social centre for the elderly and a community centre. The work was completed in 1982, but the low level of interest in the project demonstrated by the council then in office prolonged the final formalities and the connection of utilities. This resulted in a delay of another three years before the residences were handed over.

Conclusions

In the first period, 1870-1924, the innovations at the level of the whole historic centre were essentially financial. Massive investments, some of a speculative nature, by national and foreign groups transformed the morphological structure of a small city into that of the capital of a great new national state. In the second period, 1924-1950, changes in the morphological structure were due above all to political motives. The 'Romanness' of the era of the emperor Augustus had to be brought out at all costs in order to enhance the city's image as the capital of the new Italian empire. Between 1950 and 1962 changes were caused above all by the transformation to a democratic decision-making process, with the morphological transformations resulting from bargaining between the various components of Roman society. The fourth period, 1962-1978, was dominated by the gentrification of the historic centre and its increasing dedication to tertiary economic activities, another instance of socio-economic innovation. The final period, 1978-1988, was dominated entirely by environmental concerns and demands for a better quality of life, by cultural innovations.

Within the Tor di Nona quarter, technological innovations had a major effect on the morphological structure between 1870 and 1933. Following this, in the period 1933-1957, a city planning instrument was produced which resulted in a change from private to public ownership of property and laid the basis for a radical transformation of the quarter. The dominant feature of the third period was the fact that historic buildings were not demolished due to cultural bargaining involving the entire city. During the last period, 1976-1988, a political event was linked with a new cultural attitude of international value which led to the development of financial and technical instruments for the rehabilitation of the quarter.

At the level of individual buildings, transformations were more subtle and more identifiable with planning instruments. This was the case with Palazzetto Lancellotti in 1957-1978, when they led to inaction, and in the period 1978-1988 when they effectively resulted in restoration activity.

References

AA.VV. (1984), *Architettura e urbanistica. Uso e trasformazione della citta' storica (Architecture and town planning. Use and transformation of the historic city)*, Marsilio, Venezia.

AA.VV. (1985), *I Ministeri di Roma Capitale: l'insediamento degli uffici e la costruzione delle nuove sedi (The ministries of the Capital Rome: office settlement and the building of new offices)*, Marsilio, Venezia.

Almagià, R. (1958), Note sul più recente incremento della Città e del Comune di Roma (Note on the more recent growth of the city and Comune of Rome), *Rivista Geografica Italiana*, 65 (3).

Benevolo, L. (1971), *Roma da ieri a domani (Rome from yesterday to tomorrow)*, Laterza, Bari.

Caracciolo, A. (1956), *Roma capitale (Rome capital city)*, Editori Riuniti, Roma.

Figà-Talamanca M. (ed.) (1960), *Roma, popolazione e territorio dal 1860 al 1960 (Rome, population and territory, 1860-1960)*, Comune di Roma, Roma.

Insolera, I. (1971), *Roma moderna: un secolo di storia urbanistica (Modern Rome: a century of urban history)*, Einaudi, Torino.

Insolera, I. (1980), *Le città nella storia d'Italia: Roma (Towns in Italian history: Rome)*, Laterza, Bari.

Montanari, A. (ed.), (1974), *Roma sbagliata: le consequenze sul centro storico (Mistakes of Rome: the consequences for the historic centre)*, Bulzoni, Roma.

Montanari, A. (ed.), (1976), *Roma centro storico: 1924-76 (Rome's historic centre: l924-76)*, Italia Nostra, Roma.

Montanari, A. (1979), 'Il centro storico di Roma tra spreco e speculatione' (The historic centre of Rome between waste and speculation), in *AA.VV. Il San Michele*, De Luca, Roma.

Montanari, A. and Petraroia, P. (eds) (1989), *Città inquinata: i monumenti (Polluted Rome: the monuments)*, Istituto Poligrafico e Zecca dello Stato, Roma.

Seronde - Babonaux, A. M. (1980), *De l'urbs à la ville, Rome: Croissance d'une capitale*, Édisud, Mondes Méditerranéens.

5. LIVERPOOL

Bill Chandler, Helen Fitton, Leslie Forsyth, David Massey,
Department of Civic Design, University of Liverpool, England.

Introduction

The method used to understand the relationship between built form, environment, land use and innovation is based on historical investigation concentrating on the last hundred years. In order to set the study in context the early development of the metropolitan area and city is described in Levels 1 and 2. The remaining levels focus only on the last century. Within each level the periods selected for investigation vary according to the important stages of development. However, the approach is consistent. A brief history of the time period is described before changes in the urban form are noted. This is followed by suggestions as to the likely causes of these changes. Finally, in the Conclusion, the influence of innovation is considered. The source material for most of the work was Ordinance Survey maps of the areas, amplified by the publications listed in the Bibliography, and occasional site visits.

Level 1: Metropolitan area, Merseyside

The county of Merseyside lies in north west England, its coastline defined by three river estuaries, the Ribble, Mersey and Dee. It has an equable climate and unspectacular landscape which contains substantial areas of high quality farm land as well as mud flats and salt marshes which constitute one of the most extensive wilderness areas in lowland Britain.

Chester, on the River Dee, was the regional centre and major port from Roman times until the fifteenth century. However, the River Dee had a weak tidal scour and Chester Haven was abandoned as the port facilities migrated downstream as far as Parkgate. Liverpool, on the east bank of the Mersey, gained its independence from the port of Chester in 1660 and began to develop the trading links which would make it the port of Merseyside.

1660-1825

By 1715 Liverpool had built its first dock with a further six to follow. During this same period communication with the hinterland was improved by river navigation, canals and turnpike routes. Steam ferries across the Mersey helped to establish settlements on the Wirral shore.

1825-1886

The first docks on the west bank of the Mersey were built by the mid-1840s but were merged with Liverpool's more established and extensive system in 1858 under the Mersey Docks and Harbour Board. Pioneering rail links to Manchester, the North West region and beyond were established between 1830 and 1840. Birkenhead, Tranmere and Wallasey underwent dramatic growth in the last thirty years of this period. The completion of the Mersey railway tunnel in 1886 opened the north shore of the Wirral for suburbanisation.

Morphological changes The structure of the area had grown into a densely built up group of towns along both shores of the mid-Mersey estuary with a commercial and retail core in Liverpool. Peripheral villages and outer townships began to develop a suburban character. Some were linked to the core by rail. Small industrial settlements lay inland round St. Helens, in the south east at Runcorn and Widnes.

Probable causes A growing inland railway network combined with the extension of the port's oceanic trade routes led to the continued extension of the docks system and its related employment. Port-based warehouses, services and industries grew alongside the docks and the labour force lived immediately adjacent. The wealthier classes moved away from the unhealthy environment of the core areas.

1886-1940

Boundary extensions to the Borough of Liverpool took place in 1895, 1901, 1905 and 1913. The influence of the railways, which had already drawn Southport and Ormskirk into the commuter hinterland of Liverpool, brought rapid growth in the Wirral. Electrification of the lines between 1903 and 1938 enhanced the growth of these towns. The Manchester Ship Canal, Mersey Road Tunnel and river channel embankments were among the series of infrastructure investments made during this period. To the north of Liverpool the East Lancashire Road provided a link to the industrial hinterland and drew St. Helens into closer contact with the metropolitan core. In the south of the area Port Sunlight provided a growth industry and 'model village', while oil refining began at Stanlow.

Morphological changes The periphery of the continuous built up area was extended on both sides of the Mersey estuary, especially in low-density housing estates. Fingers of suburban development also reached out from the core towards the towns in the rest of the metropolitan area, such as St. Helens, Widnes, Southport, Ormskirk and along the north shore of the Wirral peninsula.

Probable causes New rings of residential and industrial expansion from the core were the result of slum clearance and new industries. Increasing possibilities for commuting generated the extensions to the dormitory suburbs. The areas of expansion tended to evolve along the routes of the electrified suburban railway network.

1940-1986

After extensive post-war reconstruction the docks system enjoyed a phase of prosperity until the late 1960s, when a container port was constructed to meet the new ship design and cargo handling requirements. The new facility failed to attract the amount of trade which had previously come to the port and together with the loss of passenger traffic, the result was a major loss in employment in the docks and in port-related industries. The whole of the south docks were closed in the early 1970s.

During the 1960s slum clearance programmes took place adjacent to the central core of Liverpool, in Birkenhead and St. Helens (Figure 5.1.). In the 1970s this was replaced by rehabilitation and improvement to the existing stock. New peripheral local authority estates were developed until 1967, but little building for owner occupation took place except in the outer commuter suburbs and St. Helens. New free standing settlements were developed at Kirkby during the 1950s and 1960s and Skelmersdale and Runcorn from the 1960s. Redevelopment schemes strengthened the regional sub-centre functions of Birkenhead, St. Helens and Southport.

From the late 1960s to the early 1980s a sub-regional motorway system, including a second road tunnel, was constructed. Other transport developments included a new airport runway and an underground rail loop in the centre of Liverpool.

In 1972 the Metropolitan County of Merseyside was designated excluding Widnes, Runcorn and South Wirral, but by 1986 the County Council, with its strategic planning and transportation authority, had been abolished.

Morphological changes Peripheral development extended the continuous core, but the inner areas became a patchwork of postwar rebuilding, clearance, dereliction and rehabilitation. The 'spokes' of development were reinforced by the construction of planned new towns and the expansion of dormitory suburbs. A new motorway system overlaid the existing radial development. A formal green belt was designated to contain the outward growth of the continuous built-up area and other towns.

Probable causes Slum clearance and industrial relocation accounted for the continuing peripheral development. Obsolete docklands, industry and railway sidings occupied large tracts of the inner areas. A new agency, the Merseyside Development Corporation took a number of initiatives to regenerate the docklands. The investment potential of residential property and improved road and rail links provided the impetus for continuing suburban expansion.

Figure 5.1. *Planning policies since 1945*

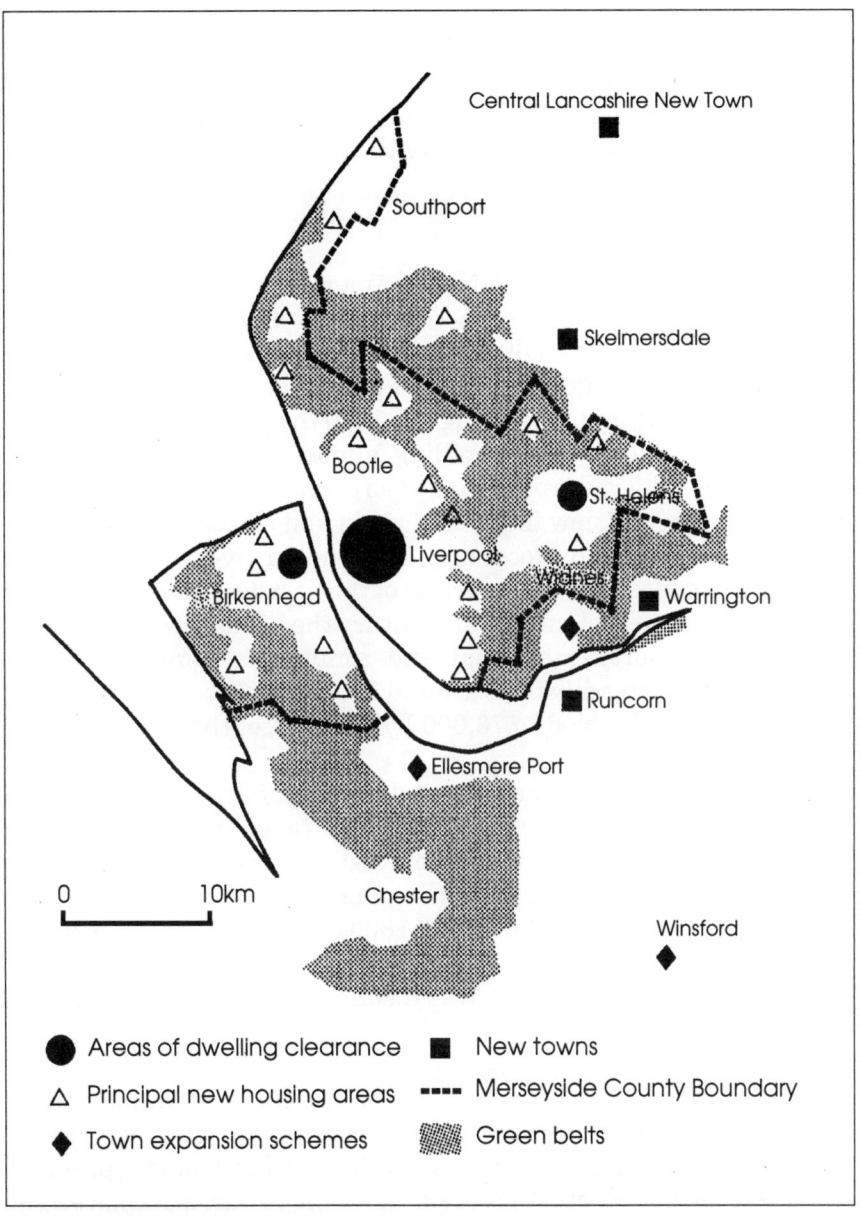

Source: Gould, W.T. and Hodgkiss, A.G., 1983, eds., The Resources of Merseyside, Liverpool
University Press, Liverpool

Level 2: Core-continuous built-up area, Liverpool

Liverpool is the regional centre for the county of Merseyside. The Borough of Liverpool received its Royal Charter in 1207 but it was not until 1660, when the port became independent from Chester, that it began to develop the trade links and port infrastructure which was to form the basis for its future growth.

1660-1825

New trade was opened with North America and the Caribbean in addition to existing links with West Africa and Northern Europe. Exports of textiles and salt and imports of sugar and tobacco then cotton provided the impetus for the port-related industries of ship building, refining and processing. By 1800 the port led Europe in the West African slave trade handling almost three quarters of the vessels involved. Trade routes had been established with Central and South America and new imports of grain and oil seed continued the growth of refining and processing. Demand for river anchorage and wharf space caused the construction of the first dock in 1715 followed by a six dock system. Later in the century the Leeds-Liverpool Canal extended the port's industrial and commercial hinterland. The population rose steadily from around 1000 in 1660 to 34,000 in 1773 then more than doubled to 76,000 by the turn of the century.

Morphological changes The medieval core expanded to the south and a series of suburban developments had taken place in outlying villages by 1800. Within the core low cost housing in enclosed courts or insanitary back to back properties were crammed alongside often noxious industries. In the early 1800s new suburban Georgian squares and grids were laid out and partially built to the south west of the core. The river frontage was reclaimed as the dock system was constructed.

Probable causes Expansion was due to the increase in population caused by migration from all areas of the British Isles. Within the confines of the medieval boundary the form of development provided the cheapest and densest solution adjacent to the workplace. The planned, residential expansion for the wealthier classes allowed them to escape the unhealthy core. Docks were essential to provide for the numbers of ships since the Mersey had a fast running and deep tidal range.

1825-1860

Trade continued to flourish with the total reaching 4.4 million tonnes. There was a corresponding growth in servicing for shipping and trading which increased the dominance of the port in the employment market. The establishment of the Liverpool-Manchester railway in 1830 provided the basis for the development of passenger and frieght rail links to the North West region and by the mid-1840s to a national network of routes. Continued immigration, inflated by the effects of the Irish potato famine, increased the population to 376,000 by 1860.

Morphological changes Twenty one docks and 16 km. of quays were built during this period under the direction of Jesse Hartley, including the Albert Dock. Despite topographical difficulties an extensive system of railway lines and marshalling yards to serve the docks was begun. Processing and refining industries began progressively to displace older industries from the waterfront. The built-up area continued to expand behind the docks causing the Borough Boundary to be extended in 1835. By 1840 the former retailing area in the medieval core had been taken over by merchant-related activities and a new shopping axis created to the south. The range and quality of the shops indicated the emergence of the city as a regional centre. The first belt of parks which ring mid-Victorian Liverpool was laid out, with land for villa developments laid out around them. Other, non-public, residential park estates were developed.

Probable causes Expansion of the docks and the built-up area was essential to satisfy the demand from increased trade and immigration. The resulting increase in merchant activity in the central core and Liverpool's emerging wealthy elite and regional significance provided the basis for a new shopping area. Processing and refining benefited from close proximity to the discharge point of their raw materials. The extent of the built-up area, the amount of overcrowding and pollution were instrumental in the creation of public parks since there were no ancient parks or open areas. The wealthier classes were able to occupy houses around the parks or moved outside the built-up area.

Port tonnage increased to 19 million by 1911. Exports were primarily cotton goods, other textiles and engineering products. Imports, although less than exports, had a much more profound effect. Wheat and oil seed continued the move to port processing but, manufacturing was limited, creating a narrow industrial base with employment vulnerable to the fluctuating cycles regulated by the port trade. The population rose to around 750,000, mostly attracted by the opportunities for work. But much of the workforce was poorly paid and the nature of port employment resulted in a sharp division between the mass of the people and a rich elite.

Morphological changes Liverpool's physical growth was reflected in the pattern of residential and industrial development, the completion of the railway network in the late nineteenth century and the northwards extension of the dock system to include three new basins in the second decade of the twentieth century. Garston docks were also built in this period, well south of the core-continuous built-up area. The industrial area behind the docks was considerably expanded and industry also developed along the Leeds-Liverpool canal and beside major rail routes and sidings on the eastern and northern fringes of the built-up area. The residential area expanded beyond parks such as Stanley and Newsham, and reached the line of Queens Drive, Liverpool's major circumferential road built in the early twentieth century. The axes of building followed the lines of the railways which provided the main means of commuting, except for the wealthiest class.

As well as expansion on the edge of the built-up area (Figure 5.2.), considerable amounts of infilling of previously open land occurred within the 1885 boundary during these decades. Between 1864 and 1914 10,000 houses were demolished. The first municipal flats in the UK were built in 1869. Two storey terraced housing regulated by bye-laws blanketed the residential areas. Wavertree Garden Suburb was begun in 1911. Right at the end of this period, Liverpool's first peripheral municipal estate was begun.

Soon after the turn of the century two significant developments took place in Liverpool's inner area, University College, founded in the late nineteenth century, became an independent university, and the construction of a large Anglican cathedral was started. The decade between 1910 and 1920 was one of little change.

Figure 5.2. *Core-continuous built-up area: stages of development*

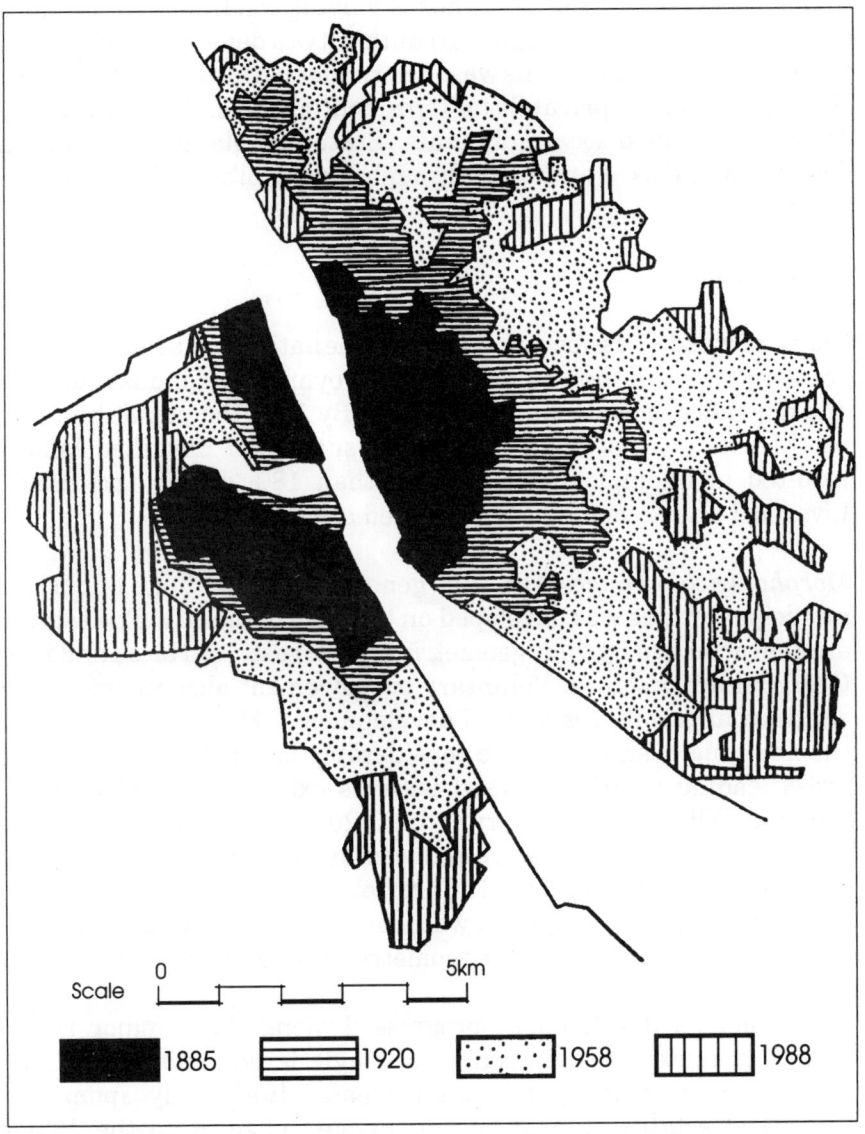

Scale 0 — 5km

■ 1885 ☰ 1920 ∴ 1958 ⦀ 1988

Probable causes Expansion of the built-up area reflected the doubling of the population and the continuing port and port-related economic activity. The pattern of the development followed transport routes giving access to the centre. Unhealthy living conditions in the older housing areas prompted the local authority to demolish and rebuild some of the worst areas, this was then reinforced by the Public Health Act of 1875 for new, privately built residential areas. The Garden City Movement gained local adherents. Statutory planning became a function of the local authority through Liverpool's 1907 Act and the 1909 Act.

1920-1940

The interwar depression and changes in the national and international economy revealed the city's narrow employment and industrial base and dependence on declining industry. By 1936 the population had reached its peak of 867,000, but the surplus of unskilled labour resulted in unemployment of more than 18% during the 1930s. Liverpool was accepted as a depressed area.

Morphological changes The first generation of the huge suburban municipal estates was developed on the edge of the city. The estates are characterised by their geometric layout, rustic red brick and neo-Georgian architecture. Voluntary outmovement also occurred into substantial private residential developments. Housing densities in the new ring of suburbs were considerably lower than in the inner areas leading to urban sprawl with its extensive use of land for building. Within the boundary of the 1920 built-up area, although in some districts densities were reduced by slum clearance, in other parts intensification occurred as previously open sites were built on. The main form of development was four and five storey walk-up flats, each block having a distinctive geometric plan surrounding a common open area.

Industrial development progressed along these major radial transport routes by the opening of single large factories and the expansion of existing industrial areas. Municipally sponsored industrial estates at Aintree and Speke, adjacent to the 1930's airport, had just begun to be developed when this period of expansion was brought to a temporary halt by the outbreak of war in 1939.

The highway network was still developing at this time. Particularly significant was the East Lancashire Road, linking Liverpool with Manchester, which opened in 1933.

Probable causes Commuting by public and private road transport was increasing reflecting the growing desire and opportunity for low density, single family dwellings in the suburbs. Low density peripheral estates were a response to the overcrowded inner city slums. Owner-occupation became a more widely available tenure and the layouts were poor imitations based on Unwin's ideas. The large blocks of inner area flats were strongly influenced by European movements in architecture. Although facing difficulties in the coordinated building of social and community facilities on its outer estates, the city council was aware of the need for a more integrated approach to housing and planning by the late 1920s, including sub-regional planning, industrial diversification and planned satellite towns.

1940-1958

The immediate post-war revival was short lived since the long-term structural weaknesses in the economic base offset any expansion. In the 1950s central government policies brought some new manufacturing jobs, but other closures and the continuing loss of jobs in the port cancelled them out. The population dropped to 750,000 by the end of the period, mainly through overspill arrangements with towns in the surrounding region.

Morphological changes During World War II parts of the dock system, the city centre and some inner areas of housing were destroyed by heavy bombing. Other buildings deteriorated through neglect. A new programme of housebuilding was undertaken from 1945 onwards, involving both reconstruction in the central areas and the continued addition of new peripheral public housing estates, generally in tandem with new industrial estates. The planned satellite housing estate at Speke was completed and substantial areas of private housing were built on the fringe, especially to the south-east of the city, towards the latter end of the period.

Very large industrial estates grew up, and in some areas a single large factory was built adjacent to a new housing estate. Industrial expansion was not limited to the periphery, but also occurred within the existing built-up area along the major lines of communications such as the Leeds-Liverpool canal, the main east-west railway line and the East Lancashire Road. The railway network began to decline with the closure of a number of stations, but the highway network was upgraded and started to carry ever increasing volumes of traffic.

The beginnings of a change in retailing could be glimpsed in this period with the construction of a large shopping centre in the north of the continuous built-up area. Previously, retailing outside the city centre was undertaken in rows of shops along the main radial thoroughfares, in local centres and small parades dating from the 1930s.

Probable causes New industries did not require a central or port-related location and moved close to their labour force. Merseyside was designated as a Development Area in 1949. Car ownership and frieght movements rose. The immediate post-war limitation of building rights to local authorities was lifted encouraging speculative housing development from the mid-1950s. This was further enhanced by mortgage availability and tax relief.

1958-1970

During the 1960s two branches of the motor car industry were established in the south of the continuous built-up area, making manufacturing a growth sector and restoring a sense of optimism. However, the docks continued to decline despite the opening of a new container terminal. The registered dock labour force fell from 23,000 in 1963 to less than 10,000 in 1972. Population fell to 611,000 in 1971 as overspill arrangements, including the designation of new towns at Skelmersdale in 1961 and Runcorn in 1964, were implemented.

Morphological changes Post-war rebuilding in the city centre, which had originally been slow, took on new forms with major retail and office projects. Slum clearance, encompassing large areas of inner Liverpool from the early 1960s, led to greatly reduced housing densities as open space standards were applied. This was the era of system-built concrete construction in the form of large tower and deck-access slab blocks of flats. Some parts of the city, such as Everton just north of the centre, became totally dominated by this type of dwelling set in a new form of city park. Falling densities in the inner city meant that the overspill population had to be moved elsewhere, some into new towns well beyond the built-up area boundary, but others into the last major public housing estates built on the periphery.

Falling densities were also occurring in industrial estates, where the extent of the inner industrial belts remained very similar, but large numbers of businesses within them either relocated to the periphery or, more commonly, closed altogether. The trend of industrial decentralisation was intensified by the large-scale expansion of the existing industrial areas in the 1960s, most notably through the establishment of vast plants by the motor industry at Halewood and on the Speke Estate.

The 1960s was also a decade of major change in transport infrastructure as large stretches of the railway network became disused and motorway construction began on the eastern fringes of the built-up area. A new, long runway was built to serve Speke airport.

Probable causes Regional policy forced the car industry to come to Merseyside. New standards for housing layouts reduced densities in clearance areas. New construction technology and central government financial incentives dictated the high-rise form of replacement housing. Car ownership continued to increase.

1970-1985

The optimism of the previous decade was not sustained. Between 1966 and 1977 40,000 jobs were lost as branch plants closed or moved elsewhere. By 1985 the rate of unemployment was 27%, double the national average, and the port workforce was reduced to 3,000. In the late 1970s almost 70% of all jobs were in the service sector and almost one half of these in the public sector. By 1981 the population had dropped to 511,000.

Morphological changes New parking structures and an inner ring road in the city centre represented the partial fulfilment of ambitious plans from the mid-1960s. Within the centre, cleared areas for office development remained undeveloped. In the first half of the 1970s local authority housing continued to be built on slum clearance sites within the built up area. From the mid-1970s the slum clearance programme was wound down and replaced by phased, small area rehabilitation programmes. During the 1980s blocks of flats from the 1930s and 1960s were demolished and replaced by semi-detached houses laid out in traditional street patterns. A major private housing estate began to be constructed on the eastern periphery. A system of culs-de-sac with primarily two storey semi-detached houses was introduced.

The motorway network was considerably extended around the city, but the railways continued to decline, although further planned electrification and city-centre tunnelling schemes facilitated suburban commuting. The docks south of the city centre were closed in the early 1970s, but in the early 1980s the regeneration process was illustrated by the International Garden Festival site, the Albert Dock complex and the Brunswick Business Park. Some docks had been filled in, others had been dredged and opened for recreation.

Out of town retail warehouse parks had been developed on former industrial estates in the south of the city.

Probable causes Central government set up the Merseyside Development Corporation to regenerate the south docks. The City Council abandoned its terraced-areas slum clearance programme and established an Urban Regeneration Strategy to improve the overall environment in areas of the worst municipal housing. New retailing requirements and shopping patterns were suited to industrial building forms with large car parks.

Level 3A: Core-centre

The definition of the core area was based on the study by Muir, 1910, City Centre Plan studies and reports, 1965 to 1987, and the work of Masser, 1970. The definition accepted by Masser and the boundaries chosen follow the technique devised by Murphy and Vance, 1954. The correlation between the areas identified and the uses distinguished in the Muir study are sufficiently close to indicate growth patterns over time.

1880-1920

Existing central area functions were strongly established with commercial activities in the old medieval core and shopping to the south east. Although port and trade-related activities were still dominant, there was a steady growth in banking and insurance, much of it also port related. Public buildings, administrative activities and transport terminals delineated a transition zone along with housing and institutional functions.

Morphological changes George's Dock was filled in and the three major office blocks which form Pierhead were constructed. Older storage structures and residential streets on the periphery were converted to commercial uses. Rebuilding took place on existing sites or consolidated subdivisions. The basic road pattern remained largely unaltered.

Probable causes The predominance of the port and related activity generated the demand for trade, shipping then financial services. Wealthy merchants and institutions commissioned architects to design buildings of 'significance', especially in the office quarter.

1920-1939

The economic recession slowed all new building work, although new commercial, traffic and housing requirements lead to considerable changes.

Morphological changes A number of very large new bank and shipping company offices were built. The new road tunnel under the Mersey discharged at two points within the core requiring new road links. New four storey walk up flats were built on the edge of the area to designs which created new, larger linear blocks in strict geometric patterns.

Probable causes Increase in frieght traffic, car ownership and commuting. European architectural influence on housing layout and the 1930's slum clearance programme.

1939-1950

War damage led to the clearance of large sectors of the core area, but post-war restrictions on building meant that little new development took place for most of this period.

1950-1970

Although some rebuilding took place on the site of war-damaged or destroyed properties, the first major proposal did not arrive until the 1958 St. John's Market scheme. In the mid-1960s this led to a comprehensive redevelopment plan for the city centre, including a new ring road, multi-storey parking, vertical and horizontal pedestrian segregation, and a conservation area for the best nineteenth century offices.

Morphological changes Throughout the area a series of individual plots, and a number of medium-sized accumulations of between two and three plots were cleared or redeveloped. Two large sites immediately north and south of the core were cleared but remained empty. Most redevelopment resulted in higher buildings and new architectural forms. A college, school and tower block were built on the north west periphery, over the street pattern and ignoring the surrounding texture. St. Johns Market was redeveloped in 1970 enclosing two city blocks and including a multi-storey car park and hotel. The first pedestrianisation scheme was completed. A flyover and road improvements took place at the tunnel entrance and exit.

Probable causes The continuing strength of city-centre retailing activity attracted proposals for shopping area renewal although there was not the demand for new office development as in London and Birmingham. Land parcels were accumulated through speculation and planned compulsory purchase for future comprehensive redevelopment. Vehicular servicing and controlled environments created new forms of retailing development. Increases in car ownership and commuting caused congestion.

1970-1988

In the 1970s the north west extension of the office area was completed. Refurbishment of larger premises took place in the main shopping street. Opportunities for redevelopment were created by the closure of two surface level railway stations. New road construction was begun as part of the proposed ring road. Since 1980 there has been sporadic refurbishment of existing major office buildings and the development of new offices and premises for Central Government agencies. Three major shopping developments have been built, one in conjunction with offices. On the eastern edge the Albert Dock has been rehabilitated for retail, commercial, cultural and residential uses.

Morphological changes The larger building complexes on the periphery of the central area have modified the texture of the built form through land coverage, height and architectural form. Older areas in the centre of the core have, for the most part, retained the nineteenth century plot division and scale. Shopping has been consolidated to the east of the core and is virtually surrounded by new pedestrian streets and public spaces. A new axis has been created to the south through the regeneration of the docks.

Probable causes Effective conservation policies in the centre of the core. Evolving retail practices and shopping habits. The creation of the Merseyside Development Corporation.

Level 3B: Ring-centre, Maghull

Maghull lies 13km. from the centre of Liverpool. It is a dormitory suburb with a population around 30.000.

1895-1935

Prior to this period the original village had been connected to the regional transport network by the Southport road, the Leeds/Liverpool canal in 1774 and the Southport and Ormskirk/Preston railway in the mid to late nineteenth century. A new, three-lane road, Northway, was built in 1933 prompting the development of new residential estates in the south west. These were accompanied by the first secondary school, public open space and public utilities. Although agriculture remained the dominant land use, three large houses and their grounds were converted to specialist hospitals.

Morphological changes The elongated village and agricultural field pattern was extended in the south by a small development of detached Victorian villas and later by the first speculative housing estates of two storey, semi-detached houses with front and back gardens. The shape of the new developments respected the original patterns of country lanes. Special hospital uses secured large areas of private open space on an east-west axis across the village.

Probable causes Movement of population away from the city. Increased accessibility through an improved road network. Land prices allowed development at low densities, and financial institutions made mortgages available. The availability of suitable properties for providing institutional care in a healthy environment.

1935-1965

Two local authority housing estates were begun in the early 1950s, but the increase in population to 20.000 was the result of additional speculative housing. A smallholdings estate and two industrial estates were designated.

Morphological changes The form of residential development continued as low density semi-detached, close to the two main roads. Areas of playing fields and public open space punctuated the new estates. In the core of the built-up area there was a triangle of open space at the junction of the two main roads with a few of the original or older village properties on it.

Probable causes Inner city slum clearance reinforced the population movement to outlying-suburbs. Planning standards determined open space requirements in new housing estates. The core area was being held for the provision of central facilities and services. Post-war tax levels.

1965-1985

At the end of the 1960s Northway was improved to a dual carriageway and in the early 1980s the M58 link was opened. Speculative residential estates continued to be constructed on the remaining open land. Park Lane Special Hospital was constructed on the eastern edge of the town.

Morphological changes The core of the town had been consolidated in a triangle at the junction of the two main roads. New residential development illustrated moves away from semi-detached housing to small estates of courtyards with more informal road layouts. The physical barriers to expansion, the railway embankments, limited development to the north and east.

Probable causes The improved road network increased accessibility and removed a portion of through traffic. The promotion of owner occupation as the main housing tenure by government, the investment value of property and restrictions on speculative house building in Liverpool. Planning legislation was used to set aside sites for centralised community services. The relaxation of standards governing the layout of roads in urban areas and design guidance for residential areas.

Level 4: District, university and hospitals

The study area runs approximately north-south parallel to the River Mersey overlooking the city centre. It covers an area of 0.7 ha. (Figure 5.3.).

1890-1938

The University was founded during the late 1870s and early 1880s and established on the site of the former lunatic asylum, close to the teaching hospital associated with the medical faculty. During the 1920s middle-class families began to move to smaller houses in the suburbs leaving the late Georgian and Victorian town houses around Abercromby Square. The properties were divided into flats and some were acquired by the University.

Morphological changes University expansion on its original limited site demanded additional accommodation and the 'red brick' quadrangle was completed in 1903 on land conveyed by the City Council. In 1890 the new Royal Infirmary was completed on the site immediately to the north of the University. It provides the first indication of encroachment into the street pattern between the University and the Infirmary. Land to the south of the University, containing the Parish Workhouse and Fever Hospital was bought by the Archdiocese in the late 1920s. The buildings were partly demolished and a Roman Catholic Cathedral was commissioned to a design by Sir Edwin Lutyens.

Probable causes The location of the area is convenient for institutions which require access to the city centre but do not have to be in the core. The establishment of links between medical services and medical education and research encouraged proximity of location. Welfare services moved to income supplement and smaller in house care for the elderly making large workhouse institutions redundant. The

Figure 5.3. *Development of the University and hospitals district*

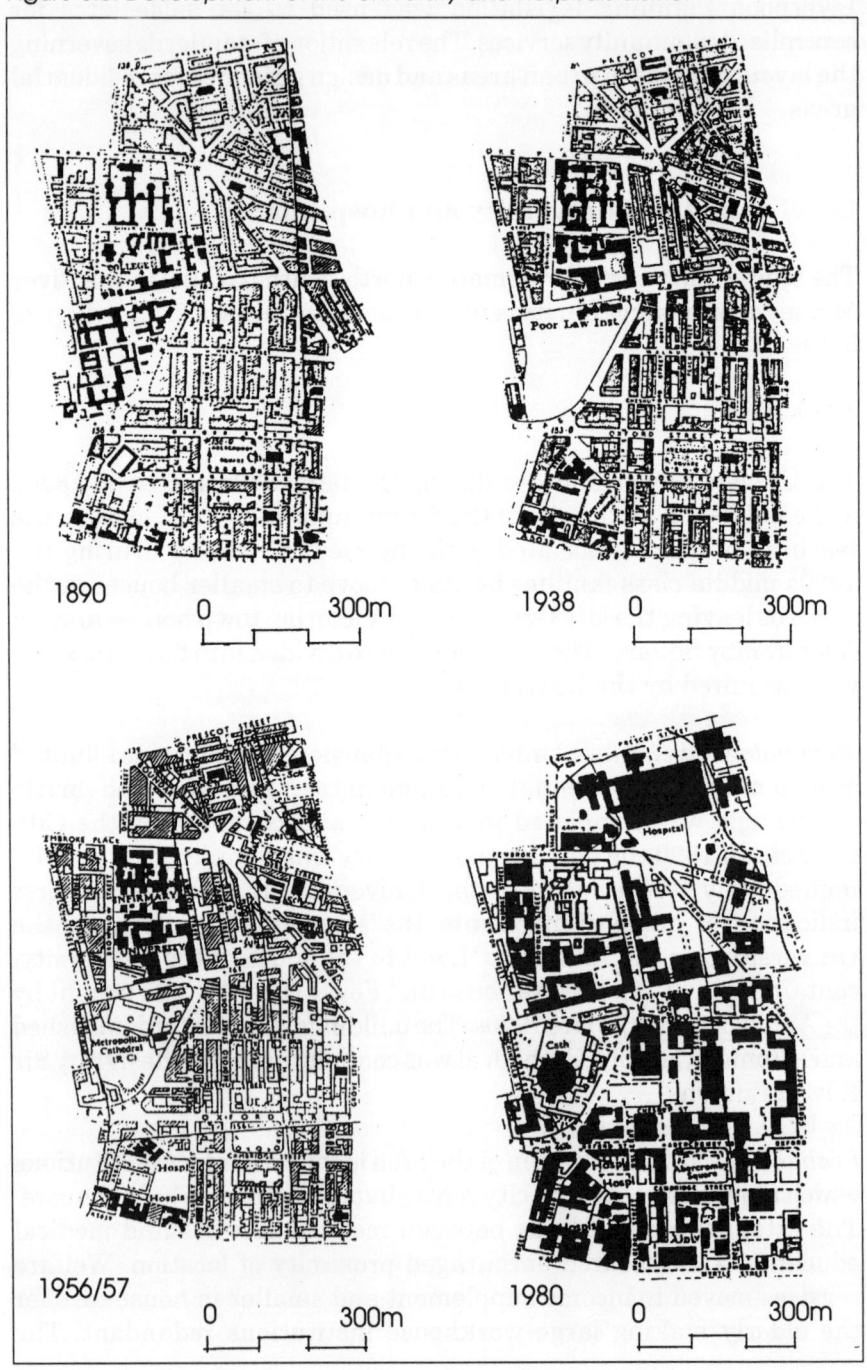

establishment of a substantial Roman Catholic population in the city required a significant centre for worship and the cathedral site dominated the city centre and competed with the neighbouring Anglican Cathedral. Central area residence ceased to be attractive for wealthy families.

1938-1957

Four voluntary hospitals were amalgated under one administration through the 1937 Local Government Act and by 1957 a decision had been reached to develop a new hospital on a single site. The University decided against a move to a peripheral site and began to prepare a comprehensive plan for future building under the guidance of (Sir) William Holford. It continued to accumulate property throughout the area. Liverpool City Council's 1947 Development Plan designated the area for university and hospital uses. Work on the Cathedral stopped at the level of the crypt in 1940.

Morphological changes Consolidation and infill around the existing blocks containing the University, Royal Infirmary and Roman Catholic Cathedral.

Probable causes Legislation uniting the hospitals ensured that a single site would be seen as advantageous, and the Hospital Board decided that it needed to remain in the district. The University rejected a move to the periphery since the potential site could not accommodate a hospital and university facilities. A joint approach to the City Council by the University and the Hospital Board resulted in the designation of the area for institutional uses. Little development was possible for the first half of this period due to World War II, but from 1947 the potential was evident.

1957-1980

The University's expansion dominated the land-use pattern from the Royal Infirmary southwards. New buildings proliferated, the majority constructed between 1960 and the mid-1970s. The new Royal Liverpool Hospital, opened in 1980, occupied the northern quarter of the district with a new Dental Hospital and Blood Transfusion Centre Immediately adjacent. The old Infirmary was closed. In 1967 the Metropolitan Cathedral was completed, built to a competition winning design by Sir Frederick Gibberd.

Morphological changes The period up to the mid-1970s illustrated the most significant changes. Substantial portions of the street pattern have been eradicated and the scale and grain of new development has increased. Single buildings occupy whole blocks or the major part of blocks with open space around them. Tower and slab blocks are the predominant building form. New outdoor spaces and courtyards have been created linked by pedestrian routes.

Probable causes The development of a deeper and broader range of academic studies and increased access to higher education demanded an increasing scale of land use and new types of building form. In-patient health care became centralised and more specialised requiring appropriate centralised accommodation. New, system-building techniques were introduced to house the specialised activities. Planning and urban design ideas promoted vehicular/pedestrian segregation and heavily landscaped, external, semi-public courtyards.

Conclusions

The methods used to categorise types of innovation are those suggested by Whitehand (1987). This approach defines four broad categories of innovation.

Construction:	building materials, legislation, technology.
Functional:	methods of manufacture and trading.
Transport:	as a very important subdivision of function.
Town planning:	as a form of public intervention.

In addition Whitehand stresses two other factors. The relationship between cycles of the economy and construction activity in causing change to built form, and the importance of the motive for development including the way it is brought into being, highlighting the role of the agents responsible.

The analysis of each level in Table 5.1 takes account of three components, economic cycle, type of innovation and development agency. The matrices which are used are not complete, but serve as a starting point for more detailed investigation. Time has not been considered in detail other than dating the period in which each innovation was perceived. If Schubert's 'Time-Phase Diagram of the

114

Table 5.1. *Innovation in each time period on each level*

LEVEL 1: METROPOLITAN AREA, MERSEYSIDE

DATE	ECONOMIC CYCLE	INNOVATION	DEVELOPMENT AGENT
1660 to 1825	Initial growth	TRANSPORT Canals Turnpike roads Stream ferries CONSTRUCTION Dock System	Liverpool Corporation
1825 to 1886	Growth	TRANSPORT Suburban railway	Railway companies
1886 to 1940	Growth then decline after 1919	TRANSPORT Rail electrification Road tunnel PLANNING Peripheral extension Model village FUNCTION Industrial location	Borough coucils Industrialist
1940 to 1986	Decline	TRANSPORT Motorways PLANNING New towns Green Belt Dockland regeneration	Central government Country Council Development Corporation

LEVEL 2: CORE-CONTINUOUS BUILT-UP AREA, LIVERPOOL

DATE	ECONOMIC CYCLE	INNOVATION	DEVELOPMENT AGENT
1660 to 1825	Initial growth	CONSTRUCTION Dock System PLANNING Georgian grid	Liverpool Corporation Liverpool Corporation
1825 to 1860	Growth	CONSTRUCTION Dock warehousing FUNCTION Processing industry location PLANNING New retail zone Public parks Private parks/estates	Merchants Liverpool Corporation Speculative developers
1860 to 1920	Growth and stability	TRANSPORT Ring road CONSTRUCTION Municipal flats Bye law housing Garden suburb	Liverpool Corporation Liverpool Corporation Private landlords Tenants association
1940 to 1958	Decline (Minor growth)	PLANNING Work/home relationship in new estates FUNCTION Shopping centre	City Council

115

Table 5.1. *continues*

1970 to 1985	Decline	PLANNING Infill and high rise demolition Dockland redevelopment FUNCTION Out of town shopping	City Council Development Corporation

LEVEL 3A: FUNCTIONAL UNIT, CORE CENTRE, LIVERPOOL CITY CENTRE

DATE	ECONOMIC CYCLE	INNOVATION	DEVELOPMENT AGENT
1880 to 1920	Stability	PLANNING Pier Head complex	Insurance companies
1920 to 1939	Recession	TRANSPORT Road tunnel CONSTRUCTION Walk up flats	Liverpool Corporation
1939 to 1950	Decline		
1950 to 1970	Decline	TRANSPORT Flyover Multi storey parking PLANNING Comprehensive redevelopment Pedestrianisation Conservation CONSTRUCTION High rise, curtain wall offices FUNCTION Indoor, controlled shopping	City Council Private developer City Council City Council City Council Speculative developers Private developer
1970 to 1988	Decline	PLANNING Main retail zone pedestrianised Albert Dock rehabilitated	City Council Merseyside Development Corporation

LEVEL 3B: FUNCTIONAL UNIT, RING CENTRE, MAGHULL

DATE	ECONOMIC CYCLE	INNOVATION	DEVELOPMENT AGENT
1895 to 1935	Stability then decline	TRANSPORT Major road construction PLANNING Victorian villas Semi detached housing estates	 Speculative developers Speculative developers

Table 5.1. *continues*

1935 to 1965	Decline	PLANNING Seml detached housing estates Public open space In residential areas Industrlal estates	Local authority Local authority Local authority
1965 to 1985	Decline	TRANSPORT Dual carrlage way Motorway PLANNING Central area redevelopment Informal road and residential layouts	Local authority Central government Local authority Local authority and speculative developers

LEVEL 4: DISTRICT: UNIVERSITY/HOSPITALS

DATE	ECONOMIC CYCLE	INNOVATION	DEVELOPMENT AGENT
1890 to 1938	Stabllity and recession	PLANNING Street bullt over	Royal Infirmary
1938 to 1957	Decline		
1957 to 1980	Decline	PLANNING Street pattern eradicated Pedestrlanisation Building occupy whole site or surrounded by open space CONSTRUCTION System building In reinforced concrete	University and hospltal University, hospltal and cathedral

Life Course of Innovations and their Effects on Urban Change'
(URBINNO Steering Committee, January 1988) is considered:
Initiation-Formulation-Enactment-Implementation-Life Course-
Effects, then the table only illustrates the period in which the
innovation became obvious due to change in the urban fabric. No
attempt has been made to trace the effects through the stages of
growth, equilibrium and removal. Similarly, no attempt was made to
trace the innovation back through the earlier stages of the diagram,
however, the inclusion of the development agency is a start to that
process.

117

References

Allan, A.R. (1986), *The Building of Abercromby Square, Liverpool* University Press, Liverpool.

Barker, T.C. and Harris, J.R. (1954), *A Mersey Town in the Industrial Revolution: St. Helens 1750-1900*, Liverpool University Press, Liverpool.

City and County Borough of Liverpool (1951), Housing Committee, *Housing Progress 1864-951*. (1965), City Centre Planning Group, *Liverpool City Centre Plan*. (1971, 1972), Planning Department, *City Centre Plan Review*.

Freeman, T.W. (1959), *The Conurbations of Great Britain*, Manchester University Press, Manchester.

Gentleman, H. (1970), 'Merseyside: Problems of Definition', in Lawton, R. and Cunningham, C.M. (eds), *Merseyside: Social and Economic Studies*, p 1-37, Longman, London.

Gentleman, H. (1970), 'Merseyside and its Region', in Lawton, R. and Cunningham C.M. (eds), *Merseyside: Social and Economic Studies*, p 38-57, Longman. London.

Hodges, R. (1953), 'The Dock System of the Port of Liverpool', in Smith, W. (ed.), *A Scientific Survey of Merseyside*, p 164-169, Liverpool University Press, Liverpool.

Hordern, R. and Miller, I. (1977), *A History of Maghull*, Hordern & Miller, Maghull.

Hughes, J.Q. (1969), *Liverpool*, Studio Vista, London.

Hyde, F.E. (1971), *Liverpool and the Mersey: The Development of a Port 1770-1970*, David & Charles, Newton Abbot.

Kelly, T. (1981), *For Advancement of Learning: The University of Liverpool 1881-1981*, Liverpool University Press, Liverpool.

Lawton, R. and Pooley, C. (1986), 'Liverpool and Merseyside', in Gordon, G. (ed.), *Regional Cities in the U.K. 1890-1980*, p 59-82, Harper & Row, London.

Liverpool City Council, (1978), *Heritage Bureau, Buildings of Liverpool*, (1987). City Centre Strategy Review.

Marriner, S. (1982), *The Economic and Social Development of Merseyside*, Crook Helm, Croom.

Massey, D.W. (1983), 'Conserving and Renewing the Built Environment', in Gould, W.T. and Hodgkiss, A.G. (eds), *The Resources of Merseyside*, p 131-146, Liverpool University Press, Liverpool.

Patmore, J.A. and Hodgkiss, A.G. (1970), *Merseyside in Maps,* Longman, London.

Pevsner, N. (1969), *The Buildings of England: South Lancashire,* Penguin, Harmondsworth.

Pickett, K.G. and Boulton, D.K. (1974), *Migration and Social Adjustment: Kirby and Maghull,* Liverpool University Press, Liverpool.

Smith, W. (1953), 'Merseyside and the Merseyside District', in Smith, W.A. (ed.), *Scientific Survey of Merseyside,* p. 1-18, Liverpool University Press, Liverpool.

Vereker, C. and Mays, J.B. (1961), *Urban Redevelopment and Social Change,* Liverpool University Press, Liverpool.

Whitehand. J.W.R. (1987), *The Changing Face of Cities: A Study of Development Cycles and Urban Form,* Blackwell, Oxford.

Erikson, J.A. and Hoggkins, A.E. (1970), Merseyside Maps. Longman, London.

Pevsner, N. (1969), The Buildings of England: South Lancashire. Penguin, Harmondsworth.

Pickett, K.G. and Boulton, D.K. (1974), Migration and Social Adjustment: Kirkby and Maghull. Liverpool University Press, Liverpool.

Smith, W. (1953), 'Merseyside and the Merseyside District', in Smith, W.A. (ed.), Scientific Survey of Merseyside, p. 1-16. Liverpool University Press, Liverpool.

Vereker, C. and Mays, J.B. (1961), Urban Redevelopment and Social Change. Liverpool University Press, Liverpool.

Whitehand, J.W.R. (1987), The Changing Face of Cities: A Study of Development Cycles and Urban Form. Blackwell, Oxford.

6. THESSALONIKI

Pavlos Loukakis, Eleni Paraskevopoulou, Maria Sfougari, University of Thrace.
Maria Kouroukli, University of Thessaloniki.
Special assistance and overall supervision George A. Giannopoulos, University of Thessaloniki, Greece.

Introduction

Context

Built on the eastern coast of gulf of Thermaikos on the site of the ancient city of Thermi, Thessaloniki is a city with continuous history since the sixth century B.C.. It is the second most populated urban centre of Greece, after Athens, and the main point of attraction in northern Greece. Among the oldest cities in Europe, Thessaloniki is built in an amphitheatric form surrounded by hills. From Roman times the town had a cosmopolitan character which remained until the beginning of this century.

The surface area of the contemporary city is approximately 61 sq.km. and its population was 702,107 at the 1981 census. Thessaloniki is a centre of administration and development and at the same time an educational, cultural and transportation centre. It is also an international trade centre with a world Trade Fare each September and many other sectorial trade fares. As a result Thessaloniki has become the focal point of development in Northern Greece, as well as an important industrial and handicrafts's centre with considerable political influence.

The urban sprawl of Thessaloniki consists of a solid and continuous built-up area formed by the fifteen local authorities. Municipalities, Ambelokipi, Eleftherio-Kordelio, Evosmos, Kalamaria, Menemeni, Neapoli, Polichni, Pilea, Stavroupoli, Sikies, Thessaloniki, Triandria. Communes, Aghios Pavlos, Nea Efkarpia, Panorama.

Methods

Within the context of the URBINNO project the intention was to define sub-periods in which there were different structures of spatial organisation, imposed by more general socio-economic rules. The changes studied were in the use of land, the type and size of the buildings, the urban road network, public transport and public services, the creation of points of attraction in the urban area such as large hospitals, large schools or hotels, business districts and shopping areas.

The study was divided into two parts level 2, the core-continuous built-up area and level 3A, the core centre. The first part deals with the general and most important stages of development that occurred in the whole of the urban settlement of Thessaloniki. In the second part the changes which happened in the central area are described in more detail. The analysis was divided into five sub-periods defined by some very important events in the recent history of Thessaloniki, 1880-1917, 1917-1922, 1922-1940, 1940-1960, 1960-1980.

1880-1917 This period illustrates how Thessaloniki was before the beginning of the twentieth century, under the dominion of the Turks, from which it was liberated in 1912 after 482 years of occupation. Soon after the liberation the Balkan War began, 1912-13. The year 1917 was chosen because of the fire which burned down the most important part of the city within the walls. At the same time this was the beginning of new period in the life of Thessaloniki.

1917-1922 The fire of 1917 gave a unique chance to plan and build the central part of the city again. The decisions about urban planning were radical and a special team of famous planners was called in to make their suggestions and find solutions. The defeat of the Greek army in Asia Minor in 1922 is one of the biggest tragedies in the history of the Greek nation. The arrival of 1,200,000 refugees in Greece had a decisive role in the definition of the later image of

Thessaloniki. Not only the quantity but also the quality of the Greek population changed from that moment.

1922-1940 During these years the industrialization of the city took place. The years between the two World Wars form a unique period.

1940-1960 After the end of World War II a considerable interest for the sciences, literature and the arts began with the construction of the University and rapid economic growth.

1960-1980 This period marks the so-called modern era of the city. Since 1970 a rapid rebuilding of Thessaloniki has taken place and at the same time considerable industrial development.

Level 2: Core-continuous built-up area

Development of the urban area

The main access to the city of Thessaloniki at the beginning of this century was through the sea and the same situation existed for a long time even after the construction of the railway connection with the city of Larisa and the rest of Southern Greece, in 1917. The ground plan of the city is of a trapezoid shape at the foot of Mount Chortiatis. Up to 1867 this trapezoid was surrounded by the Byzantine walls. Gradual demolition of the seaside section of the wall continued until 1912. In this way the new contemporary seashore made its appearance.

The city's heritage, as revealed through various ancient manuscripts and documents is reflected in its current urban form. The significance and the importance of the city during the years of the Roman Conquest is shown by the ruins of the Palace of Galerius. Some churches are left from the 11 centuries of Byzantium, while several buildings remain from the 480 years of Turkish domination. The boundaries of the city did not change for twenty two centuries although inside the walls, the city became more or less dense in various phases of history.

Different nationalities, languages and religions existed separately in this city, in different neighbourhoods for Jews, Christians and Moslems. That is why Thessaloniki had a European and an Eastern face at the same time. It had always been a big cosmopolitan city and this situation continued until 1912, the year when the city was liberated from Turkish rule. The population of the city doubled between 1830 and 1910, but after 1912 its population gradually changed from mostly Turkish and Jewish to mostly Greek. However, during the first World War, the arrival of European troops and their allies again gave the city a varied and rather confused form.

After the end of the war the new city of Thessaloniki was built upon the ruins of the fire of 1917, which had burnt down the central area. The new plan and style of Thessaloniki was totally different and completely free of any eastern influence. The special character of the city, given by different nationalities living together in the same place for about 400 years, had disappeared forever. The old city of Thessaloniki which was destroyed by the fire of 1917 looked like a city of the East. There were narrow streets made of stones and many mosques with high minarets. At the same time the neighbourhoods by the seashore in the eastern part of the city were quite European. There were buildings of European style, large streets, big shops, and all the elements that together make a picture of a western city.

In 1920 the population of Thessaloniki was 170,321 of Greek, Jewish, Turkish and European origin. At that moment, it gave the impression of an ethnological mixture, even though almost half of the population, 47%, was Greek. This situation gave rise to potentially very serious problems. However, in 1922 the Greek refugees from Asia Minor, Pontos, Ionia, East Thrace, and many other places, arrived after the fall of Asia Minor to the Turkish army. The arrival of all these refugees changed the ethnic mixture of the city towards a predominantly Greek population.

After the end of World War II a huge expansion of the city and its population began, because of the national internal migration of the population towards urban centres. Thessaloniki quickly changed to an important industrial centre with the construction of large factories, especially in the western part of the city. Today Thessaloniki is the most important industrial city of northern Greece, and the second most important in the whole of Greece. The main stages in Thessaloniki's evolution are shown in Figure 6.1. and can be summarised as follows.

Phase one ends in the year 1430, when the city was still limited within the walls built during the Roman Imperium. The city extended to an area of 330 ha., and the population was somewhere between 20,000 and 50,000 inhabitants of various nationalities.

Gradual expansion took place towards the east and by 1917 the area was 900 ha. while the population had reached 160,000. As previously mentioned the fire of 1917 marked the beginning of a new period for the city of Thessaloniki, in which there was a sudden increase of population, especially after 1922, when the Greek refugees from the near east arrived.

Figure 6.1. *Core-continuous built-up area: stages of development*

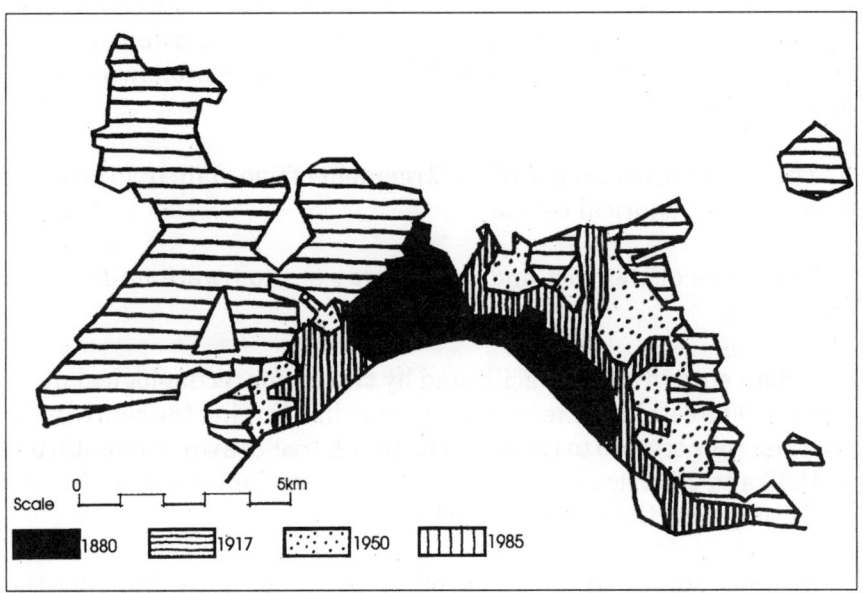

Up to 1940 the city continued to expand especially towards the east, but due to the influence of the two World Wars urban growth was very slow.

Since the end of World War II there has been rapid population growth, particularly in the last 15 years, reaching 702,107 inhabitants in 1981. During this time the city has been expanding towards the north-west, along and near the road which goes from Thessaloniki to Athens, while the expansion towards the east and north-east still continues. The expansion towards the north-west has happened as a result of industrial and port-related development, because those industries are the main point of attraction for the workers of the city. That is why the low income groups live in the west of the city, while the more wealthy inhabitants prefer the east.

Town planning

The beginning of any serious attempts at town planning in Thessaloniki took place in 1866 when the wall along the seaside was demolished. The object of these attempts was the historical centre, traditional place of all the economic activities. A very noticeable boost to the efforts at town planning was the first election and appointment of a town council and mayor in 1869. The main actions following these innovative events included:

The creation, opening up or enlargement of five main road arteries through the historical centre.

The areas of the port and the railway station were unified and enlarged.

Urban sprawl began, facilitated by two major technological innovations. The establishment of a tramway line linking the centre with the then countryside to the east, the first horse-drawn trams started in 1893 and electric ones in 1907. The further demolition of the city walls to the north-west and south-east.

Various administrative and other buildings were built by the Turkish administration of the city. Notable examples are the Central Administration Building (Dioikiterion) still used as the Ministry for Macedonia-Thrace, the railway station, banks and industrial buildings.

Various infrastructure networks and public services began to be created. Examples include, 1890 gas and sewage, 1893 water and public lighting, 1901 fire brigade, 1905 port health control.

The first serious efforts at town planning were made due to the town planning act of 1864, which set regulations designed to control the urban sprawl and building stock. These efforts were intensified after the fire of 1890. At the beginning of the twentieth century the city walls had been demolished, except for a few pieces lasting until today. Sanitary and other controls had been organised, but development was still taking place in a largely unchecked and disorganised way. Five principal innovations formed the basis for the development of town planning in the city during the twentieth century.

The necessary procedures for intervening in the organisation of the physical space of the city and its development were understood and it was recognized that planning should be the work of specialists.

The local authorities assumed greater responsibilities for town planning.

Social sciences made significant progress and the role and responsibilities of the authorities towards achieving social equality and public amenities through urban planning became better understood and accepted.

The first citizens organisations and special interest groups began to appear.

The value of property for homes and shops began to increase in the centre. The higher income groups were to be found here leaving the lower income ones in the suburbs where the various services, including transportation, were much poorer.

Population

The population growth curve for Thessaloniki has been an upward one ever since the Turkish occupation ended in 1912. The period between 1923 and 1940 was the most significant. After 1922 the influx of refugees affected the city's population both in numbers and composition. Most of the 1.2 million refugees settled in Thessaloniki, Athens and Pireaus. However, as they were in special camps outside the cities, population statistics for the actual cities may show a decline during this period because the refugees were not included.

During the period 1940-1971 the population in the greater Thessaloniki area increased steadily, from 278,399 in 1940 to 302,635 in 1951. In the two subsequent decades it increased by 25.7% and 42.4% respectively. The greater increase was for the outlying areas while in the actual municipality of Thessaloniki the increase was lower. The variations in the rates of population increase by municipality or commune were due to changes that took place in factors such as, allocation of employment, building activity, and socio-economic conditions which affected internal and external migration levels. The high percentages of population increase in the northern and western suburbs were mainly due to the increased industrial employment that took place after the allocation of industry to these areas.

In other areas such as Kalamaria and Panorama, to the east and north-east, the population increase was due not so much to the employment opportunities but mainly to the quality of these areas for residence attracting the most well-off sectors of the population. As employment opportunities seemed to affect the choice of place for residence, the rates of increase of the population in the western parts of the city were higher than those in the east especially for the period 1940-1971. Later this difference evened out. At the height of this increase, the period 1961-1971, the data shows that 63%, 112,000, of that increase was due to internal migration and only 37% to the natural increase of the population.

Level 3A: Core-centre

The Thessaloniki centre, like many other central areas in old cities, had narrow streets and the houses were built very close together. As a result of this the fires, which occurred quite frequently, caused considerable destruction. However, they also signalled a new planning era in the development of the area. The first fire of 1890 was a disaster in which 2000 houses were burned as well as the main church of the city and the precious records of this church. The inhabitants of this area were mainly Jews. They built this part again after the fire following a new plan. Long after the fire the area was known as 'Kamena' meaning burnt.

In the period after 1890, the eastern aspect of the old city combined with a western one, which already existed in the neighbourhoods near to the sea as well as those in the east. In the neighbourhood called Frangomahalas near Vardari Square, where the residences of most of the rich Greek and European inhabitants were, the outlook of the city was purely European. The neighbourhoods of the eastern part were similar as they belonged to the rich Jews of Thessaloniki.

The fire of 1917 which destroyed all of the central part of the city, 1.12 sq.km., was a very important event in the urban evolution of the central core. The city was temporarily paralysed and only the part round the White Tower, which was not burned, remained of the old glamour of Thessaloniki. The reason for this disaster was never discovered. However, a new city centre was to appear soon planned by the French engineer Hebrard.

Unfortunately only a little part of Hebrard's plan was to be carried out, Aristotelous Square, Aristotelous Street, Elefthorias Square and Diikitiriou Square. The intensive building reconstruction went on until 1935 and started again in 1956. However, the construction of multi-storey buildings changed completely the beautiful, old aspect of Thessaloniki. Today the centre, where all commercial, administrative and recreational land uses are concentrated, extends between the White Tower and the railway station and between the coast and Aghiou Dimitriou Street.

The wholesale trade is mainly around the port as well as along Monastiriou Street. Retailing takes place along Vasileos Gheorghiou, Vasilisis Olgas and Leoforos Stratou. There are also many local centres for the various sections of the city. In the south-east end of the centre there is the International Trade Fair, the University of Thessaloniki and the military school and installations. In the north and in the north-west there is the port and the passenger railway station.

In spite of the intensive growth of the centre during the last 15 years, most of the fundamental characteristics of the structure of Thessaloniki are still recognizable: the same industrial area in the west, the importance of the port, the difference between the centre and the other parts of the city. But today the old city is the centre of an urban area with 800,000 inhabitants and considerable influence over northern Greece.

Change

The centre is the part of the city in which most of the administration, commerce, and other economic activities take place. Also, some of the architectural and urban projects, which were developed after the fire in 1917, remain the most attractive places in the contemporary city. In spite of the various catastrophes that have struck the city in the last 100 years a large number of buildings have been preserved. Of special importance are the famous early Christian and Byzantine monuments through which one can see the evolution of architecture and painting over 10 centuries and more. More modern buildings have a variety of morphologies and styles.

Unfortunately, for the period before World War II quite a lot of historical data is missing, especially regarding building condition and land use distribution. The building stock consists mainly of pre-1970 construction, much of which was built before the war. Around 20% of the stock is estimated to be in good condition, 30% reasonable, and 50% in medium to poor condition. The high density of buildings is reflected in the height and site coverage indices for the core. The recent trends are towards renovation and a change in the traditional land uses with a shift towards more commerce and offices. The private sector is by far the main agent of building activy.

Almost all kinds of land uses are to be found in the centre with the major ones being residential, services and commercial. In the last 15 years residential land uses have declined in certain areas of the central core while offices and services have increased. The only stable residential area in the core is from Egnatia Street towards the north, the Old City. The only area of the core where no residential land uses exist is between the port and the railway station in the west. The prevalent land uses there have been transport-related activities and wholesale. The building stock is rather old with small trends towards very slow renewal.

The principal road axes existed before the war, so the structure of the road network remains virtually unchanged. The major additions in recent years have been the realisation of the inner peripheral road, which has been under construction for the last 15 years and is not yet fully completed, the extension of Egnatia Avenue which follows the same alignment of the ancient Egnatia Street first built by the Romans, and the eastern route towards the airport and the Chalkidiki peninsula. As is the case for most urban areas, major road extensions or road construction have initiated development along the road with positive and negative effects when, as was the case in Thessaloniki, no special measures were taken.

In the central road network there is evidence of changes in the volume of traffic flows, but their relative distribution to the various axes is the same. This signifies the fact that there were no major changes in the importance of each road within the hierarchy of the network. For the last 10 years traffic volumes have remained virtually unchanged in the centre, which shows that the road network in the centre and in the areas around it is working at capacity levels.

Conclusions

The case of Thessaloniki illustrates the development of a medium to large metropolitan centre in a very short period of time. The growth rates were quite spectacular until the 1970s, but they have slowed in the 1980s and this had a very notable effect on northern Greece and the country as a whole. The driving force behind this growth seems to have been the strategic location of the city at the crossroads of major transport routes. For the time period examined in this study, the urban area has been the focal point of transportation flows to and from the other Balkan countries with all modes of transport. Migration to the city and mixture of population has been a deciding factor and the major cause of population increase. As a result Thessaloniki became a focal point for commerce and services for the whole of northern Greece.

Population

Internal migration has been the major factor behind Thessaloniki's increase in population all through the period considered. It appears that the saturation population level of the city will be somewhere around 1,100,000, to be reached by the year 2000. The rates of population growth by area within Greater Thessaloniki have changed dramatically in the last 15 to 20 years. The western areas near the industrial park and similar other developments have traditionally grown fastest, now the opposite is true and these areas are in decline in favour of the eastern side. This trend is further accentuated as public transport links become more efficient and the construction of the so called interior ring road bypassing the centre is nearing completion.

As regards the characteristics of the population it can be seen that the main origin of modern external migration is the Central Macedonia region, while a major flux of immigrants from Asia Minor and the Black Sea regions of today's Turkey, occurred in a short period after 1922. The basic demographic and social characteristics of the population remain the same in each of the regions. These have remained unchanged practically for the whole of the last period of consideration, 1960 to date. However, income and professional activities tend to segregate by region. On the whole it can be said that Thessaloniki is a city without serious social differences and frictions. In this sense the city is closer to a medium-sized provincial city rather than a large metropolitan centre like Athens.

Land use

The main land uses in the central core are retail and offices but a large proportion of residential and administration still remains. Residential uses have slightly reduced in recent years but not to the degree evident in other large metropolitan areas. The expressed policy of the Thessaloniki Planning and Environmental Protection Agency is to keep the present residential uses in the centre. Outside the central area, the predominant land uses are residential followed, along the major road axes, by consumer retailing and other commercial activities. Major concentrations of land use, for example, hospitals and education, are dispersed throughout the urban structure.

Transport

The existing road network remained virtually unchanged for all the time periods considered. This was due to the modern town centre built after the fire having sufficient space for transport needs up until the last period of study. During the years 1960 to 1980, three notable additions have had and are still going to have serious effects on the city's development. These are the inner peripheral road, the extension of Egnatia Avenue, and the eastern seaside exit of the city towards the airport and the Chalkidiki Peninsula. Public transport is by a network of buses which follows the expansion of the city by extending existing bus lines and occasionally creating new ones. The tramways that existed in the city disappeared just after the war with the increase in other traffic which made their operation very cumbersome and expensive. The average trip lengths for journeys to work are relatively short, around 15 to 20 minutes, with the lack of parking space being the main problem for residential and commercial land uses both in the centre and outside it.

Core-centre

A considerable amount of building stock renovation was observed in the last period of reference. The increases of the permitted building factors and densities initiated this renovation in the 1960s but unfortunately no provision was made for building adequate parking spaces, a fact that has serious consequences today in the function of many parts of the city, especially the centre. Major service, administrative and retail functions continue to concentrate in the central area. This strong centre situation is observed throughout the 100 year period examined, ever since the demolition of the city's walls. The intense expansion of the city in the 1920s and 30s as well as in the 1960s and 70s failed to create any serious peripheral centres although they are beginning to emerge. Wholesale commercial land uses are still concentrated in the centre around the port which remains in its original position, at the edge of the centre.

Innovation

On the whole innovation, in its widest sense, including all kinds of major actions that changed the existing status quo at any given point in time, has had a major effect in shaping Thessaloniki's current urban form and development. The major innovations that have been noted and correlated with major shifts in the development of the urban area are described below.

1869 The creation of the municipality with a mayor and a town council.
1893 The establishment of a tramway line linking the centre with the eastern areas of countryside, which was electrified in 1907.
1864 The first town planning act.
1870 The beginning of the demolition of the city walls.
1890 Gas and sewage networks.
1893 Water supply network.
1917 A new plan for the city following the fire.
1922 The sudden influx of Asia Minor/Black sea immigrants.
1950 The removal of the tramways and their replacement by a

system of bus lines during the decade.

1960 The introduction of new higher density and building factors for the central area over the decade.

The creation of the few new road arteries notably the inner periphery and the extension of Egnatia Avenue.

Although innovation has undeniably played a decisive role in fostering development and shaping Thessaloniki's urban form, other factors both exogenous as well as endogenous have also been decisive. These were the particular socio-economic conditions that existed in the city, notably its mixture of nationalities and cultures, its strategic geographic location at the crossroads of north-south and east-west routes, and its hinterland that extended beyond the Greek borders to the neighbouring countries which wanted access to the Mediterranean Sea via the port of the city.

References

Demetriades, B. (1983), *The Topography of Thessaloniki during the Turkish Occupation 1430-1912,* Foundation for Macedonian studies, Thessaloniki.

Gerolymbou, A. (1985), *Planning and Rebuilding Thessaloniki after the Fire of 1917,* Thessaloniki University Press, Thessaloniki.

Greek Industrial Development Bank (ETBA) (1987), *The Beginnings of Industry in Thessaloniki 1870-1912,* Greek Industrial Development Bank Press, Thessaloniki.

Loukakis, P. (1987), *Some Aspects of Development and Planning in Metropolitan Areas: the Case of Thessaloniki,* National Centre of Public Administration, Thessaloniki.

Moskof, K. (1973), *Thessaloniki 1700-1912: a Cross-Section of the Manufacturing City,* Magazin Stochastis Press, Athens.

Papagiannopoulos, B. (1982), *The History of Thessaloniki,* Rekkos, Thessaloniki.

Technical Chamber of Greece, (1981), *Study of the Building Shell Development of Thessaloniki,* Technical Chamber Press Section of Central Macedonia, Thessaloniki.

Triantaffilides, I.D. (1966), *Town Planning Study of Thessaloniki,* Ministry of Public Projects of Thessaloniki, Thessaloniki.

7. AACHEN

Gerhard Curdes, Andrea Haase, Lehrstuhl und Institut für
Städtebau und Landesplanung, RWTH Aachen, Germany.

Introduction

This contribution is a summary of the contents of a more detailed
study of Aachen which was undertaken as part of a national project.
The work used the recommended programme as a basis for considering
appropriate time periods, the type and location of significant aspects
of urban change, and the phenomena of the effects of innovations on
urban structure. With regard to the whole time period of more than
150 years, the selected sub-periods which were investigated accord to
the important stages of development within each level. The methods
of analysis which were used are introduced at the beginning of each
level. Spatial units of investigation of the levels one to three have been
distinguished in relation to the existing urban structure and its
historical development.

Level 1: Urban area

The distinction of relevant time periods of investigation refers to an
overview of the main periods of innovations (Henckel, 1986) which
proved relevant for the stages of urban development of the Aachen

region, 1785-1842, 1842-1897, 1897-1940, 1940-1988 (Table 7.1.).
The last period has been subdivided into two 1940-1965, 1965-1988.
The survey of urbanisation processes is based on empirical data
analysis and on a model of different stages of urbanization, established
in the relevant literature. As urban structure change mainly concerns
the extension of built-up areas, respect is given to this aspect in every
time period, while densification, as a continuously occurring effect of
urban structure change, is reviewed for the whole time period of
investigation.

Figure 7.1. *Urban area*

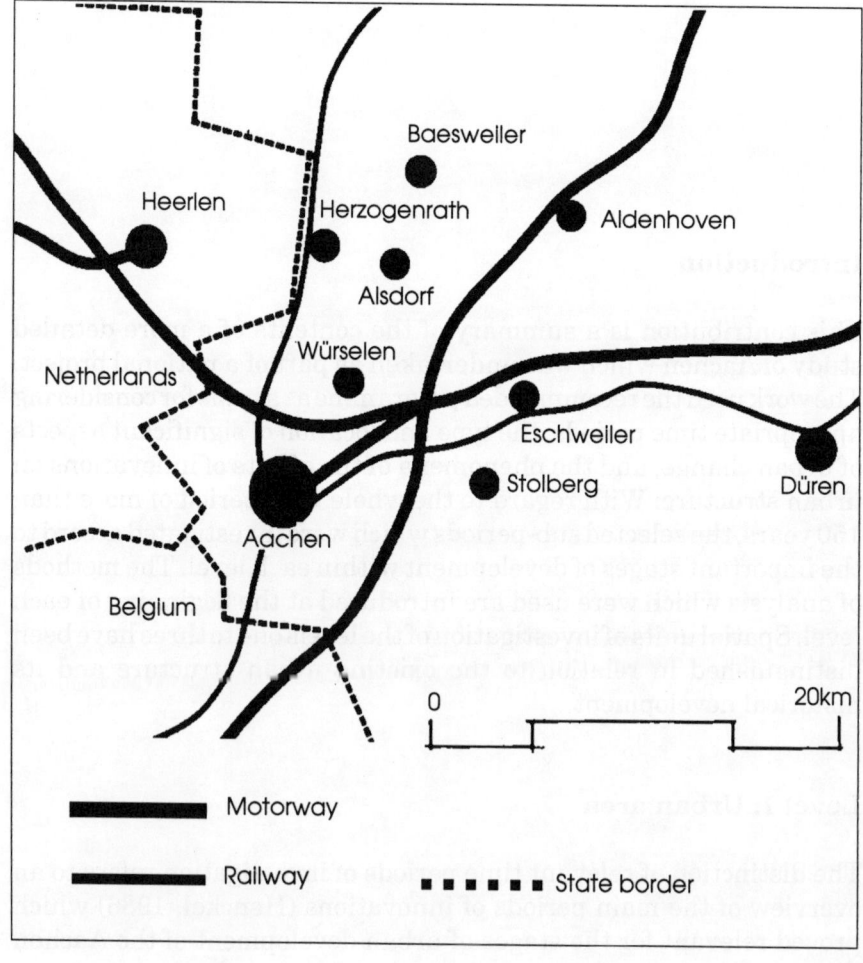

Table 7.1. Outline of urban development, 1874 - 1977

Period	Politics	Branches of economic growth	Population and land use	Infrastructure and transport	Social processes and urban forms	Innovations
1785-1842 Centre	1807 'continental barrier' of the French government. 1815 Prussian government	Bathing, cure, trade and service. Textile and needle industry.	1827 Population increase starts. Densification of trade, industry, residential uses.	Representative public buildings, theatre and station.	Urban centralization, workers residences adjacent to factories, medieval areas around town gates become residential. Defortification of the second wall.	Product: glass, iron, brass, mining. Production: machine. Transport: railway. Planning: axes.
Region		Industry: textile, needle, brass, iron, glass, mining, quarrying.	Densification of industry and residential uses around village cores, new axes.	Junctions to the south east, axes, railways, stations.	Industrialisation of villages, urbanisation of villages with trade functions in the south east.	
1842-1897 Centre	1870-71 French Prussian war, reparation payments from France. 1871 foundation of 'Das Deutsche Kaiserreich'.	1875 Technical University. Industry: wagon, steel.	1890 Main population increase begins. Densification/expansion of residential uses.	Boulevards (2nd ring). street network. Horse tram. 1896 electric tram.	Urban centralisation, decline of factories, migration of industry and residences into 'inner expansion areas'. Development of new quarters on both sides the 2nd ring. Function separation and social segregation begin.	Production: machine. Transport: horse tram, electric tram. Planning: boulevards, new quarters and public space, private development agencies.
Region		As above. Mining in the north. Leisure in the south.	Population decrease. Specialisation/expansion of industry, densification of residential uses.	Street and tram network. Public institutions for health and education.	Linear enlargement of industrial and residential uses along axes and railways. Early forms of suburbanisation, residential/industrial in villages without an urban core. Function separation begins.	

137

Table 7.1. continues

1897-1940 Centre	1914-18 World War I. 1919 Weimar Republic. German reparation payments. 1930 occupying forces leave the Rhineland. 1929 economic depression, USA. 1933 National Socialist government. 1939 World War II.	Economic crisis 1900 food industry and insurance, Aachen-Münchener, Aachen-Leipziger. Closure of the steel industry.	1900 Main population increase ends. First enlargement of the University.	Completion of the street network. Stagnation of the tram network. Elimination of railway stations. Public institutions.	Stagnation of urbanisation, residential uses move to the outskirts in settlements of the 1920s and 30s. Function separation, social segregation, public spaces, parks. Green fingers in outer areas.	Transport: motor car. Planning: 1919 green fingers as an element of the plan for the functional and spatial development of the whole town. New settlements, separation of uses.
Region		Economic crisis. 1910 china industry, chemical industry.	1910 Population increase begins. Densification and expansion of residential uses.	Completion of the street network. 1939 motorway reaches Aachen.	Urbanisation of industrialised villages without an urban core, suburbanisation in those with a core. Further concentration of industry, separation of uses.	
1940-1965 Centre	1945 end World War II. Mining in the Rhur area under UK authority. 1947 currency reform. 1949 constitution of the Federal Republic of Germany.	Increase in the tertiary sector, University, banks, insurance, offices.	Low population increase. Second enlargement of the university, expansion of the service sector.	Street widening, rings and radials. Pedestrian areas, subways, public parking, motorway, decline of boulevards. Bus system, reduction of the tram network.	Dereliction of 60% of the core-centre. Reconstruction of inner city areas. Substitution of housing by offices. Growing linkages between the University, core centre and residences in the outskirts (new settlements in the 1950s and 60s).	Transport: individual motor car traffic. Planning: suburban settlements, function separation, early inner area renewal.
Region			1960 Population increase ends. Densification and expansion of industry and housing.	Motorways, completion of the street network. Bus system.	Suburbanisation of villages. Tertiary industrial uses at overland junctions. Function specialisation and social segregation within villages. Inner area tertiarization of old villages with urban cores.	

Table 7.1. *continues*

Period	Politics	Branches of economic growth	Population and land use	Infrastructure and transport	Social processes and urban forms	Innovations
1965-1989 Centre	1973 oil crisis. Growth of the economic influence of western Europe, USA, Japan. 1989 demolition of the wall between the two Germanies. Reduction of border controls.	Tertiary sector, University and new technologies.	Stagnation/decrease of population. Third enlargement of the University including the new clinic and technology centre.	Pedestrian areas. Park & Ride. 1975 Tram becomes obsolete. Spas move to outside the 2nd ring. Radials widened, 3rd ring, traffic reduction.	Centralization of tertiary uses. Renewal of inner residential areas. Reuse of old factories for culture and housing. Expansion of the University into the city centre. Social segregation.	Transport: Discussion about the TGV link between Paris and Cologne and bus system between station and region.Environmental ticket. Reduction of motor car traffic in inner city and housing areas. Planning: restructuring of inner areas for housing and open green space, discussions about the car free town. Environment protection, reuse of areas and buildings.
Region		Implementation of new technologies starts.	1970 Stagnation/increase of population. Decline of heavy industry. Closure of the mining industry.	Completion of motorways, village bypasses.	Suburbanisation of villages. Aachen remains the main centre. 10-20,000 commuters living in Belgium and the Netherland. Recycling and recultivation of former industrial areas, high technology centres. Environmental improvement and protection.	

139

Pre-1785

The urban area of Aachen was the most western part of the economic district of the Rhine-Ruhr area which was part of the federal state of North-Rhine-Westfalia. Traditional political borders were the state borders to the west, Netherlands and Belgium, economic borders were traditionally given by the river Rhine, 70 km to the east and were actually defined by towns of regional- national importance, Düren-Cologne in the east, Mönchengladbach-Düsseldorf in the north, Maastricht-Liege in the west and by the rural landscape in the south, 'Eiffel' (Figure 7.1.). The town was Aachen was extended over a 'kettle-like' valley in which the core-centre was exposed on a hill. Fertile soil in the surroundings of the town and the existence of various hot springs even in the core-centre have basically influenced the economic life of the town of which the origins are a Roman settlement, 0-400 AD. The use of natural water sources was basic to very early industrial development in textiles and brass. Overland junctions between the Rhine and the North Sea provided location advantages for international trade. The function of a cure and bath centre, flourishing during the eighteenth century, supported the central importance of Aachen over minor regional towns.

1785-1842

Growth The industrialization of the core centre, textile industry, and of villages in the north-east of the region, mining and iron industry, showed prosperity until the end of the eighteenth century. Although the 'continental barrier' excluded innovative influences from England, the waggon industry was founded in 1838. The density of trade, service, leisure, industry and international transfer of goods and information, made the city not only a regional centre but also a national marketplace. The local foundation of an industry and trade company, 1804, supported efforts of economic independence and enforced the consciousness of a local community within the region.

Urbanisation process The duality of urbanization, originated by the town of Aachen and by the old eastern industrialized villages, caused urban development without the effects of urbanization in the northern mining areas.

Urban structure change: extension Enlargement happened around old settlement cores, around the core centre of Aachen and around industrialized villages in the south-east. It also followed regional overland junctions by developing single, settlement units. From 1827, population rates increased continuously in the core and region.

1842-1897

Growth The railway line, central station 1841, provided new location advantages and supported prosperity in the waggon industry, but could not compensate for the lack of transport advantages by ship which led to a migration of the iron and chemical industry into the Ruhr area. The Technical University, 1875, was given the task to compensate for a threatened decline of industry. The polarity of economic growth between the core-centre and eastern villages was still maintained.

Urbanisation process Apart from the enforced urbanization of the town of Aachen, there were processes of decentral urban growth within the east and the north of the region, supported by subcentral railway stations, Stolberg, Eschweiler, Langerwehe. Early forms of suburbanization in the town, housing and industry, and in the region: in the north, industry and housing; in the east, housing.

Urban structure change: extension The extension of the core of Aachen benefited from large areas of land cheaply available in the east of the town which was not available in the east of the region. Population rates of the town of Aachen increased continuously from 1850, parallel to those of the region until 1890, then the town's population increased more than the region's.

1897-1940

Growth The production of heavy industrial goods did not reach full prosperity. But glass, chemical, food and the old traditional branches, like mining, textile, needle, iron and metallurgy and the waggon industry contributed to a continuity of the industrial sector within the region. The foundry decline in Rothe Erde, 1924, introduced an industrial crisis.

Urbanisation process The core dominated, suburbanization reached the nearby villages which became more closely bound to the town by incorporation. Further suburbanization enforced the subdivision of the town in functional and socio-spatial units. Thus, the town reflected the separation of functions in the region, agriculture and industry in the north, industry and trade in the east, residences and recreation areas in the south, agriculture and trans-national exchange of goods to the west. Suburbanization in the region enforced its existing character.

Urban structure change: extension While the core enfolded its dominance by centralization and concentric urban growth, the development of the region was steered by a concentration of mining in a few locations, Herzogenrath, Kohlscheid, Alsdorf, Baesweiler and of a mixture of trade and industry in Eschweiler and Stolberg. Villages in between lost central functions. A differentiation became obvious between the town with a core-centre and extension areas on one side and single industrialized villages within the region on the other. Population increased in the region from 1910 while the population of the town area increased until 1900.

1940-1965

Growth The development of individual traffic contributed to the growth of industry, especially mining and glass. But competition with oil restricted the chances of growth in the mining sector. The tertiary sector, banks and insurance, increased, having a local tradition of more than 100 years.

Urbanisation process The importance of the tertiary sector and of individual traffic for the town increased corresponding to the time dependant idea of 'urbanity'. The service sector took the new urban conditions as location advantages and contributed to a concentration of economic forces within the town. Binding an increasing number of working places within the tertiary sector, the town of Aachen showed urban growth. Suburbanization was directed to the west, residences, and to the east, industry. Obsolescence of inner area residences started to become obvious near locations of a change of morphological structure. Tertiarization took place in central areas of the region but turned to a kind of tertiary, industrial character, especially in the north, discount-markets. In the east, Eschweiler and Stolberg succeeded in maintaining a growth of the tertiary sector within central areas but could not avoid negative effects on morphological structure due to traffic and building structure.

Urban structure change: extension Enlargement was enforced by function separation starting from the town of Aachen, then, successively reaching regional towns by a large, spatial development of junctions. While the mining areas in the north became closely linked up to the core-centre by individual traffic, the old centres in the east could not maintain their relative centrality, despite being supported by a variety of urban functions. The western enlargement of the core centre affected areas even outside the national borders, Vaals, Kelmis. A continuous increase of population showed a time lag of 5 years between the town and region.

1965-1988

Growth Electronic, chemical, and information technology have become the youngest branches of industry in the region. A large number of self-employed professionals originating from the University, has widened the spectrum of a new tertiary influenced industrial sector which is still dominated by the traditional tertiary sector. The decline of the traditional mining industry began by the closing of the main mines in Kerkrade, 1970. Further decline is projected for the north of the region until 1995.

Urbanisation process Enforced suburbanization concerns all parts of the region, especially the west and the south-west. While Eschweiler and Stolberg have successfully improved the old town quarters and restricted through traffic, the northern communities, especially Alsdorf and Baesweiler, have suffered from location disadvantages in consequence of having exhausted not only coal but also landscape resources and urban qualities. The villages in the outer enlargement areas have shown the first signs of 'disurbanization'. The first signs of 'exurbanization' have been found in towns outside the region. The core-centre of Aachen has shown the initial signs of 'reurbanization' by increasing residential densities on former industrial locations and by specialization in the service sector.

Urban structure change: extension Even during the threatening decline enlargement is still going on concentrating on the west of the core-centre, caused by the new clinic. Residential expansion in the region has appeared near the old trade centres, Eschweiler and Stolberg, less in the northern areas and most in the south, Roetgen. The demand for housing in the border states and in the privileged outskirts of the town has increased with building taking place primarily on former agricultural land: Richterich-Laurensberg, Kohlscheid, north; Walheim, south; and Haaren, east.

Urban structure change: densification

1785-1988 Like processes of enlargement, densification activities within the region have been introduced by a linear extension of settlement and have also been formed by filling up areas between cores, linear development and first enlargement. Densification has, in principle, supported concentric growth by being oriented on to the respective (sub) central area(s).

North   The old industrialized mining villages remained influenced by agriculture and, for a long time, had no densified core. Settlement completion was more addition than densification, following the needs of industry and of residences.

East   In consequence of a variety of land uses and a relative independence from the core centre, Aachen, some villages developed urban density by service functions, Stolberg, Eschweiler. Densification there is comparable to that of Aachen, based on an early densified mixture of use.

South-west   Small agricultural villages and free-standing settlements have either a traditional urban core, Kornelimünster, which was difficult to densify or have been extended by addition of detached residences, Roetgen, Netherlands and Belgium.

Urban innovation

Two main periods are distinguished for the spatial implementation of inventions, 1785-1940 and 1940-1988.

1785-1940 Having its origins in medieval times, the industrialization of the region is related to a 'long wave' period which was prosperous for the early iron industry of the south during the first partial period,

144

1785-1842, for the mining industry of the north during the second partial period, 1842-1897, and for a mixture of machine, metallurgical and chemical industry of the north-east during the third one, 1897-1940. Product innovations have had the most important influence, production innovations relatively less, because natural resources of energy remained the basic potential of industry. The sequence of prosperity and decline became spatially established by settlement development, moving from the south-east to the north-east. Agriculture was an appropriate basis for the introduction of product industry into social and spatial systems. Transport innovations supported the local traditional industries of the east but did not guarantee the prosperity of the north.

1940-1988 During this period the effect of innovation can be distinguished in four areas.

North   Transport and production innovations caused a further specialization and a concentration of skilled employees. The resulting homogeneity of industrial uses and residential areas signifies the difficulties for implementing social and technological innovations, for example, growth-promising branches of high technology. Nevertheless, the first planning and investment activities for technology parks have taken place in Alsdorf and Kohlscheid. The spatial effects of transport innovations, individual traffic, are the most important factors of investment retardation, though differentiated infrastructure is a main demand.

East   Production specialization did not extend. The development of main roads and motorways did not really affect these towns as much as the northern ones. The early location disadvantages of Stolberg, in a narrow valley, become an incentive for attracting service industry and housing through the improvement of the historical centre during the 80s.

South   Having maintained ecological resources and having increased attraction for leisure activities and residences, mainly during the first post-war period until 1965, the south has importance in compensating for the disadvantages of the eastern agglomerations spatially and funtionally.

West   Innovations inside the state borders, the new clinic and the motorway, 1970-1984. Outside the state borders, there have been recent initiatives to support the economy through tourist development within the 'Three Country Point' of FRG, Netherlands and Belgium.

Level 2: Core-continuous built-up area, Aachen

At this level the main subjects of survey are morphological change and urban innovation. The selected periods illustrate significant development phenomena. The subject of morphological change is distinguished by the categories of enlargement (Figure 7.2.), densification, and actual trends of development.

Figure 7.2. *Core-continuous built-up area: stages of development*

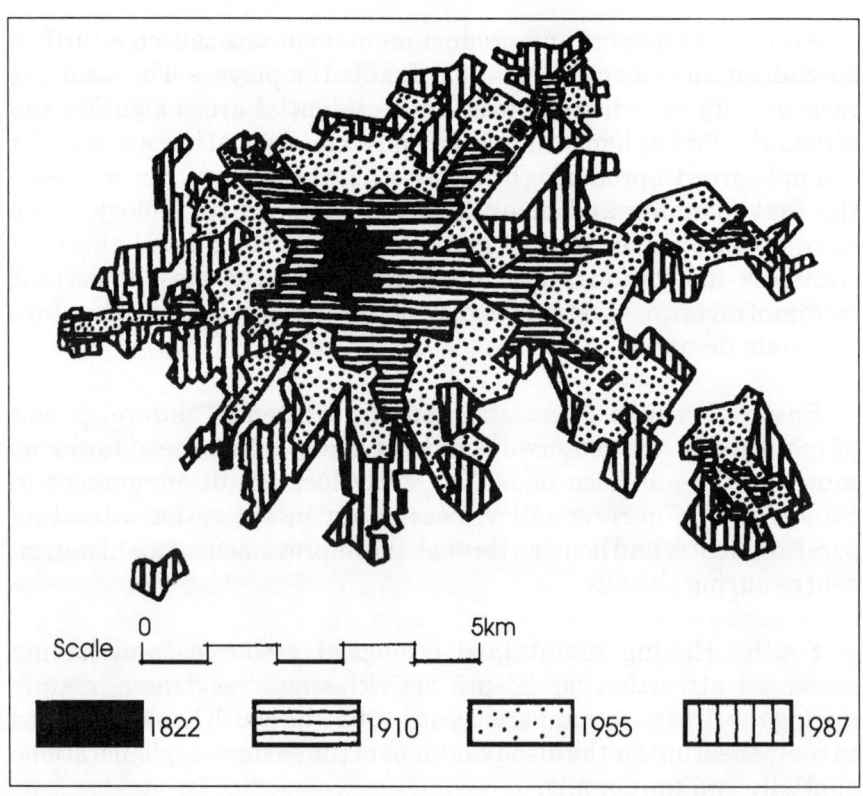

Scale 0 ⊢———————⊣ 5km

■ 1822 ▤ 1910 ⦂⦂ 1955 ‖‖‖ 1987

Morphological change: enlargement

1850-1930 (inner areas / east) Private investment generates the early enlargement along the axes developed by Napoleon. Small streams and wet areas cause low land prices but provide advantages for industrial location. The settlement extends with mixed use to the north, Rehm and Steffensviertel, and mainly residential in the southern and newest part, Frankenbergviertel. The existing street system was completed during this period and the 'Gründerzeit' pattern taken for the urban structure during the 1920s.

1950-1980 (outer areas / west) The major expansion began at Hanbruch in the 1950s and is still not finished. The expansion of the University and later of the medicine faculty with the new clinic, was responsible for the enlargement of residential use. The proximity to the Netherlands' and Belgian borders are seen as advantageous for a central public institution within an area of adjacent nations about to intensify their economic relations. This period mainly added spatial units of a different system of socio-spatial organization, 'flowing' space.

Morphological change: densification

Central areas of the core-centre Signified by a network of streets still similar to their medieval origins. Different dimensions of plots and different heights of buildings have developed there under investment pressure in the post-war period, causing a discontinuity of urban structure framed by the function of central services.

Nineteenth century inner areas and 'Gründerzeit' quarters The regularity of street development and of deep plots has created a densification corresponding to the respective individual needs of users and to the single plots. This kind of densification guaranteed a high degree of urban structure continuity which is significant for most parts of the inner town areas. Influences of the post-war period have changed the use and form of the block fronts, from vertical to horizontal structures, and also within the block yards, from industrial to residential use, to the tertiary or cultural sector.

Inner areas from the beginning of the 1920s Densification did not take place there except for additions of similar building structures within the surroundings, Junkersmühle, Rolandplatz, Talstrasse.

Morphological change: actual trends

Extension and densification continue. Areas vacated by a migration of uses, public institutions, factories, are reused by the tertiary sector, insurances, high technology.

Urban innovation

The two periods in which the main influences occur are 1820-1920 and 1950-1975.

1820-1920 The actual face of the town became structured by various innovative factors. Water, as a natural resource, served for bathing and cure and for industry. Transport innovations, railway and tram, supported the importance of central trade and service. The network of trams ensured the employees' transport to their working places. Production innovations caused a migration of traditional industry, textile, needle, into inner and outer extension areas because of expansion needs. The Technical University, as a functional innovation of the public educational sector became the symbol for centralizing public functions within the core-centre. The face of the town had become formed by representative elements of the function of bathing and cure, the idea of leisure was implemented very early. The 'Generalbebauungsplan' of Henrici, Schimpff und Sieben, 1911-1919, intended to establish 'green fingers' reaching from the south into the centre, signifies the local beginning of environmental consciousness and the concept of socio-spatially, ecologically and functionally well-organized development.

1950-1975 The main effects on urban structure were caused by transport innovations, the 'Neuordnungsplan' of 1950, introducing a fundamental change of urban structure function and form by widening inner rings and radials, closely linked up to the influences of the tertiary sector which itself had strongly been affected by storage and transport innovations and by a concentration of financial capital into companies. These influences dominated the subsequent reconstruction activities.

1975-1988 The cooperation between the University and the youngest branches of the tertiary sector is a revival of a traditional field of innovation within the town and has given a boost to the region. Technology parks have begun to occupy areas between the central core and outer areas of extension. The railway network is going to become influenced by the 'TGW' connection between Paris and Cologne.

Level 3A: Core-centre

History

The central area of the core accords to the Roman settlement. The actual core of the town is extended over two thirds of the area within the second ring, oriented towards the east, having followed the economic forces of the eastern extension of the town during the nineteenth century (Figure 7.3). It is nearly identical with the medieval town area, originally fortified by two walls; the 1st wall around 1300, the 2nd wall at the turn of the 13th/14th century. The first one was maintained in most parts as an additional fortification while outside new buildings with only a minimum plot depth were added by Napoleon between 1799 and 1807. Outside of these plots a ring street was symmetrically built up. Since the second wall, except for some gates, was destroyed a second ring street, boulevards, was established, linking the town up to future extension areas and providing a new linearly-developed public space. By 1812-1820, the built-up areas outside and even partially within the inner ring and in the south, Burtscheid, still had not been completed. Inside the second ring town development was primarily oriented on the ancient gate streets and on some public buildings (Komphausbadstraße, Elisenbrunnen), the theatre being located in the 1st ring. Territory potentials inside and between the rings were successively filled up by new functions and streets. The medievally pre-structured core has shown a high stability of main structural elements, streets and ranges of blocks. During 1648-1910 the main street did not change in width or form. In 1950, the street grid was still the same, but many of the demolished buildings have not been rebuilt.

Figure 7.3. *Development of the core-centre*

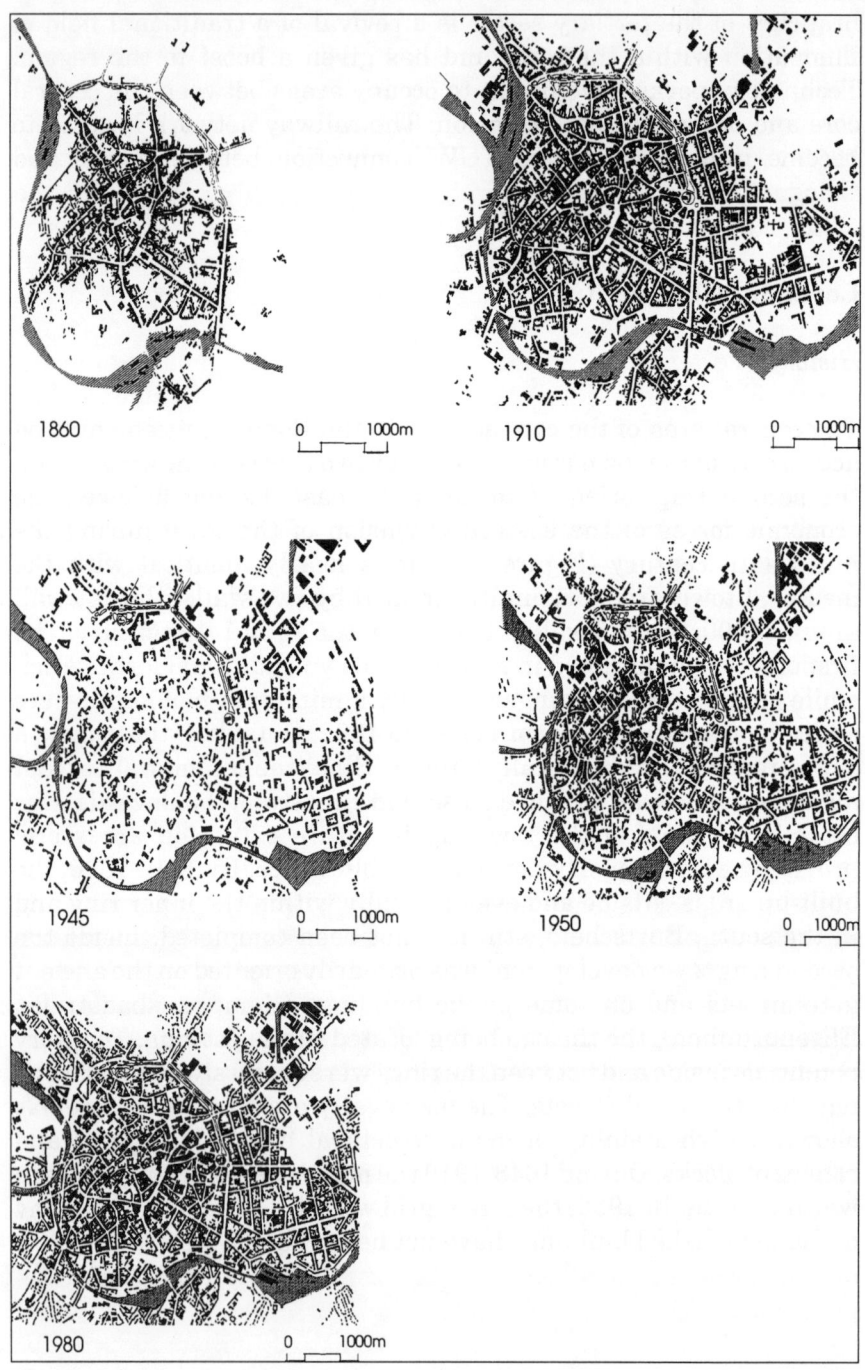

1860   0   1000m

1910   0   1000m

1945   0   1000m

1950   0   1000m

1980   0   1000m

Morphological change

The main changes have taken place in the periods, 1950-1964 and 1964-1980.

1950-1964 This period resulted in a fundamental, discontinuous change of form, function and use/users by opening the city for the motor car. A proposal for a new west-east axis through the city, discussed since 1919, would have preserved the main structural elements by leading the axis through back areas in the south of the first ring. Comparably, in 1950, a north-south connection was planned, creating a double ring system with a crossing on the market place, this time not giving any respect to existing structural elements. The final decision, the 'Neuordnungsplan' 1950, basic to its implementation during 1950 1960, left out the north-south connection but concentrated on a widening of the rings which still affected the medievally and neo-classically influenced surroundings as if the scale and design elements of a motor-way had been brought in. Since the 60s, a network of pedestrian areas and of parking spaces have been developed.

1964-1988 Individual traffic continued to be supported by additional offers of parking space and by further corrections of the width of the rings. The biggest change took place in the east, the break through of the new north-south connection as part of an outer ring. The tram network was closed by 1975. Over time the current discussion for individual traffic capacity restraints developed. A big warehouse transformed the southern part of the city core dramatically and is to be extended. During the last 20 years, more and more tertiary and secondary functions have left the core area. Since 1965, the central area has begun to change. Obsolescence near locations affected by a fundamental change of morphology during the preceding period (Peterstr.), effects of individual traffic. Substitution of mixed uses by residences, between the city centre and University. Modification of building and population structure within the old quarters of factory workers' residences (Rosviertel, Annastrasse, Rehmviertel), the effects of urban renewal and improvement. The changes of that period are not definitely discontinuous, although the use and users have been exchanged partially, the old function of the city becomes revived. New influences of built form pick up main elements of the old structure, public space in narrow streets, common or private space in block yards (Deliusstrasse). The ranges have changed only in some areas,

151

but the inside of most of the blocks has changed, many of the old back buildings have been demolished or have not been rebuilt.

Urban innovation

The main innovations occurred in medieval and neo-classical times, in public space and buildings. Individual traffic and service concentration tend to be regarded less favourably. The actual development is innovative by returning to the qualities of narrow space, open places and high quality specialist shopping. The tradition of a main market is maintained by temporary free and trade markets beside the weekly food market.

Level 4: District, university

History

The university was founded in 1870 by the Prussian state as an element of regional policy to encourage science education. During the first 35 years the number of students was less than 1000. In 1920, the number increased, still restricted by the economic crisis and the war. Since then, the total number of students has grown exorbitantly, 35,000 students in 1988, and 8,700 employees in 1987.

Morphological change: extension

The influence of the respective planning ideals on the university morphology is obvious (Figure 7.4.). In 1870, the first nucleus became located north of the first and nearby the second wall. In 1910, the university had filled the free space between the 1st and the 2nd ring and had its second nucleus north of the railway station. In 1928, this station was given up and replaced by buildings expanded at the old and the new site. In 1950, a third location west of the railway was opened. In 1964, expansion proceeded rapidly at all the three places, most in the core area, the university crossed the first ring and moved to the city core. In 1978, the expansion was still continuing, a new campus west of the 2nd ring had been opened. The medicine faculty migrated from the former hospital to the new clinic and was opened in 1982. Building complex and landscape transformation have created an obvious break of scale with the surroundings.

Figure 7.4. *Development of the university*

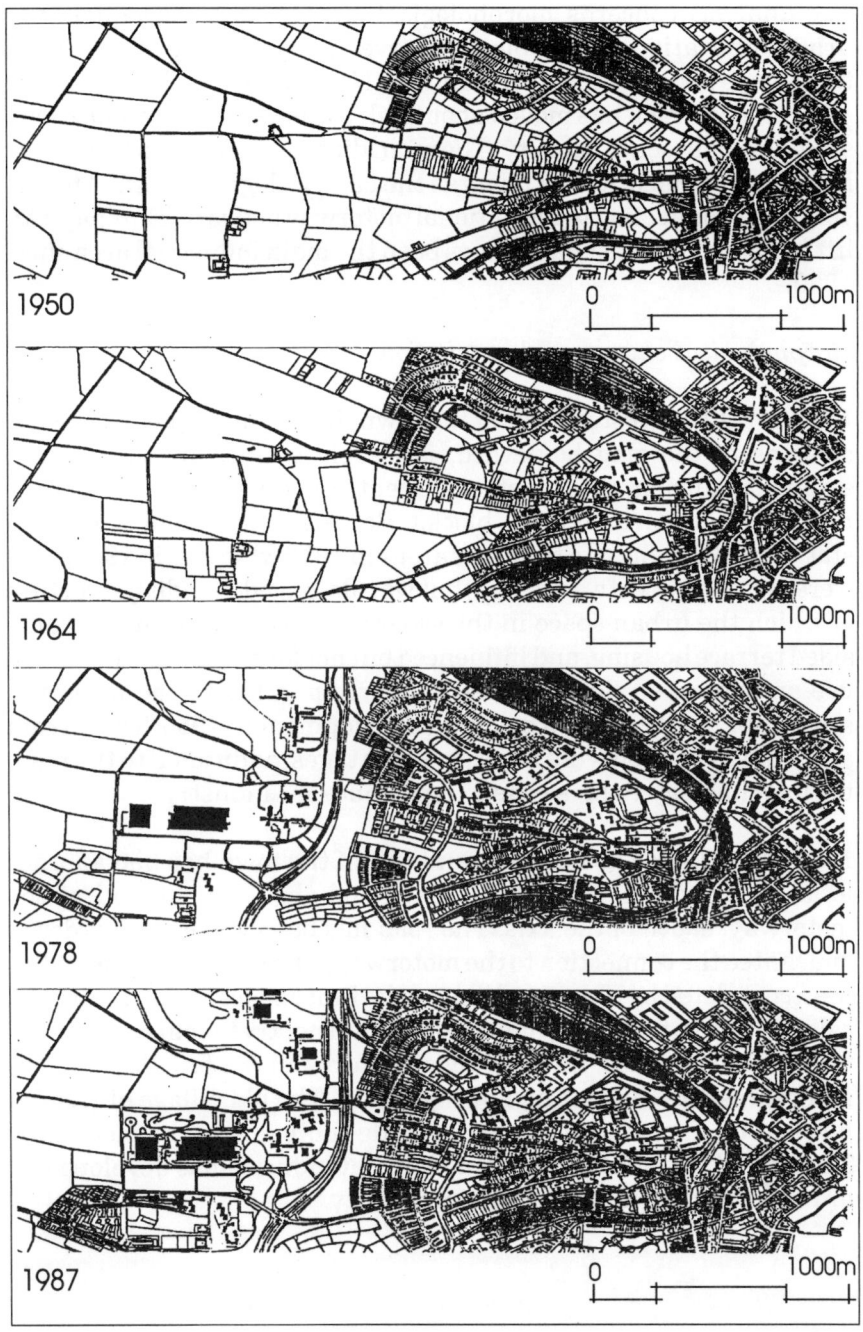

1950   0   1000m

1964   0   1000m

1978   0   1000m

1987   0   1000m

Level 5: Specific area, main east-west axis

The analysis concerns morphological change as a consequence of urban innovation. The subject of survey is part of one of the oldest tradeways between the Rhine and the North Sea. It still borders the central core which has moved about half a kilometre to the south east since the eighteenth century. The main direction of exchange of goods has always been oriented towards the east, Cologne, signified by the very early connection to the national motorway network, Europaplatz, in the 1950s. The east has always been the main industrial area since 1850.

Morphological change and innovation

1910-1950 The western exit of the town, linked up to the core by the crossing of the ring boulevard, is clearly defined by the eastern extension during the nineteenth century, consisting of workers' residences in 'Gründerzeit' blocks. Some factories and a railway station still belonged to the outer areas in 1910 whereas, in 1950, they were already part of a densified, but war-damaged, building structure of which the urban space in the streets was mainly formed by one-sided terrace housing, and influenced but not formed by the industrial use on the other side. A stream, nearly parallel to the axis, was already partially covered by urban development in 1950. The stream and railway station were location advantages, additional to the road, until technology reduced their importance for industry.

1950-1978-1987 The entrance and exit of the town has had another location, resulting from the large round-about at the end of the motorway, south of the axis. The road has been widened in order to guarantee the connection to the motorway. The row development has been completed by ribbons, filling up the inner space of the residential side. The densification of industrial use did not try to contribute to the form of urban space, its building structure follows functional needs and represents the idea of 'flowing space'. The old village of Haaren has lost its importance as a main trade route due to the motorway. The availability of cheap land means that industry has developed all around and that this use is given priority.

Level 6: Blocks

This survey concerns change of use and basic building types, and influences of dimension and form on stability. In Aachen, different morphologies represent the implementation of new ideas for built form.

Medieval form, Pontstraße and Driescher Gäßen

Location At the old arterial road to the north, crossing the 1st ring. *Change* On the west side, buildings have been continuously renewed whereas on the east side, the very small medieval types of houses have been conserved. One plot depth was removed due to the widening of the first ring. The immediate urban connection towards the southern side has been lost. Since 1950, most of the plots have been extended, new buildings have replaced the former vertical structure by a horizontally-oriented architecture. Only the narrow buildings in the middle still have the original scale.

Medieval form, Jakobstraße and Annastraße

Location Between the main east-west axis and an ancient rural footpath at the west end of the medieval town.
Change The block has had four front sides and had been subdivided by a minor way until, during the 80s, new residences and underground garages have filled its southern part, replacing industry and defining a new public space in between. The renewed part has become subdivided by a new minor way, linking up its yard sides. One remaining factory, Barockfabrik, has become reused for culture. One corner, mainly formed by a cloister, has lost one plot's depth by street widening. The width of plots has changed since the seventeenth century, the original minimum width of medieval times does not exist any longer but was substituted for the last examples, in the middle of the block, until 1950. The block is an example of inner area renewal maintaining its edges.

Mixed medieval and nineteenth century, Deliusstraße

Location Between the main east-west axis, the inner ring and the railway, here nearly identical with the line of the second wall.
Change In 1910, the area was subdivided by two streets, developed

155

by housing at the edges and some factories within the inner areas. The newly developed inner part has been subdivided again, during the 80s, into three parts, after the industrial use has left the area. One factory has been reused and subdivided into flats. The inner areas have been densified by housing.

Gründerzeit 1870-1900

The block structures proved to be very stable during the 100 to 120 years of their existence. The size of plots, narrow but deep, allows every kind of densification and inner area reuse for housing or environmental improvement.

Rehm-Viertel, ca.1860

Location Eastern enlargement, south of the main east-west axis.
Change Taking back of one plot depth by street widening of the axis. The high density of various functions within front and back buildings and most of the buildings themselves, except for those of the axis, have been maintained.

Adalbertsteinweg, court and prison

Location Between a south-east axis and minor streets of later parts of the eastern extension area.
Change Block developed during a whole century, housing at the edges, only interrupted by the court, the prison within the inner area. While planning ideas of 1910 still gave respect to a stream, the urban development of the following time covered it. Now, the prison is moving to the outskirts. This block is a typical example of the later extension area blocks; housing forms the front, industry or in this case public infrastructure is given space in the block inner area.

Reform, Talstraße, ca.1925

Location Near the first post-war motorway entrance into the town, south of the east-west axis.
Change Neither of use, housing, nor of building structure. But some shops, originally located at the corners, have been closed for a long time. The image of the area suffers from traffic.

Row structure 1925, densification 1960-75

Location Between the main east-west axis and the motorway entrance.
Change Successive closing of buildings along the roads, except for the inner areas, industrial uses in the west and wet areas in the east. Infilling of ribbons of housing and public buildings since 1965.

Artistic period of town design, ca. 1930

Location East of the eastern extension.
Change No change of the main use, housing, but of the shops at the corners. Still attractive example of a very early inner area densification.

Garden city type layout, ca. 1925

Location South of the railway line and of the latest parts of the eastern 'Gründerzeit' extension.
Change Because of privatization and of no guidelines, gradual demolition of the unique ensemble. Inner space oriented and distinctively formed structure with places, sequences, terraced houses, private gardens, modest architecture.

Medieval with ribbon infill, Hirschgraben

Location Beside the first ring.
Change The replacement of war damage by ribbons is one of the early forms of inner town densification using this type of building structure which originally had been designed for extension. The space of the street has been altered, the flats suffer from traffic.

Level 7: Plots and buildings

Buildings and plots develop either within planning rules, corresponding to major elements of the building structure, the block, or they follow the economically-oriented ideas of investors or fulfill the local specific demands of users. Types develop, if there is a standardisation of uses independent of a specific location. Their addition produces a specific structure. The traditional example for a standardized type in Aachen, mostly built between 1830 and 1925, is the Rheinish three-window house, with and without back buildings. This very solid type of building and its ground plan with rooms of about the same size allows a variety of uses and subdivision. The brick walls fulfil modern demands of sound absorption and insulation. This stage of the project did not regard building types as an isolated subject of survey but only within the context of the preceding levels.

Morphological changes and innovations, nineteenth century buildings

Subdivision of originally one-family houses and flats into flats or apartments, one or more per floor. Exchange of shops, ground floor, into flats and of housing into offices, elimination of back buildings or reuse of industrial (back)buildings. Extension of houses by balconies, or reuse of the space under the roof or by adding additional stories. Reuse of roofs as roof garden.

Conclusions

The contents and methods used in this stage of the project have proved to be appropriate in relation to the goals. Conclusions are drawn for the fields of innovation and their effects on sequences of form, use and function for levels one to seven. The type and location of those innovations concerned with transport and town planning are shown in Figures 7.5 and 7.6.

Production, levels 2 and 3

Old factories within the inner town area and in the centre maintained their original use and their built form until the first movement of expansion, specialization and migration supported a change of use, 1850-1910. This change concerns the kind of industrial use, for example, from textile industry to chocolate production, and the built form is only modified. A change from the secondary to the tertiary sectors began in 1960 and caused a second, internal modification of building structure for trade, service, culture. Since 1980 changes of form and function in the centre where residential uses and garages are replacing tertiary uses have begun to appear.

Figure 7.5. *Innovations in the transport network*

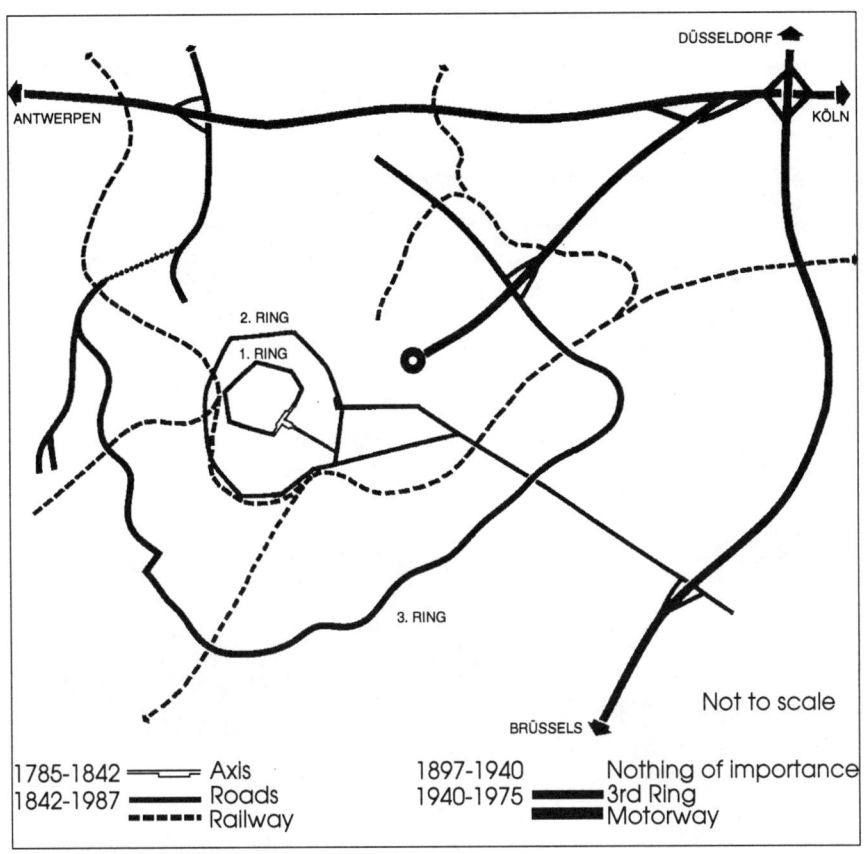

Transport, levels 1, 2 and 3

Change of function of the railway line: loss of importance for industry, increasing importance for individual transport, obsolescence of ancient goods stations near old factories, 1930. Change of the tram-function: superseded by individual traffic during 1950-1970. The development of the second and third 'rings' and the development of parking places and car parks as individual urban functions.

159

Built form, levels 2 and 7

Change of the use of blocks in consequence of the emigration of industry and of the tertiarization of the inner town, 1950. Initial obsolescence of solitary buildings within suburban settlements, because of a stated loss of life quality, 1975.

Figure 7.6. *Important innovations in town planning*

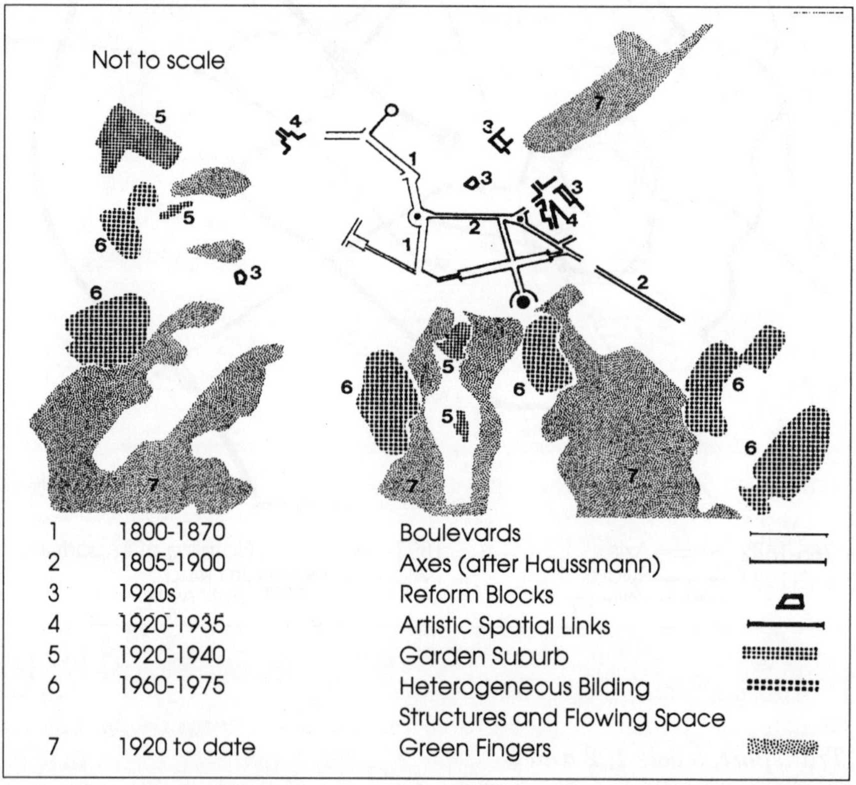

Not to scale
1 1800-1870 Boulevards
2 1805-1900 Axes (after Haussmann)
3 1920s Reform Blocks
4 1920-1935 Artistic Spatial Links
5 1920-1940 Garden Suburb
6 1960-1975 Heterogeneous Bilding Structures and Flowing Space
7 1920 to date Green Fingers

Infrastructure, levels 2 and 3

Obsolescence of building substance or location within inner town areas for institutions of the period 1890-1930, since 1975. Improvement of building substance, exchange of use within the infrastructural sector by a mixture of public and private use. Insurance companies benefit from the image of the old built form or from location advantages, for example, the use of the old clinic within the privileged situation of a green park, near the southern residential areas. The decrease of birth rates has caused a translocation of offers within the education sector.

Green areas and public space, levels 2 and 3

Boulevards and public space suffer from individual traffic, outer areas are protected by a 'landscape plan', 1978. Pedestrian areas and measurements of civic design do not compensate for the disadvantages in inner areas.

Environment protection, levels 1 to 6

Since 1975, an increasing awareness of the need for environmental protection has been caused locally by traffic problems and by pollution of industrial areas in the region. The increase of consciousness still has its centre of origins within the town.

References

Aretz, J. (1986), *Kohlscheider Bergwerke,* Aretz, Herzogenrath.

Balchin, P.N., Kieve, J.L. (1977), *Urban Land Economics,* Macmillan, London.

Curdes, G., Haase, A. and Pasternak, S. (1988), *The Development of the Urban Structure and the Influence of Innovations: Aachen, Volumes I, II,* unpublished.

Curdes, G. (1987), *Innovation and Urban Development (URBINNO) Working programme,* unpublished.

Curdes, G., Haase, A. et al. (1987), *Morphology and Innovation, Pilotstudy Aachen, Urban Area Structure Analysis,* unpublished.

Curdes, G. Haase, A. (1989), *Aachen: Case Study Summary,* unpublished.

Fricke, W. (1987), *Die Entwicklung des Deutsch-Belgischen Grenzraumes bei Aachen,* Städtebaulicher Werkstattgespräche Institut für Städtebau und Landesplanung, RWTH Aachen.

Friedrichs, J. (1981), *Stadtanalyse: Soziale und räumliche Organisation der Gesellschaft,* Westdeutscher Verlag, Opladen.

Gemeinde Kohlscheid (ed.) (1971), *Kohlscheid, Beiträge zu seiner Geschichte und Entwicklung,* Kohlscheid.

Haase, A. (1988), 'Die 'Raum-gewordene-Geschichte' der Stadt Aachen, in Phasen und Faktoren der stadträumlichen Entwicklung-Beispiel Aachen, Gemeinsame Einführung Städtebau', *Materialien zur Vorlesungsreihe der Kooperierenden Lehrstühle der Abteilung Architektur,* unpublished, RWTH Aachen.

Hamm, B. (1977), *Die Organisation der städtischen Umwelt,* Hubert, Frauenfeld.

Institut für Städtebau und Landesplanung und HMS: Helmer, Meier, Seiler (1988), *ZAR Zukunftsinitiative im Aachener Raum,* unpublished, Aachen.

Koebe, H. (1970), *Neuere Veränderungen im industriellen Gefüge des Raumes Aachen-Lüttich-Maastricht,* dissertation unpublished, Köln.

Krohn, H. (1988), *Auf der Schiene: Die Geschichte der Reisezug und Güterwagen,* Jubiläumspublikation der Waggonfabrik Talbot 1838-1988, München.

Kujath, H.J. (1986), *Die Regeneration der Stadt: Ökonomie und Politik des Wandels im Wohnungsbestand,* Christians, Hamburg.

Meyer, L.H. (1989), 150 Jahre Eisenbahnen im Rheinland, Entwicklung und Bauten am Beispiel der Aachner Bahnen, Bochem, Köln.

Peterek, M. (1981), *Das Aachener Rehmviertel: Eine typologisch-morpholo gische Studie,* Vertiefungsarbeit unpublished, RWTH Aachen.

Schmidt-Hermsdorf C. & G. (1984), *Stadtlesesbuch Aachen,* Lehrstuhl für Planungstheorie, RWTH Aachen.

Stadt Aachen (1965, 1972, 1976), *Statistische Berichte. (1980, 1985, 1987), Statistische Jahrbücher. (1956-71), Statistisches Handbuch,* Stadt Aachen, Aachen.

Wynands, P.J. (1986), *Kleine Geschichte Aachens,* Cobra, Aachen.

8. BARI

Domenico Di Bari and Francesco N. Nitti, Institute of Architecture and Town Planning, Faculty of Engineering, University of Bari, Italy.

Introduction

Context

In a regional study of the Bari area, it is expedient to start from the time of National Unity, 1861, although, in the past forty years the area has achieved full industrialisation and considerable tertiarisation with consequent broad, regional transformations. Bari, the main city, grew and developed within this context (Figure 8.1., Table 8.1.). Traditionally a merchant and maritime centre, it soon became an important railway and motorway junction, then an administrative centre. This development produced a phenomenon of urbanisation until twenty years ago. Today the Bari metropolitan area (A. M. B.) is second to Naples in southern Italy.

The A. M. B. has been defined to include all twenty municipalities connected by means of public and private transport to the centre of the city of Bari within thirty minutes, in normal traffic conditions. The area can be seen as two 'belts', the first including the ten municipalities which border on Bari, the second made up of the remaining nine. The territory has a surface area of 2670 sq.km., 20% of the province of Bari and 5.3% of the region of Apulia. The population is 654,955, 1981 census, 44.7% of the provincial and 16.9% of the regional population. Bari itself covers an area of 116.14 sq. km. and has 371,022 inhabitants. It is located in a perfectly central position, provincially and regionally. Until 1927 the territory was made up of 24 municipalities, but by 1937, unification and absorbtion had reduced this to 20 (Figure 8.2.). During this period the municipal territory of the city of Bari increased from 73.84 sq.km. to its present size and shape.

Figure 8.1. *Core-continuous built-up area: stages of development*

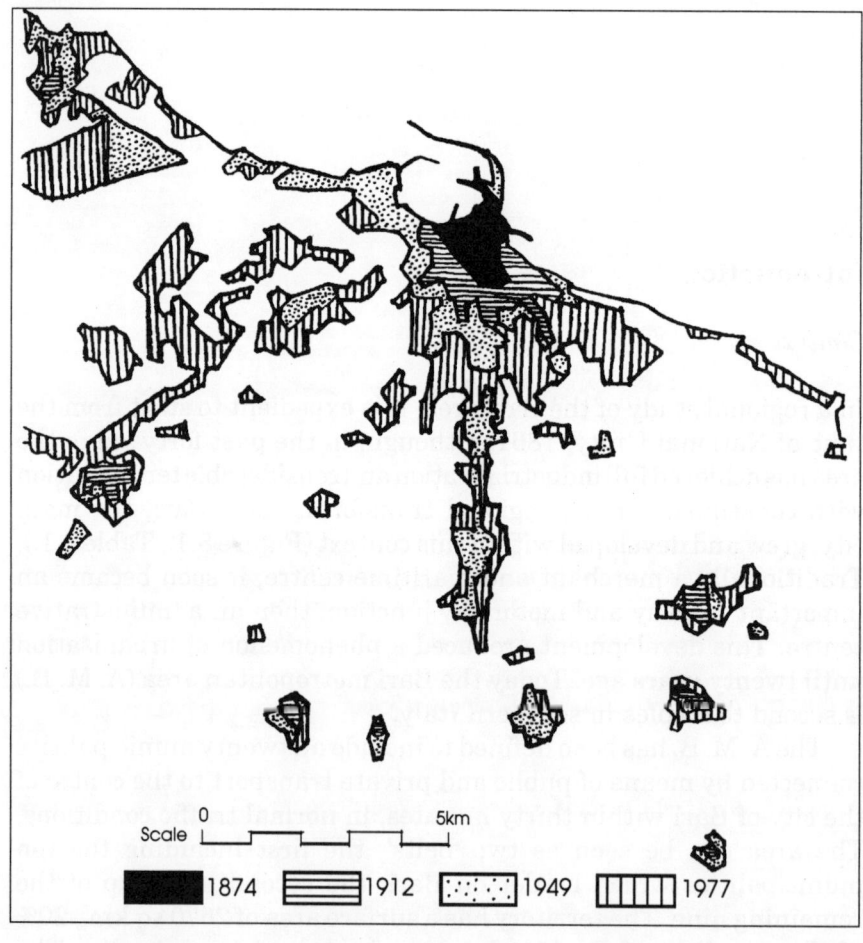

Scale 0 ——— 5km

■ 1874   ☰ 1912   ⋯ 1949   ▥ 1977

Methods

Maps, aerial photographs and census information were used in coordination to study the regional transformations over the 120 years between 1861 and 1981. The research used data from the census closest to the years of the four series of maps (*Fogli*) issued by the Istituto Geografico Militare, 1874, 1912, 1949 and 1977. The census data from 1871, 1911, 1951 and 1981 was supplemented by information from 1961 and 1971 to outline the evolution of the industrialization process over the last thirty years. Maps of inhabited areas were those prepared by the municipalities and approved by the IGM before publication, consequently they have different dates of origin from the *Fogli*.

Figure 8.2. *Boundaries of the Town Councils (after 1934)*

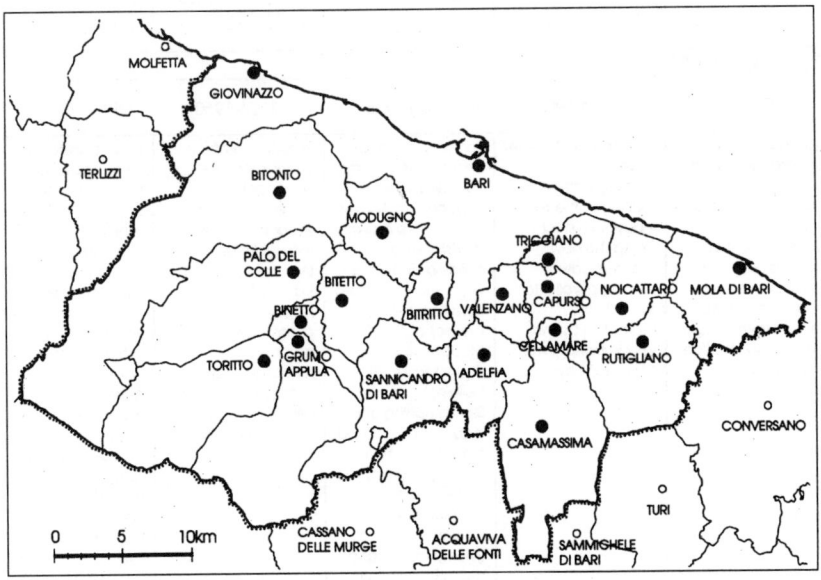

165

Table 8.1. *Outline of urban development, 1874 - 1977*

IDEALS

BEFORE 1874	1874-1900	1900-1926	1926-1950	1950-1977
Spreading of the new town (ex borgo murattiano) to match the demographic growth and to improve housing conditions.	Government's expansion proposals through new plans replacing the 1888 version.	New proposals for expansion beyond the railway, provision of zoning and containing the urban density within 200 inhabitants/ hectar.	Proposal to move the railway station to favour the southern expansion of the town. Lowering of building densities in the old town.	Planning of the whole city territory and town expansion without moving the railway. Further extension on account of sudden demographic, building, industrial and tertiary growth resulting from new proposals for the railway.

LEGAL TOOLS

BEFORE 1874	1874-1900	1900-1926	1926-1950	1950-1977
Murat Statutes (1814). Act 2359/1865 about expropriation for public utility with the first town planning standards of Italian legislation.	Proposal to apply the law used in Naples for slum clearance.	Second plan for town expansion according to paragraph II of law 2359/1865, approved in 1926.	Application of the fisrt paragraph of law 2359/1865 in the clearance plan for the old town (Bari Vecchia) approved in 1931. First Italian organic town-planning law (1150/1942).	First Master Plan according to the new town-planning law, approved in 1954 and in force for 22 years. Second Master Plan and town-planning standards forecast, approved in 1976.

RESULTS, BUILT FORM

BEFORE 1874	1874-1900	1900-1926	1926-1950	1950-1977
Building closed in quadrangular and rectangular blocks delineated by public ways. Development of the new harbour built before Italian Unity and arrival of the railway from Foggia and its continuation to Lecce and Taranto.	Persistence of former town planning and building formalities. Making of the first industrial district along the new circuit-roadway running parallel to the railway, promoted by the very important role of the railway station.	Fan-form expansion beyond the railway following some important radial and half ring-like roads. Persistence of building standards and initial proposals from linear models (houses of Madonnella) and single family buildings. Strengthening of the industrial district and formation of two other new areas running along the new south-east railway (Mungivacca) and along the railway junction with the harbour (Marisabella).	Persistence of the block in private buildings and development of linear standards in the public sector. Development of rail transport, the beginnings of automobile transport and the initial formation of aspects of the metropolitan area including some communes in the first belt.	Block standards were followed by linear ones then by single family housing. A new intercommunal industrial area meant that earlier settled districts were vacated. Large satellite district created (St. Paulo quarter). Automobile transport supercedes the railway. Metropolitan area develops beyond the communes of the first belt.

166

Level 1: Metropolitan area (A. M. B.)

Growth and density of inhabited areas

The survey was carried out by measuring the built-up areas and calculating the regional and urban population densities, expressed as inhabitants/sq.km., of municipal and urban surface area respectively. In spite of the lesser graphical accuracy of the first two series of maps, and the unavoidable approximations, the results are appropriate for this study.

Regional densities have increased in all towns, due to population explosion, whereas urban ones have decreased. This has been especially relevant in recent decades as a result of the expansion of urban areas and consequent increase in dwellings and residential surface areas caused by higher expectations and living standards.

The population of the territory increased by 231% in 110 years, from 197,684 in 1871 to 654,955 in 1981. Built-up areas have increased by 644%, from 6.28 to 46.75 sq.km. Consequently, the average regional density has more than tripled, 191 to 635 inhabitants per sq. km., whereas the average urban density has dropped by more than half, approximately 31,000 to 14,000 inhabitants per sq. km., the density of the residential area of the city of Bari has dropped from approximately 400 to 135 inhabitants per ha., and the town of Grumo Appula from approximately 510 to 170.

Construction of dwellings

ISTAT figures illustrate a 278% increase in the total number of rooms between 1951 and 1971, 214,166 to 818,242, whereas the population increased by 35.6%. During the same period the phenomena of the second home contributed significantly to the increase in uninhabited rooms, 10,783 to 97,089. The room occupation indices for all of the towns in the area reflect this increase. The ratio between members of families and numbers of rooms occupied was 2.0 in 1951, with exceptions in Triggiano and Toritto. In 1981 the mean values for the area were 0.9 for twelve municipalities and only two had an above average index, Grumo Appula and Toritto, 1.1.

Most of the dwelling stock in the towns in the area is recent, with a considerable amount of facilities. However, many of the residential areas have been built without following planning regulations. At the end of World War II single-family dwellings were the norm in the towns, 'casa a schiera' in particular, which was really only adequate for the rural areas. Multi-family houses had been built in Bari much earlier, but since the war this type of housing has become prevalent, despite its unsuitability in small towns and some residential areas. This diffusion of multi-family housing is due to speculators, the Istituto Autonomo delle Case Populari (low income housing authority) and other agencies working in the field of subsidised construction.

Economic indicators

After the 1951 census, the twenty towns of the area appeared, as the Apulia region, to be agricultural. Yet, between 1951 and 1981 the situation changed completely, as can be seen from the evolution of assets percentages in the three economic sectors. Those of the primary sector dropped from 59.2 to 25.7%, but the secondary and tertiary sectors increased respectively from 22.0 to 28.9% and 19.8 to 45.4%. By 1981 the area and the province were closest to national percentages as both were greatly affected by the situation in the capital city, 2.4, 25.7, 71.9%, and Mudugno, 2.8, 44.1, 53.1%.

A. M. B. is now an industrialized and tertiarized region. This is confirmed by other economic indices such as the percentage composition of the sectorial structure of the per-capita income (value added to the cost of factors). These were computed for all the municipalities of Apulia in 1981. Mean regional values for the three economic sectors were, 3.95, 29.88, 66.17%.

Bari appears to be the most tertiarized city, 71.9% of assets, 74.85% of income. Modugno is the most industrialized, 44.1% of secondary assets, 68.3% of income. Rutigliano and Sannicadro continue to appear rural with over 40% of assets in the primary sector.

Transport and utilities

Roads In the eighteenth century, Apulia had very bad roads, as all the peripheral regions of the Kingdom of Naples. Nevertheless, it had a considerable number of harbours and, therefore traded by sea with Venice, Trieste and other Mediterranean ports, much more than by land with the rest of the kingdom. Towards the end of the century, the new 'royal' road was under construction, connecting Naples to Salento. It reached Bari by 1790, when the first project for a 'borgo' (suburb) was being discussed, due to the expansion of the city outside of the walls. Subsequently, many other roads were built, to the extent that

Bari soon found itself in the middle of a very rich crossroads.

In the 1960s all the roads of the network around Bari were linked together, with crossings at different levels, and with a 16 km. by-pass that freed the city from excessive traffic. The latter, perhaps the most important public work of that decade, was also linked by the two junctions which lead to the highway paytolls. The A16 highway reached Bari in 1965, from Naples, the southern city where the Autostrada del Sole ends. A few years later the A16 reached Taranto. It connects Bari and all of Apulia to the national highway system.

Harbour Some years before the Unity of Italy, construction of a new harbour was started, as the existing one was totally inadequate. This bold initiative took up the old idea of a westerly harbour in the bay between the city and the S. Cataldo peninsula. As years elapsed, the harbour was enlarged, connected with the railroad, 1916, equipped with an oil wharf, 1937, and cereal silos, 1960. More recently, 1980, it was equipped with a new passenger terminal for the many tourists who take ferries to the opposite shore of the Adriatic Sea, Greece and the Near East.

Railways Only after National Unity did the railway system reach Apulia. It reached Bari in 1864, coming from Foggia, Pescara, Ancona. The Adriatic line was built very rapidly as it represented, at that time, the only possible railway connection between Torino, capital of the new Kingdom, and the south. The Tyrrhenian line to Naples, and from there to Bari via Foggia, was not viable due to the existence of the Pontifical State. Therefore Bari was connected to its former capital almost fifteen years after the first proposal for a Naples-Bari-Lecce railroad was made by an engineer from Bari. The railroad reached Lecce, via Brindisi, in 1866, and two years later it reached Taranto. Thus, in 1869 Bari became the first railway junction of the Apulia region.

In 1882, besides the littoral Foggia-Lecce line and the connecting line to Taranto, an internal one to Barletta via Bitonto, Terlizzi, Ruvo, Coratoand Andria, was built. It was made up of a steam tramway, narrow gauge, almost always along the edge of the road. In spite of the very simple technical characteristics, this 'economical railway' built with foreign funds, achieved the goal of connecting Bari to centres other than those along the littoral line. The reconstruction of this line began in the 1930s with normal gauge, new track, not connected with roads, and new railway stations in towns, yet, only in

1965 were the works finished. A positive consequence of the delay was the transfer of the station, envisaged in the western suburban area of the city, to beside the national railway station.

The Sud-Est railway was built at the beginning of the century, with an ordinary gauge. It started off from Mungivacca, in the south-eastern suburban area of the city, with two branches directed to Conversano and Putignano. This line was also connected to the central station. Finally, in 1915, the narrow gauge Calabro-Lucana railway became operational. It connected Bari to Matera and Potenza, via Gravina. This line ran parallel to the Bari-Taranto line, for the first 30 km., and it also had its terminal station next to the national station. The urban section is made up of a 1.5 km. long reinforced concrete viaduct. In 1915, out of the twenty towns of the Bari metropolitan area only Bitritto and Cellamare did not have a railway station, whereas Modugno and Grumo Appula had two, National Railways and Calabro-Lucana. The Bari junction embraces the four lines, FS, FSE, FCL, FBN.

The 1864 railroad layout was built at a good distance from the expanding 'borgo'. As the Municipal Council considered it to be too far away, the location of this 'countryside station' was disputed. Only at the end of the century was it reached by buildings, and the square in front of it took up the aspect it has now. Already at the beginning of the twentieth century, the city had expanded beyond all forecasts, especially with the construction of the San Pasquale, Carassi and Picone neighbourhoods. This is how the issue of the railway station first appeared and it is a problem which remains unsolved.

In 1919, the first proposal was made to move the station further south and lay the track parallel to the existing one. A project of the 1938 town plan suggested a closer station and lowering the track cuttings, but this was not approved. In 1954, the new general town plan was approved, which rejected both the idea of moving the station and of transforming it into a terminal station. The connection of the two parts of the city split by the station was solved by planning bridges. For the most important one, which follows the axis of Corso Cavour, a monumental architectural solution was suggested, four tall buildings at the two ends and along the sides, a row of shops just as on Ponte Vecchio in Florence, in order to conceal the view of the railway. At the end of the sixties, instead of that project a simple one way steel bridge was constructed. Yet, the problem of the railway still worried the city of Bari, and several alternative suggestions were also presented: the transit station in the western outskirts, San Giorgio;

the terminal station moved westwards; a transit station embanked in the present area; a terminal transit station with the tracks of the Brindisi-Lecce line and of the Ferrovie del Sud Est underground.

At present the cargo area is being restructured between the western outskirts of the city and the industrial area, at Ferruccio, where the new cargo and container depot is located. It is connected to urban and suburban roads by means of the by-pass road, which also links it to the airport and harbour areas.

Airport Regular flights with Rome became operational at the beginning of the fifties. Until 1975 planes used the runway built before World War I, in the military airport of Palese, a littoral, outward-lying in the west of the city. At the end of the sixties, the construction of the civil runway began, located not far from the military one. It became operational in 1975. It is 2,600 metres long, of which only 2,200 are presently used, and 45 metres wide. The airport terminal, however, remained in the small premises of the military building until 1983, when an improved but still inadequate facility was built.

The airport of Bari is busier than the other Apulian airport, Brindisi. It is only 12 km. from the city centre and occupies a surface area of approximately 200 ha.. It has a total traffic of 6,400 planes and 420,000 passengers per year. Its share of the total Italian airport traffic is 1.3% of planes and 1.1% of passengers. It has inter-regional importance, however, as it attracts traffic also from the eastern part of the Basilicata region and, obviously, serves all of the northern and central part of Apulia, including Taranto.

At present from Bari Palese there are daily regular flight connections with Rome, four, with Milan, two, with Bologna, Catania, Cagliari and with Titograd, capital of the Republic of Montenegro, Yugoslavia, besides some charter flights. In a few months time regular flights with Paris and Frankfurt will become operational.

Water Surface water courses are almost non-existent in Apulia, in spite of the approximate 500mm. per year of rainfall, due to the geology of the territory. For the same reason, underground water is abundant. Only in the past twenty years has it been drawn from the water table by means of artesian wells and employed mainly for agricultural purposes, at times for industrial uses, and recently, even for civil uses. In the past, agricultural crops and the inhabited areas of Apulia have been constrained by the scarcity of water.

From 1914, however, the situation changed when Bari received water from the longed-for aqueduct from the Sele source, a river of the Campania region which flows into the Tyrrhenian Sea south of Salerno. These enormous works, already proposed at the end of the reign of the Bourbons, were accomplished only half a century later. It was 1937 before the Acquedotto Pugliese, at that time the longest in the world, reached all the inhabited areas of the region. Obviously, population growth and the expansion of inhabited areas, as well as the development of industries and recently, also of tourist resorts would not have been possible without the water, although as a per-capita supply it is very modest. Today, the aqueduct also uses water from other sources or watersheds, even from the Basilicata Region, within the framework of the agricultural irrigation programme designed immediately after World War II.

Irrigation water, and that of artesian wells, have already made a positive contribution to Apulian agriculture. Besides traditional crops, mainly shrubs and trees, industrial crops, such as sugar beet, sunflowers and maize, have been grown.

Energy Coal unloaded in the harbour was the first energy resource of the area and Bari was the first city in Apulia to have gas pipelines for home use. In the second half of the century, electricity produced in the mountains of Sila, in Calabria, reached the area. After World War II, the thermo electric-oil-power plant started producing energy. Together with other plants, oil, coal-powered, or multi-functional, it supplies the electricity required in the area. By means of the 380,000 KV power line, the Bari area is linked to the national electricity supply network.

It is almost twenty years since another energy source became available, natural gas, previously obtained from the Basilicata and the Dauno sub-Appenninic fields of the province of Foggia. Today it comes also from Algeria and northern Europe thanks to the link-up with the national methane network.

Industrial and tourist sites

The first industries were established shortly before National Unity. They processed the products of local agricultural crops, olives, grapes and wheat, few produced consumer goods. The advent of the railway system, 1864, called for the establishment of a small metal and carpentry factory, which started the industrial area along the suburban road, at the back of the railway station. A second industrial area was formed at the beginning of the century in the south-eastern periphery, along the new Ferrovia del Sud Est, parallel to the road to Capurso and Taranto. A third area was established in 1916 in the north western suburb, not far from the harbour, along the junction track that links the harbour to the station.

A few years before the last war, a big oil refinery was built in the western suburban area, along the road to Modugno. The site was chosen as it enabled the connection of the plant with the new oil wharf constructed in the harbour. At the end of the war, other industries were established in the surroundings, to the extent that very soon it became the most important industrial area. Expansion also took place in Modugno and Bitonto, consequently, in 1962, this area was selected as the main conurbation of the Industrial Development Area (ASI).

The old, obsolete areas were featured as residential areas in the general town planning scheme (PRG) of 1954. The first one, along the suburban road at the back of the central station, has already been completely reconstructed. At present, the land of the second area is being utilized along the tracks of the Ferrovie del Sud Est. Only the area facing the harbour, no longer with factories, awaits a new use. The last was defined as an out of use area by the 1954 town planning scheme but by the present scheme, 1976, it is characterized as an area for the construction of public buildings and important communication infrastructures. Many small industries were established in other areas of the Bari municipal territory, beyond the ASI conurbation, mainly in the towns of the first belt.

The main tourist attraction of the area is the sea. At the beginning of the nineteenth century, small seaside resorts sprang up along the coastline, Santo Spirito and Palese in the north-west, San Giorgio and Torre a Mare in the south-west, next to the railway stations. The diffusion of automobiles and the failure to comply with the 1954 town planning scheme have led, in the past thirty years, to a continuum of tourist resorts which go from Giovinazzo to Mola. However, there are few hotels as the seaside tourism is made up mainly of residents.

The hinterland does not have any tourist attractions but, in the lower slopes of the Murgia, in the town of Toritto, the Quasani village has been organized, 350 m. above sea level. Bari has a good potential hotel capacity, over 5,000 hotel beds, almost half of all those of its province, as it is a place to which people come in transit, or for business or for cultural purposes.

The present regional structure

The present structure of the area, 1988, does not correspond to the one recorded in 1977. In 11 years, the regional transformations which have taken place have been many. To understand them it is expedient to bear in mind the division of the twenty towns in the two belts surrounding Bari.

Compared to the past, the city has a greater industrial importance and plays a more qualified tertiary role. Industries are mainly located in the western part of the area, the site of the new airport and the future railway cargo area now under construction. This area, already equipped with many transportation infrastructures, is also located next to the highway and will soon have better connections with the harbour.

Bearing in mind the ideal north-south axis, which starts out in the old city and passes across the southern and 'murattiani' neighbourhoods, to Carbonara, Ceglie del Campo, Adelfia and Casamassima, the first belt of towns appears to be split up in two arcs of a circle. The western one, Giovinazzo, Bitonto, Modugno and Bitritto, has a mainly secondary productive backbone, while maintaining a good tradition of agricultural activity.

The main ASI conurbation embraces not only the Bari surroundings, but also the regions of Modugno and Bitonto. Many small factories and storehouses have been developed along the Bari-Modugno-Bitonto axis, thus forming an industrial continuum. The eastern sector, Valenzano, Noicattaro, Triggiano and Mola di Bari, is not as industrialized, but more advanced from an agricultural point of view. There are some factories around Triggiano and between Noicattaro and Rutigliano. The situation is constantly changing in this sector, due to the establishment of some new industries, east of Rutigliano and south of Capurso, and also tertiary activities, Tecnopolis in Valenzano, as well as north of Casamassima along state road 100 to Taranto, where large warehouses, wholesalers and offices are located. Inhabited areas of the first belt west of the axis are much more populated than the ones in the east. A good example is Modugno. Its population almost tripled between 1951 and 1981, 13,421 to 33,830 inhabitants, whereas in the previous forty years it remained stable, at the 1911 census it had 12,464 inhabitants. The towns of the second belt are rather small and only recently have become interested in town-planning processes. This organization was accomplished without any supra-municipal directive, still missing today. The ASI

Scheme is the only regional, although sectoral, document drawn up at this level.

An Inter-municipal Plan (PIB), started ten years ago, is still in its cognitive phase. Studies for a Special Project for the Bari Metropolitan Area, expanding it to the whole province, have been carried out and deserve further consideration at the regional level. Until today the region has been subject to municipal plans only because a 1967 Act called for town planning schemes to be drawn up in all Italian towns before development was permitted. In this context, the city of Bari is an exception as it has had several plans throughout the years. Two expansion schemes, 1868 and 1926, and a reclamation scheme for the old city, 1931, drawn up according to Act 2359/1865, as well as two more drawn up according to Act 1150/1942, 1954 and 1976. The present scheme, undoubtedly oversize, envisages the urbanization of 67% of the city's territory and a population of 628,000 inhabitants by the year 2011. The regional organization which took place spontaneously up to the sixties, conditioned all town planning documents subsequently produced. Each one of these has borne in mind the existing regional situation and tried to rationalize it within each context. Yet, the overall result has not been good. It is impossible to foresee when the A. M. B. Regional Plan will be drawn up but this appears essential to rationalise the situation and to avoid future problems.

Conclusions

Of the innovations which can be found throughout the history of transformation in the metropolitan area of Bari few are unique. For the most part they are the same as those which occurred in other regional contexts where the industrialisation process has been completed.

One of the most important is the 'Statuti murattiani', the town planning and building norms annexed to the 'Borgo' map which regulated construction from 1813. The map itself is of secondary importance compared to the statutes. The statutes established the principle of alignments along the roads, from 14 to 16 m. wide on average, as well as that of the perimeter construction of blocks limiting them to two thirds of the surface area, the remaining portion paved or covered with grass for public uses. These introduced into Apulia the block formed by multi-family buildings. Another innovation is represented by the expropriation and transfer, to the municipal administration, of the area for the construction of the Borgo, most of it belonging to the State and to religious bodies. The municipal administration gave the lands, in concession, to private individuals who requested a license to build, upon payment of a tax. The availability of building lands, the characteristics of the Borgo map, together with the compliance to the norms of the 'statuti murattiani', permitted Bari to expand in a very advanced way, for those years. This supported its population explosion and social and economic growth.

Innovations common to other regional contexts relate to progress in the field of infrastructure and transportation as well as energy supply due to the importance they have in the formation of metropolitan areas. The construction of the Acquedotto Pugliese, instead, represents a specific and very important innovation. Without water supply, how could the population and non-agricultural production increase?

The industrialization process, especially after 1945, was facilitated by the population and the availability of infrastructure. It does not show any specific characteristics, except for the direct intervention of the government with the establishment of the ASI (Industrial Development Area). The ASI Consortium, in compliance with an appropriate Act, expropriated and built the Bari-Modugno-Bitonto industrial area in a few years.

The very fast growth of private transport is an innovation common to all industrialized countries. However, in Bari the increase in numbers of automobiles did not go hand in hand with an adequate increase in roads and, above all, of parking areas. This is a drawback today, felt not only in the central part of the city of Bari, where most tertiary, commercial and managerial activities are concentrated, but also in central parts of the outlying inhabited areas. Municipal town planning schemes, with the introduction according to Act 765/1967, of divisions into homogeneous regional areas, did not constitute a true innovation, if compared to the theoretical and practical instances already mentioned. Innovative, instead, were the town planning standards and the maximum building limits established by them which, sometimes, have also had negative effects.

References

Acquarone, A. (1961), *Grandi Città e Aree Metropolitane in Italia,* Zanichelli, Bologna.

Ancona, G. (1984), *Pendolarismo e Bacini di Traffico in Puglia,* Cacucci Edit., Bari.

Autori vari. (1975), *Puglia: Documento Programmatico Pianificazione Territoriale,* Assessorato all'Urbanistica della Regione Puglia, Bari.

Cafiero, S., Busca, A. (1970), *Lo Sviluppo Metropolitano in Italia,* Svimez, Roma.

Casmez (1986), *Studi di Fattibilità per il Progetto Speciale Area Metropolitana di Bari,* Roma.

Di Bari, D. (1984), *Urbanistica dell'800 in Puglia,* Ediz. Full Books, Bari.

Di Ciommo, E. (1984), *Bari 1806-1940,* F. Angeli Editore, Milano.

Di Comite, G., Di Comite, L. (1987), *Evoluzione Demografica e Redistribuzione Territoriale della Popolazione: il Caso dell'Area Metropolitana di Bari,* Ricerca M.P.I., Pescara.

ISPE Centro Piani (1986), 'Progetto '80: Proiezioni Territoriali', *Urbanistica,* No. 57, Torino.

Marchese, U. (1981), *Aree Metropolitane e Nuove Unità Territoriale in Italia,* Ecig, Genova.

Petrignani, M., Porsia, F. (1982), *Bari,* Editori Laterza, Bari.

9. KECSKEMÉT

Katalin Korompay, Hungarian Institute for Town and Regional Planning (VATI), Budapest.
Coordination, Nora Hörcher, United Nations HABITAT East-European Information Office, Budapest, Hungary.

Introduction

Kecskemét is situated in the centre of the Great Hungarian Plain between the Danube and Tisza Rivers, on the highway E5 which links Scandinavia with the Balkans. With a population of nearly 100,000 it is one of Hungary's medium-sized towns. Kecskemét is a dynamically growing county-seat with industry based on the excellent agricultural potential of the surrounding region. It is a centre of cultural life and tourism.

The settlement emerged at the crossing of three important European and national roads. The first settlement, established during the twelfth century was destroyed by invading Tatars in 1241. The town's resettlement and consolidation took place in the fourteenth century. During the 150 years of Turkish occupation in the sixteenth and seventeenth centuries the population grew four-fold and the built-up area tripled. During the eighteenth and nineteenth centuries growth was far less spectacular. The medieval fortifications of trench and palisade were demolished in the late eighteenth century. The railway was built in 1851. Important development started in the years 1860 to 1870 and gained enormous momentum after 1880. This momentum was broken by World War I. After long years of stagnation the town became the county seat and has been developing dynamically since the 1960s.

This study examines the evolution of the town at three levels, the core-continuous built-up area, the core centre, and specific areas adjacent to the core. There were three characteristic stages of development which correspond to clear changes in the style of townscape. The stage of intensive growth between 1880 and World War I was followed by a period of less dynamism. Then in 1950 the town became the seat of the county. It was at that time that the recently ended phase of the town centre's further development and of the construction of new housing projects started.

Level 2: Core-continuous built-up area

1880-1914 Economic prosperity, reconstruction and expansion

The late nineteenth century was a period of great economic prosperity all over Hungary. The political relationship with Austria was consolidated permitting the bourgeois transformation of the country. Thanks to the ideas of famous agronomists and the organisational skills of excellent economist-mayors agricultural development that started was suited to the town's potential. From among the plants that could be grown, given the conditions of soil, relief and climate, vine and fruit growing, as well as horticulture helped to fix the wind-blown sand. In a few years apricot and melon orchards and vineyards were planted on 8,000 cadastral acres around the town. Pigbreeding also gained importance. The town was able to settle its debts. Agricultural industries were established, mills, a canning factory, a distillery and an abattoir.

Horticulture reinforced the town's systematic pursuit of urbanisation. In 1869 a new map was drawn and the large-scale reconstruction of Main Square, bringing about a fundamental transformation of the centre's morphology, was begun in the eighteen eighties.

During the Turkish period the town's structure, inherited from the Middle Ages, became crowded. Small blocks were inserted between the elements of the irregularly-shaped Main Square, but the layout consisted almost exclusively of single-storey houses. This unplanned, rural town centre was not worthy of the town anymore. It was decided to create a huge square which could provide sufficient space for the fruit and melon market. A broad avenue was planned to the railway station with four rows of trees and a promenade. To improve transport it was intended to lengthen the avenue to make it a main road crossing the town. Other plans were formulated to widen and straighten roads. The square's new contours were formed by newly erected houses which '...authorities insisted should express the newest and most typical trends of Hungarian architecture'. As a first step a national competition was announced in 1887 for the design of the townhall. The old town hall and a small block had to be demolished to make way for the new building. The new headquarters of the town council was completed in 1896 designed by Ödön Lechner, an architect of European fame at the turn of the century.

The first idea was to extend the main square in the shape of a regular octagon but, due to the value that old churches represented in cultural history the town insisted on preserving the church. Therefore, the irregular rectangle of the square, almost 150 metres by 300 metres, was located between and around the existing buildings. First, two blocks of houses were demolished in 1895 at the site of today's Szabadság Square, the impressive synagogue that had been built not long before was selected to be the new frame of the square. Széchenyi Square was created when the narrow block which partly concealed the baroque church was demolished. The present contours of Main Square and of the adjacent squares were formed by construction started at the turn of the century with slight adjustments to the blocks' limits. The development yielded a high-quality, interesting architecture. As the first stage of implementing the main road cutting across the town it was decided, in 1893, to create Rákóczi Street by street widening and straightening. The 42 metre broad, 500 metre long promenade was not completed until 1950-60. The Rákóczi monument and the new railway station planned at the end of the street were never built, because World War I broke out.

The ensemble of Main Square and the head of Rákóczi Street were surrounded by noteworthy public buildings and mansions. The theatre, secondary schools, banks and casinos were constructed. The huge building of the Academy of Law became not only a dominant feature of Main Square, but also an institution of higher education that enhanced the town's cultural importance. Administrative institutions received worthy positions, the Palace of Justice was built and the establishment of the tradesmen's centre took place. Commerce was installed in business houses. On the lower storeys of public buildings and apartment houses, shops were established. Main

Square was reserved for the market, and landscaping was soon completed both here and along the main roads. Streets and footways were paved.

Cultural progress took shape in buildings erected in other parts of the town. Primary schools, trade schools, a secondary school for modern languages and sciences, a conservatory, a high school of agriculture and vine-growing, suburban primary schools and an orphanage were founded. A museum, a library, an exhibition hall, a post office and a hospital were installed in new buildings. In 1909 a colony of artists was founded on the periphery of the town.

Industrialisation, the railway and construction by the army transformed the zone surrounding the town. The establishment of a suburb of detached villas, the new residential quarter meant a further expansion of the town.

The impressive course of development that had transformed the town's morphology was interrupted by World War I. By the last quarter of the nineteenth century the crowded settlement consisting of single-storey houses, with its centre marked visibly by five to six spires had overstepped the ring of the seventeenth century moat in the south-east and west only. By the time development had been completed at the turn of the century the spacious area in the town centre and the main road were flanked by two, three and four storey buildings and the pattern of single storey residential areas was also transformed into a development of terraces, composed of eclectic and art nouveau buildings. In the outskirts a specific urban structure emerged, colonies of vine-dressers. These colonies, Helvécia, and Katonatelep, were composed of single storey, rural houses designed and built in an uniform way on regular plots for hired workers.

1914-1950 Period of stagnation

World War I broke the enormous momentum of urban development that had started around the turn of the century. Some plans were never implemented. With war, Hungary's losses and the economic crisis in the background, the forces of urbanisation received no stimulation in Kecskemét.

From among the goals of previous plans the construction of the rows of houses along Rákóczi Street continued in the twenties, but it was never completed. Only a few important public buildings were built, the new secondary school of the piarists, the school mistresses' training school and the Central Post Office. This period was characterized more by reconstruction than by new construction, existing buildings received new functions, for instance the former summer casino was turned into a museum, a new boarding school was added to the Academy of Law and the baroque County Hall was reconstructed and expanded.

Previously defined aims of urban planning that had materialized in the master plan of 1913 could not be implemented. Building activities dropped to one tenth of their former level and large-scale demolition accompanied by reconstruction could not be undertaken. The avenue leading from Main Square, towards the south-west that had been planned as the continuation of Rákóczi Street was not constructed, nor were plans for widening and reconstructing the road that connected the suburb of villas to Main Square.

The reconstruction of Main Square and of the main road, at the turn of the century, yielded historically valuable areas with architecture typical of Kecskemét. But other areas were left behind by that momentum of urbanisation and still remain today. During the period under discussion housing needs were satisfied mostly in the area of detached houses on Vasi-hegy, opened up by a partially curved road system, where the layout pattern, interspersed with green spaces, was supplemented by an up-todate town plan. In order to make up increasingly acute deficiencies in the infrastructure, sewers were built, streets were paved and parks were established.

World War II did not cause serious damage in Kecskemét. Since, however, the country had other tasks of reconstruction to accomplish, Kecskemét remained at the fringe of major developments after the war. It was only in the fifties, when the town became county seat, that it started to recover.

1950-1980 Development and growth

Thanks to social, economic and technical progress achieved during the last few decades Kecskemét became one of Hungary's dynamically growing towns. The rate of growth between the fifties and the eighties is indicated by a demographic increase of almost 65%, the strengthening of the industrial base, a considerable growth in tourism, the quantitative and qualitative improvement of housing supply and services, and the gradual development of public utilities.

Due to its geographical situation and favourable location in the national transport network, to unlimited possibilities of urban growth and to its radio-centric urban pattern, Kecskemét's traditional central

183

organisational and supply functions were consolidated. The transformation brought about radical changes in the settlement's morphology, structural pattern and layout (Figure 9.1.). Extensive demographic growth was induced first of all by industries and services installed in the town. Contiguous industrial zones were formed in the south and east, The necessary increase in the supply of housing was ensured, primarily, in new housing projects constructed on previously unbuilt areas. Thus, between 1960 and 1970, on the site of an eighteenth century cemetery in the south-east, a medium-sized housing project was constructed of prefabricated blocks. Later, Széchenyiváros was built in the north with panels provided by Kerskemét's housing factory. Presently about 10,000 people live there in mostly ten storey blocks of flats. In the northern and eastern parts of the town two large, contiguous areas were allotted for the construction of one and two storey detached houses. Two to four storey apartment houses were also built here, so that the ring surrounding the town is composed of a great variety of residential areas of different technologies, layout patterns and organisational structures.

Figure 9.1. *Core-continuous built-up area: stages of development*

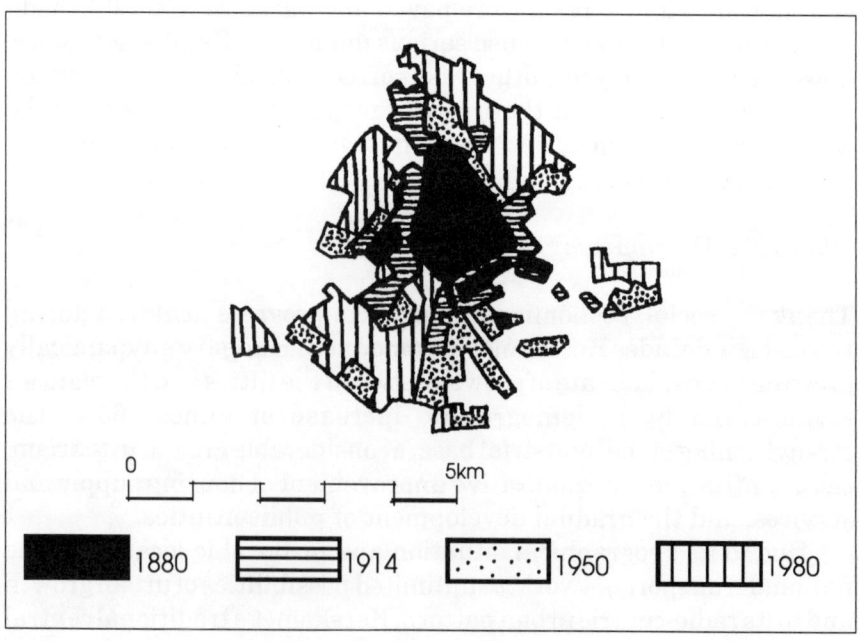

The predominantly single storey town in which two, three and four storey buildings had been built at Main Square and along major streets, has been surrounded by four to ten storey housing projects since the 1960s. Given the lowland character of the settlement, this change is clearly reflected in the urban skyline.

The reconstruction of downtown areas also started in the late fifties and early sixties. First, the row of houses along Rákóczi Street was completed and vacant lots at Main Square were filled in. The new hotel, built after extensive professional debates, called forth a structural change. The perpendicularly located, six storey building created a small bay, thus separating Main Square from the adjacent square that had joined it in a funnel shaped form.

The construction of the main road to cut across the town that had been foreseen by the urban renewal programme of the late nineteenth century was also implemented in the sixties and seventies. The avenue, designed as the continuation of Rákóczi Street to replace the south-west oriented, zigzagging main road, was meant to serve as an extension of Main Square. The idea, dating from the turn of the century, to straighten the road was re-evaluated. A new main axis was linked to the existing ensemble of Main Square, capable of housing the buildings for the central functions demanded by the explosive urbanisation.

As a first step the road was cut through and two enormous eight storey linear, housing blocks were built to form a 350 metre long frame. The huge buildings of county administration and cultural life were built on the side that is directly connected to Main Square. Shops and services, three big department stores and a service centre were installed on the other side. The ground floors of the housing blocks also contain shops.

The modification of the urban pattern was extended to link to a residential zone. The mostly single storey, partly deteriorated houses built during the second half of the last century were in poor condition. Arpád-város started in the 70s with total reconstruction. The mostly ten storey prefabricated panel houses were built in a new system, perpendicular to the next old main street of the town. This meant that a segment was carved out of the town's area. Subsequently this process was not continued as planned, urban renewal today is based on other, spatially differentiated principles. The reconstruction of the former residential area near Main Square in Kecskemét is a typical example of Hungarian urban development in the 60s and 70s.

Besides major reconstruction, small-scale intervention was also undertaken at certain sites in the town centre. First, three to four storey apartment houses were built in a rather haphazard way between single storey houses. Building permits were issued in accordance with the town planning concept of the late sixties which stated that the mostly obsolete single storey layout could be gradually replaced. But the extremely slow pace of reconstruction has called the validity of this concept into question.

At the end of the seventies an important decision related to transport had a considerable impact on the morphology. Traffic crossing Main Square was diverted so that Szabadság Square became a pedestrian zone. The one-way traffic diverted to the roundabout system of the existing road network was intended to be a transitory solution only. The construction of the new element of the radio-centric urban pattern, an inner boulevard, was soon begun. The new road that more or less follows the existing track is being constructed according to the master plan of 1978.

The town's most acute problem is the high level of the water table caused by its lowland location. The completion of the downtown sewer system is a precondition of reconstruction. Where the problem of sewerage has been solved, residential areas are being renewed. One and two storey, high density layouts and three to four storey terraces are being built preserving the existing road network and valuable buildings. The aim set by the present master plan is to carry out an up to date renewal that takes account of the need to preserve values, based on a study of the town's historic, architectural and townscape values.

Development and change of the morphological structure

By the end of the nineteenth century the town's territorial extension overstepped the route of the boulevard that followed the former line of the medieval fortifications. In the south-east and south-west rural type residential areas joined the outer side of the boulevard. In the western part of the town a garden city consisting of one to two storey, detached buildings, called Villatelep, Colony of Villas, was built before World War I, encouraged by the attraction of the new large public institutions built at the turn of the century. This type of development was a new phenomenon in the town, where the typical layout had been either a gable or L shaped low rise development, or development in unbroken rows of two to three storey buildings. It expressed the new way of life of the town's upward moving, wealthy middle class.

The hills to the north of the town were inhabited after World War I. Less wealthy citizens built colonies of one and two storey detached houses in more distant areas from the centre. Buildings were not on the plots' side-borders, these houses were detached, similar to villas, although in the case of small plots it meant a wasteful use of land.

The most crucial change in the morphology of housing construction was brought about in the early 1960s by the appearance of housing development projects, without specific boundaries, that consisted of four to five storey buildings. The first houses were built in Leninváros, on both sides of the boulevard, in the southern part of the town. The previously unknown solitary masses of tower and slab blocks appeared together with the lower 'pies' of public institutions. Only the technology used for the construction of the precast panel housing project in Széchenyiváros in the western part of the town was changed. Several phases of Hungarian housing are represented in the architecture of the five stages of construction of the new quarter planned to accommodate 45,000 people. From the initial, parallelly located, ten storey linear blocks, to courtyards, to a residential quarter of varied layout that includes houses of two to ten floors.

During the 1960s and 70s colonies of small condominiums were built in other parts of the town. The task of greatest urgency, however, is the renewal of the town centre. Organized actions to reconstruct the main square and its immediate surroundings are needed along the inner boulevard. The renewal of the historic quarter situated to the west of the main square has already started and has respected all the existing valuable elements. The one, two and three storey high buildings of the downtown area can be renewed. Either the house's interior is renewed behind the valuable eclectic or art nouveau facade, or a new house adapted to the scale of the environment is built profiting from the possibilities offered by the existing network of plots. Since the late seventies this process has started all over the town in residential areas freed from the former aim of the town centre's total reconstruction. The present master plan envisages the stability of the morphology in the downtown area. Important transformations are permitted only in the peripheral zone where the quality of rustic buildings is poor and reconstruction that increases the number of floors by one or two has already started.

Level 3A: Core-centre

Transformation of the core by traffic

Around 1880 the road network of Kecskemét, whose urban structure had followed a mostly natural pattern of growth, consisted of tortuous, overcrowded, unplanned streets and squares despite the basically circular-radial structure. The first urban planning decision outlined a 42 metre wide and 500 metre long new avenue opening towards the railway and envisioned large-scale demolition in order to create a monumental main square. It was planned to align the dangerously zigzagging main road leading south-west out of the town and transform it into a new highway crossing the town, connected to the new avenue. The zeal of rebuilding, broken in 1914, was sufficient for the development of the main square and of the avenue only (Figure 9.2.). The idea to break through a road for transit traffic had been kept alive and was realized in the 1970s, but for a short time only. By the end of the seventies the main square and its immediate surroundings became a pedestrian zone. The one-way traffic, diverted to the roundabout system of the existing road network, was intended to be a temporary solution. The construction of the new element of the radio-centric urban pattern, an inner boulevard, was soon begun. The new road that more or less follows the existing track is being constructed according to the master plan of 1978.

Transformation of the core by building

Planning decisions induced by traffic coincided with functional, social and aesthetic reasons for rebuilding. The square's new contours were formed by newly erected houses.

Between 1896 and 1913 eclectic and art nouveau buildings, representing the period's architecture, were constructed around the main square and along the inner part of the avenue. The first two to three storey edifices built in an unbroken row were public buildings, but they were soon followed by a new type of urban residential buildings. Until the end of last century the inhabitants of this town, whose main occupation was in agriculture and related industries had lived solely in one storey detached houses. The overcrowded layout, in small plots, resulted in dense residential areas, while multi-storey tenement blocks with their many flats represented a new phenomenon in the Kecskemét of the 1880-1890s. In accordance with the town's

Figure 9.2. *Development of the core-centre*

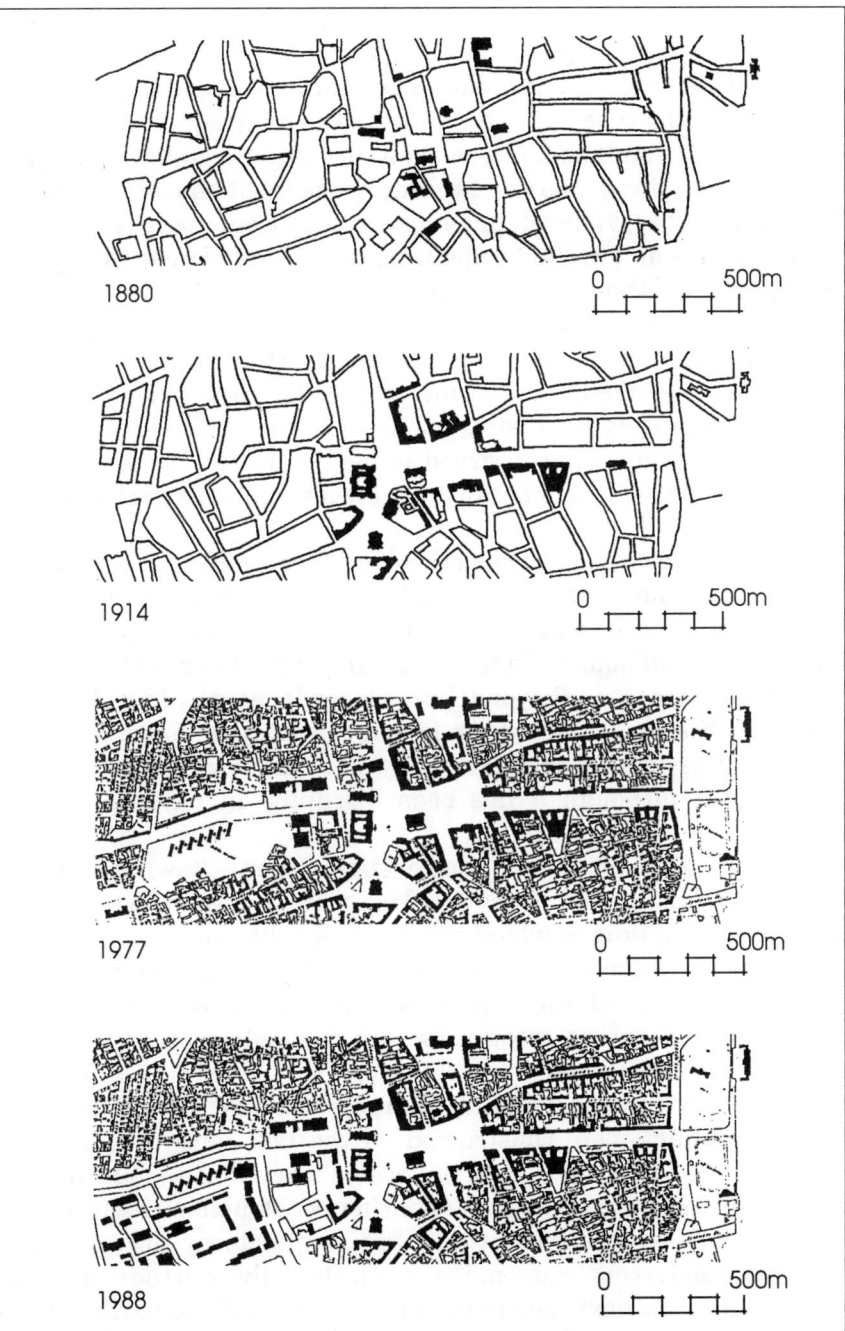

1880   0   500m

1914   0   500m

1977   0   500m

1988   0   500m

growing importance in trade, shops occupied the ground floors of residential, and even of certain public buildings.

The new public buildings raised around the main square at the turn of the century demonstrated the increased importance of the town. The redevelopment of the surrounding residential areas used several building types. In the vicinity of the main square, in some of the adjoining small squares, as well as along the old road leading towards the railway, one and two storey tenement houses were built. The single storey development was renewed along the other thoroughfares and in the residential areas situated near the centre. The former rural house that had stood perpendicular to the street was replaced by a form that was parallel to the street so that an almost unbroken row could be developed, although the original network of plots was preserved. In certain places the earlier development pattern, consisting of rustic, porched houses built in the early nineteenth century was conserved, either because the house satisfied the needs of its owners or because the owners did not have the means to rebuild their home.

Thus the town's characteristic townscape evolved at the turn of the century. The unbroken row of high-rise buildings in the centre with eight towers rising above them is surrounded by a partly unbroken row of houses. In the outer ring of the town centre as well as in the adjacent areas beyond the boulevard the typical development pattern is the rural, gable-form layout consisting of low rise, single storey houses. This characteristic layout of the town centre can still be detected, although it has been disrupted in many places by subsequent developments.

Following the interwar period of stagnation the town centre was transformed in the years 1960-1970 in accordance with its role as the county seat. Buildings which could not be inserted among edifices around the main square were erected along the south-west main road which had to be rebuilt. This is where the new centre of county administration, representing a new vertical emphasis, was developed together with new department stores and shopping centres destined to meet the increased demand for trade and services. Eight to ten storey buildings were constructed on the two sides of the road's outgoing section. Their uniform design over 200 to 500 metres drastically disrupted the scale of earlier construction (Figure 9.2.). The aim was to modernize the entire single storey town by total demolition and reconstruction. This zeal lasted only until the residential area reaching the next main road had been rebuilt. Although architects sought to adapt the outward appearance of buildings to the situation of the town centre the ten storey panel houses that replaced the mostly single storey development recall the dreariness and disproportion of peripheral housing projects and form an appalling contrast with the old environment. At the time of planning and construction it had become evident that this was not the way to follow in the renewal of the town. Therefore the inner, valuable development of the main road that borders the area of renewal from the east has been preserved and new and old are in harmony here.

In the sixties and seventies rebuilding did not leave the main square and the avenue untouched. The row of buildings lining the main square was completed by a few new elements and the row of houses along the avenue was also filled up. The new hotel also resulted in further structural change to the Main Square and its immediate surroundings.

Transformation of public space by new modes of use

The will to renew the town by the end of the nineteenth century was focused on the construction of a monumental main square. The aim was to provide a market place of national importance for the produce of vine and fruit growing. This aim was realized at the turn of the century after the main square's size had been increased to 300 metres by 150 metres. Several times a week, from the early morning till the afternoon, merchants filled the market place according to a determined order. This lasted until 1930s when the market was removed from the centre and the main square was converted into a park.

Simultaneously with the growth of its importance the role of traffic has been radically reassessed during the period under study. Around 1880 it was planned to carry out large-scale demolition and construction in order to align and widen the road and build a national thoroughfare across the centre. After the 80 to 90 years it took to realise this, at the end of the 1970s it was decided to divert transit traffic from the centre. The idea of a pedestrian town centre was born and implementation started by closing off the main square to traffic. The street that connects the main square to another near by was pedestrianised and gained new life as a shopping mall at the end of the seventies.

The way the notion of public space was understood when the new centre was extended differed from the old approach and this has caused the problem of harmonious connection between the two. No streets and squares were created. It is likely that in order to create harmony between these and other public spaces, consisting of traditional squares and streets, it will be necessary to develop zones of transition in the future.

Transformation of the morphological structure by new modes of design

In the 1880s the morphological structure of Kecskemét was rather homogeneous. The first structural change was due to the rebuilding of the centre and the erection of multi-storey edifices. Extensive areas were opened in front of the new, higher buildings, in most cases newly divided plots permitted that the width of the buildings be proportional to their height. Despite the preservation of existing elements the realisation of this systematic urban planning concept resulted in a modified scale in the centre's structure. The looser design of buildings, that was typical in that period and development in unbroken rows, created a smooth transition towards the adjoining residential area whose scale was not modified by reconstruction.

In the recent period of large scale structural transformation, when the new centre adjoining the main square was developed, architects and urban planners used entirely different elements than in the past. Together with the already mentioned change in the understanding of traffic, the design and layout of buildings also differed from the approach that had resulted in development in unbroken rows along the two sides of the street. In the zones of transition peculiar 'pie' shaped low rise wings of building represent an effort to link the new architecture to the old. To re-establish the harmony that has been disrupted will be a task for the future.

Level 5: Specific areas of change

Characteristically, market towns of the Hungarian plain have a ground level morphology. The basis of life for centuries has been agriculture and associated trades. These determined the life of the settlement until the middle of the twentieth century. The effect of urbanisation on Kecskemét meant that the agricultural way of life was pushed further and further from the centre with accompanying transformations in the morphology. These effects can be clearly seen in the areas surrounding the centre of the town.

Figure 9.3. *Specific area of change adjacent to the core*

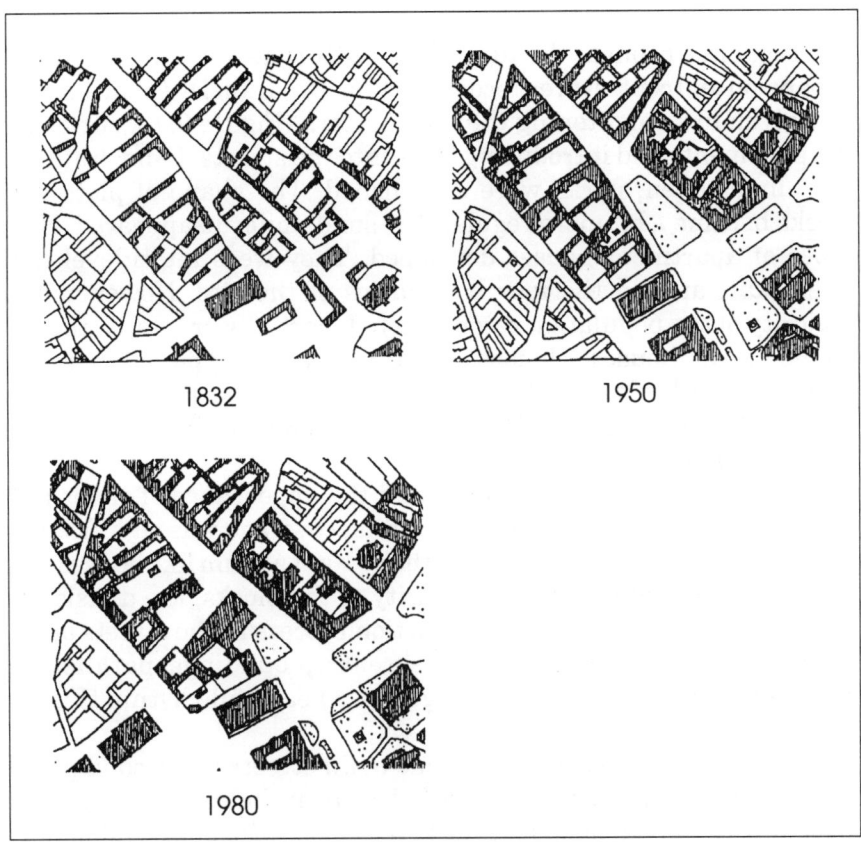

1832   1950

1980

1860-1880

The road system developed as a result of the organic growth of the town. The shape of blocks and land divisions were totally irregular. Small mills, operated by draught animals, often occupied the triangular recesses in streets. Ten metres was the average width of the roads (Figures 9.3., 9.4.).

Buildings were of the village type, beginning with one unit at right angles to the street. They were enlarged by extending on the street boundary forming an L shape. In some cases, wealthy citizens added open corridors and verandahs. At this stage the town was made up of single storey houses with the exception of some around the main square.

1914-1950

The economic boom at the turn of the century resulted in territorial expansion and central area reconstruction. In parallel to this attention was given to straightening the road pattern (Figure 9.4.), surfacing the pavements and introducing trees into the streets. Land drainage was undertaken, parks were established. The shape of plots and blocks became a bit more regular, the small mills disappeared. New types of apartment houses developed. They were multi-level and often built around the four boundaries of the site. However, the majority of the population reconstructed, partially or totally, the old single storey houses. Houses of the mezzanine type, often richly decorated with plaster stucco, were rebuilt in the Romantic Style around 1870-80, in Eclectic Style from 1880 to 1910 and in Art Nouveau between 1910 and 1925.

With the growth of the house and the appearance of shops and workshops, land utilization and building coverage was significantly increased and an intensity close to the present atrium like houses was formed. As a result of the irregularity of the blocks, the division of land, and the incomplete reconstruction process in the streets of the residential area surrounding the centre, buildings parallel, or perpendicular, to the street changed in a beautiful rhythm (Figure 9.4.).

On the main square and in some main streets the second phase of reconstruction was taking place. Significant mezzanine type, one floor houses were demolished due to the effects of urban development and increased land values. On the newly-formed lands a new type of multi-level town apartment house appeared with open-sided corridors, in many cases built round all the four sides of the site. But these were the homes of people who had given up farming therefore there was not enough demand for this type of building in Kecskemét as compared to large industrial towns at the turn of the century. In the course of the rebuilding of the main square several important public institutions were built, which together with the former buildings and the new apartment houses result in a completely new morphology in this area.

1950-1980

State developments starting in the 1950s at first aimed to continue the earlier, beginning of the century, urban development actions. One of the aims was to finish the building out of Rakóczi street. This meant traditional, three-level, brick-wall, high-roof houses built on lag lands.

Figure 9.4. *Specific area of change adjacent to the core*

1832

1950

1980

A more significant intervention in the morphology of the settlement was the location of a new hotel in the entrance to a side square (Figure 9.3.). The building represented a totally new urban architectural characteristic not only to Kecskemét, but to a whole era. The more modest, lower intensity living area was left out of consideration. In the spirit of the ideology of that period it was declared by architectural means that the ground level town, comprising 80 per cent of the 160 ha. city centre should be totally reconstructed. Such interventions came one after another between 1960 and 1970, all of them having the severest morphological consequences.

The 1968 plan envisaged the reconstruction of the town in the form of three to four level collective houses on each site. On the basis of the approved plans, free-standing four to six flat houses were built into the closed row, or still comb-like streets, about a dozen on the whole territory of the city centre (Figure 9.5.). However, on the two sides of a small park a private development succeeded in finding a harmonious form which complemented the urban structure.

The first housing estate appeared on the outer areas of the city centre, on an old cemetery. The scale, quality and position of this layout detracted further from the town's morphological coherence.

In the 60s and 70s the town started to show its increased importance by building two large, ostentatious buildings in place of two beautiful buildings on the main square, which were declared dangerous. However, there was a need for space for the county administration. The site of the huge new buildings was chosen along the axis planned at the turn of the century, behind the most significant buildings of the main square, the church and the town hall. Even with respect to the larger buildings of the main square these ten and twelve storey bulks meant a structural and dimensional change. This reflected the practice of urban construction in the early 70s (Figure 9.4.).

The plan, prepared at the end of the decade, on the basis of historical, aesthetic and living values and considering real possibilities of reconstruction or rebuilding, evaluated the low-rise buildings around the city centre in a different way and envisaged preserving the greatest portion of it. After the approval of the new plan, the new, low-rise, atrium-like, construction method on certain areas provoked national interest. Tiny restaurants, fashionable shops, and popular service activities found places in the territory formerly almost solely for housing purposes.

Conclusions

The levels of analysis and the time periods considered reveal, in broad terms, morphological responses to the influences which affected the development of Kecskemét. These influences, or challenges, are grouped in categories of social, economic and technical along with their implications for built form, environment and use. In each category the date of the influence is shown on the left and the dates of the results, where known, follow their description.

Social

1880-1914 Significant improvement in the material conditions of the population.

New public institutions are built, 1880-1914.
Inner living areas of the town are rebuilt, 1890-1930.
Villa quarter is built, c. 1900.

1887 Increase in the town's authority level.
Construction of new town hall, 1896.

1951 The town becomes capital of the county.
More government finance for housing results in peripheral estates and inner area reconstruction, 1960-1980.
New county administration centre built, 1978

Economic

1870-1914 Increase in role as market centre and the subsequent spreading of working capital
The extension of the central space to accommodate a huge market.
Appearance of factories on the edge of inner area and small workshops in housing estates.
Banking and insurance buildings in the inner area.
Apartment houses by the town and church.

1950-1980 Dynamic industrialisation.
New industrial plants and large companies in the south-east.

1950-1980 Increase in the turnover of goods.
Development of shopping network, 1960-1980.
Building of department stores, 1970-1980.

Technical

Transport system

1851,1867 Railway construction.
Reconstruction of road to station, 1860-1890.
Opening of new traffic line to station, 1880-1914.
Concentration of industry around goods station.

1960-1970 Increase in car traffic.
Road straightening and international motorway, 1975.
Increase in capacity of ring road, 1978.
Exemption of personal traffic from main square, one
way system and pedestrianisation, 1978.
Widening of inner ring boulevard, 1982-1989.

Urban development

1869 to date. Demand for a regular planned road network and
avenue.
Four lines of trees in Rákóczi Street, 1900-1913.
Varying regulations.

c.1890 Prescriptions for a 'town-like' appearance.
Up to date town centre, 1892-1914.
Formation of a closed outlook town, 1880-1930.
Rows of multi-storey houses.

1968-1978 Demand for the continuous reconstruction of the old single
storey areas.
Partially realised plan of building four storey
collective houses, 1968-1978.
Regulation for the construction of one and two storey
houses, maintaining the street system and plot
boundaries, 1978.

1964-1988 Ideas originating from the 'Charta of Athens'.
Peripheral housing estates, 1964-1988.
Inner area housing renewal, 1978-1985.
Low rise supermarket in housing estate, 1977-1987.
New architectural form for renewal,1970-1984.
New architectural form for infill, 1968-1977.

Building type, construction technology, investment systems

1875-1900 Demand for larger, private family houses
L shaped houses, fronting the street with 2 to 4 rooms, glass corridor and mezzanine.

c.1900 Demand for apartment houses.
One to three storey apartments with open corridor access, occupying two to four sides of the site.
State construction of varying types:
brick, high roof, three or four storey infill, 1950-1965.
concrete block, line and point, three to five levels, small estates, 1960-1975.
reinforced concrete frame construction, individual houses, 1970-1982.
panel construction systems, large estates in the centre and on the periphery, 1975-1989.
Private and cooperative apartment houses, detached two to four stories, or in rows, 1962-1989.

References

Bácskai, V. (1965), *Magyar mezvárosok a XV. században (Hungarian market towns in the 15th century)*, Akadémiai Kiado, Budapest.
Baczo, P. (1977), *Jelentés a Kecskemét-Kossuth téren végzett ásatásókrol (Report on the excavations carried out on Kossuth Square, Kecskemet)*, unpublished.
Bagi, L. (1896), *Kecskemét multja és jelene (Kecskemét past and present)*, Kecskemét Város Közönsége, Kecskemét.
Borovszky, S. (1910) *Pest, Pilis, Solt, Kiskun: vármegye monográfiája (A monograph of four counties: Pilis, Pest, Solt, Kiskun)*, Országos Monográfia Társasag, Budapest.

Csánki, D. (1880), *Hazánk kereskedelmi viszonyai I. Lajos korában* (*The position of commerce in Hungary at the time of Louis I*), Weisman Nyomda, Budapest.

Csánki, D. (1896), *Magyarország történeti földrajza a Hunyadiak korában* (*The historical geography of Hungary at the time of the Hunyadis*), Akadémiai Hornyánszky Nyomda, Budapest.

Entz, G. Genthon, I. Szappanos, J. (1961), *Kecskemét*, Müszaki Könyvkladó, Budapest.

Förds, L. (1934), *Kecskemét települési és gazdaságföldrajzi képe* (*The settlement, economic / geographic statistics of Kecskemét at the end of the 18th century*), Elsó Kecskeméti Nyomda, Kecskemét.

Ger, L. (1975), *Vár-épitészetünk* (*Our castle architecture*), Müszaki Kiado, Budapest.

Hornyik, J. (1860), *Kecskemét város története oklevéltárral* (*An archival history of the town of Kecskemét*), Gallia F., Kecskemét.

Hornyik, J. (1927), *Kecskemét város gazdasági fejldésének története* (*The history of the economic development of Kecskemét*), Kecskemét Törvényhatósági Jogu Város Muzeumának Kiadványai, Kecskemét.

Juhász, I. (1977), *Kecskemét története* (*The history of Kecskemét*), unpublished.

Kecskemét and Pest County (1939), *Document and data collections.*

Korompay, Gy. (1954), *A magyar falu településtörténete és településfatjtái* (*The history and typology of Hungarian villages*), Magyar Épitmüvészet, Budapest.

László, Gy. (1944), *A honfoglaló magyar nép élete* (*The life of the conquering Hungarians*), Magyar Elet, Budapest.

Major, J. (1960), *Telektipusok kialakulásának kezdetei Magyarországon* (*The beginning of the formation of plot types in Hungary*), Településtudományi Közlemények 12, Tanknyvkiadó, Budapest.

Makkai, L. (1963), *A magyar városfejldés és városépités történetének vázlata* (*A sketch of urban development and the history of towns in Hungary*), Tanknyvkiadó, Budapest.

Maksai, F. (1971), *A magyar falu középkori településrendje* (*The medieval structure of the Hungarian village*), Akadémiai Kiado, Budapest.

Mendöl, T. (1936), *Alföldi városaink morfológiája* (*The morphology of towns on the Great Hungarian Plain*), Közlemények a Debrecen Tisza István Tud. Egy. Földrajzi Intézetéból, Debrecen.

Mikes, K.B. (1977), *Kecskemét város Tanácsa a XVI-XIX században* *(The City Council of Kecskemét from the 16th to the 19th century)*, unpublished.

Rupp, J. (1872), *Magyarország helyrajzi története f tekintettel az egyházi intézetekre (The topographical history of Hungary with special regard to churches)*, Eggenberger, Pest.

Szabó, I. (1966), *A falurendszer kialakulása Magyarországon (The development of the village system in Hungary)*, Akadémiai Kiado, Budapest.

Szabo, K. (1938), *Az alföldi magyar nép müveldéstörténeti emlékei (Cultural-historical monuments of the Hungarian people living on the Great Hungarian Plain)*, Országos Magyar Törté neti Muzeum, Budapest.

Zádor, A. Rados, F. (1943), *A klasszicizmus épitészete Magyarországon (The architecture of classicism in Hungary)*, Franklin, Budapest.

10. TROMSØ

Halina Dunin-Woyseth, Gunnar Ridderström, Oslo School of
Architecture, Norway*.

Introduction

Tromsø was selected as the Norwegian case study for professional
and practical reasons. It is the administrative capital of one of the
northern counties of Norway, Troms county, and an important social,
economic and cultural centre. It is representative of the many coastal
cities which illustrate specific features of Scandinavian urban
development, the presence of the sea, long distances between human
settlements and a small population in relation to the size of the
territory. The practical reason for choosing Tromsø was the opportunity
to carry out an analysis of urban development as part of a graduate
course in urban design and planning with Tromsø as the reference point.

---

* The empirical data which forms the basis of the Tromsø case-study and the
drawings presented, were provided by twenty students from the two Norwegian
Schools of Architecture: Faculty of Architecture, Norwegian Technical University
in Trondheim, and the Oslo School of Architecture, who participated in the above
mentioned joint course. The course took place in the Spring semester 1988 and the
teaching staff responsible were: Halina Dunin-Woyseth, Karl Otto Ellefsen and
Dag Tvilde.

The analysis was a by product of the course, which aimed to produce proposals to improve the built form of the town. The North Norway Architects Associaton provided financial support to organise the material into the URBINNO format. The study was also supported by the Department of Town Planning, Tromsø City Council which gave access to important background material.

The study uses three steps for each of the seven levels. A historical review of development, establishing general features of development and changes in urban form, then tracing and 'reading' innovations from these changes. Conclusions are drawn in the form of a table of innovations from the seven levels. The innovations are categorised as 'hard' or 'soft', the former relating mainly to technological progress whereas the latter encompass those of lifestyle preferences and ideals.

Level 1: Region, Troms County

When investigating a city in north Norway it is difficult to suggest strict boundaries for the hinterland and it is impossible to discuss its metropolitan area. The whole coastal area of the country can be regarded as a single region, since it has similar characteristics, a varied coastline, long distances between settlements and low, overall, population densities. This makes the situation different from cities in continental Europe. For the purposes of this study the administrative boundaries of the Troms county were used to define the region. The city of Tromsø is the administrative centre of the county. The only other city is Harstad, with 26,000 inhabitants. Troms county is located between the Nordland and Finmark counties, Finland and Sweden (Figure 10.1.). Offshore islands and deep fjords combine to form a long, broken and dramatic coastal strip. The county has an area of 26,000 sq. km., 251 of which had been cultivated by 1979. Troms constitutes 8% of the total area of Norway, with only 147,000 inhabitants.

The sea has always been the main source of life and income in this region, shipping and fisheries are the main trades. Industry in Troms is less developed than in other parts of the country. Only 13%, compared with an average of 28%, of the total labour force is employed in industry (Moe, p.217). The most important branch is food production, primarily seafood.

The most influential factors in the development and change of the region have been technological developments in shipping, communications and the refinement of food production. Sea traffic between south Scandinavia and Troms has a tradition of 2000 years (Eriksen, 1979, p.187). First by 'jekter', small cargo boats with half decks and sails, then by 'fraktskuter', frieght boats and long ships and later by steamships, 1837. The first modern liner began to commute between southern Norway and Hammerfest in 1893 and this route became vital for the whole coast. But the real breakthrough was made by the motor boat at the end of last century.

Figure 10.1. *Troms County*

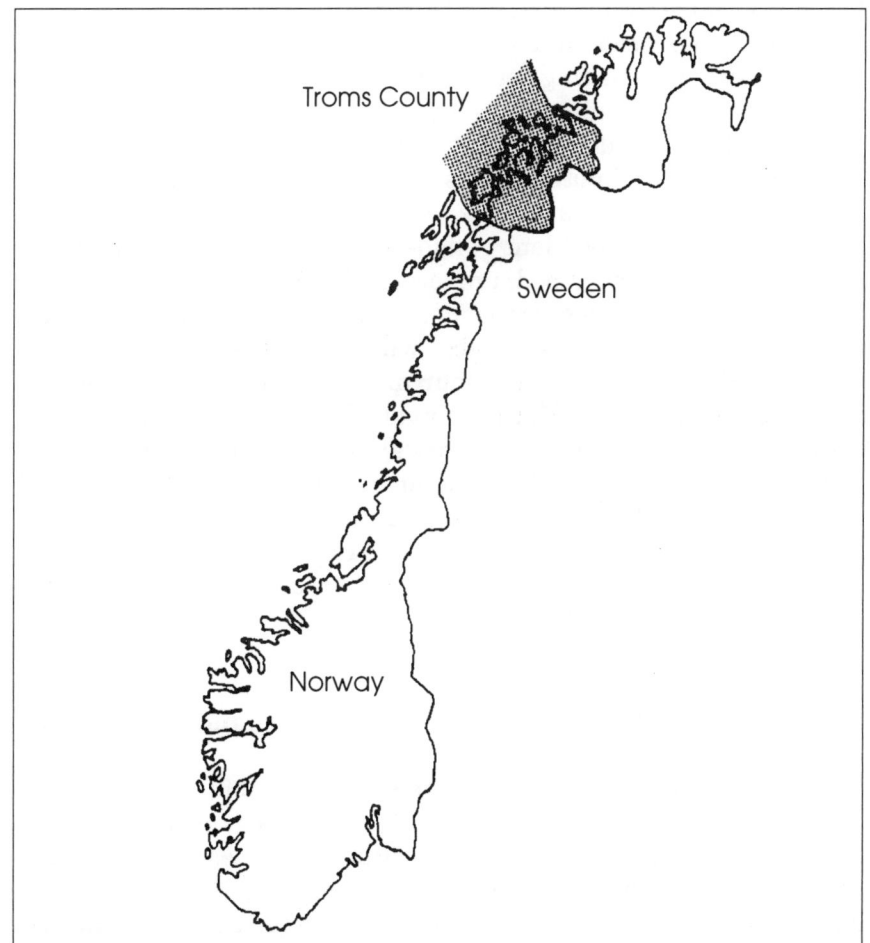

Traffic by land began slowly in 1857. At the end of the century there were less than fifty Norwegian miles of passable roads. Riksvei 50, an inland state highway was the first which reached Tromsø in 1936 (Eriksen, 1979, pp.188-9). In 1950 there were about 2000 km. of roads, more than 55% of them built between 1920 and 1950 (Troms, p.136). The real breakthrough for traffic by land happened in the 1960s when the road network was extended to follow the coastline. Collective transport was provided by bus service combined with a ferry boat service; the railway has never reached the county. The first scheduled service by air appeared in 1938 at Skattøra, and in 1964 a new, modern airport was opened at Langnes, Tromsø, which was to become a centre for north-south traffic.

Industry has been bound to processing fish or the servicing and development of the fishing fleet. However, new technology in industry has had a similar effect on settlement patterns as the changes brought about by developments in communications. These are examined in detail in other parts of this study. Since the end of the 1960s there have been attempts to establish new sources of income and new trades independent of fisheries and shipping. New service trades have been developed, university teaching and research have been established and modern communications have helped to reduce the effect of the physical distances within the region.

From these developments four main time periods can be seen within this level. Pre-1900, the point at which the motor boat became an everyday reality in northern Norway. 1900-1940, characterised by the development of traffic by land and new technology in food processing. 1940-1960, economic reinforcing of the technical infrastructure after World War II. 1960-1989, new communications, air traffic and information technology, new trades, services, university and research.

Level 2: Core-continuous built-up area, Tromsø

This level represents the city of Tromsø and its suburbs as they have grown over the last century (Figure 10.2.). The basis of analysis is the main periods of development of the city and its closest surroundings, expressed by extensions to the city boundary and the growth of the city and city community area. 'The boundary problems in a quantitative study of urbanisation are not merely problems of relatively arbitrary statistical classifications, but reflect historical changes.' (Flora, 1983,

Figure 10.2. *Core-continuous built-up area: stages of development*

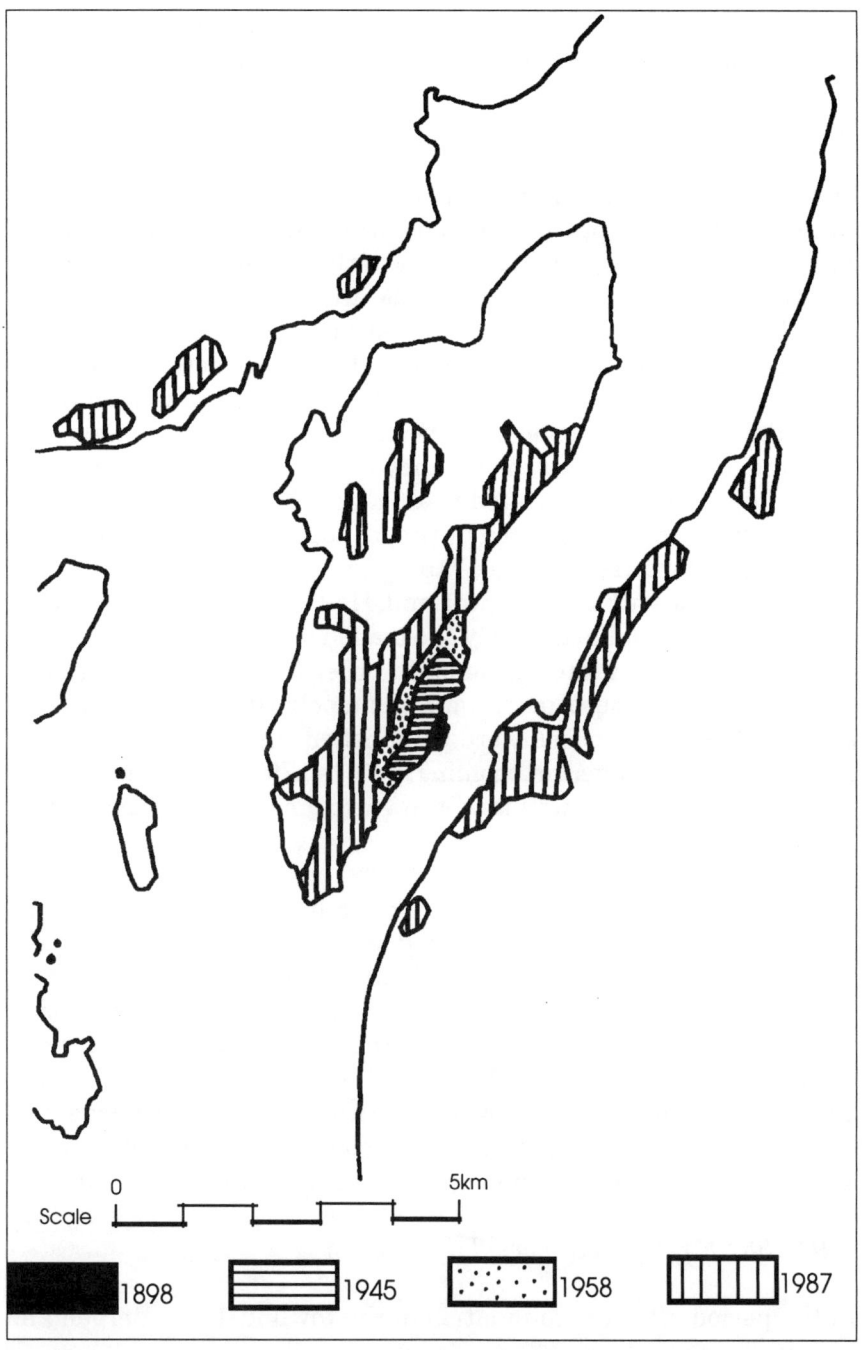

Scale

0   5km

1898   1945   1958   1987

p. 247). There have been seven major periods of change in the history of Tromsø. An investigation into the changes of urban form during these periods was used as a basis for 'reading' the influence of innovation at the scale of the town and its hinterland.

1250-1794 Before the foundation of the town

In 1250, by order of king Håkon Håkonson, a church was built for Norwegians and Lapps on the place which was to become Tromsø. The church was supposed to be part of the defence system and the basis for introducing Christianity to the area north of the Polar Circle. The church domain became a Hanseatic trade centre in the later middle ages. A letter of privileges from 1361 and another from 1560 ratified the monopoly rights of merchants from Bergen and Trondheim to trade with the northern parts of Norway. Bergen, Oslo and Copenhagen in turn, had the tenancy rights to the region, and they were allowed to run privileged trade through their companies. But the supply obligation required by the monopoly was not fulfilled and the monopoly was reduced in 1715.

Illegal trade with France, Scotland, Holland, Russia and England existed from 1500. Particularly the trade with the Russians caused great losses in the state's income. In order to solve the question of supply and that of customs income, monopoly rights were abolished in 1789. Customs houses were established in Tromsø, Vardø and Hammerfest (Lorange, forthcoming, p.7). In 1794 Tromsø received its town rights and became a market town, exempted from taxation for 20 years.

Morphological changes The original church settlement was a social and economic meeting point, located strategically in relation to the sea traffic connected with taxation. This was the reason for designating Tromsø as the main town and customs' centre of the province. Other communications were by roads from the north and the south, along the coast, and the King road from the west. The roads met at the Smørtorget (Butter Market) where the kirkestuer (church lodgings) were situated. The building complex, consisting of the church, the churchyard and the church lodgings, were situated on Prostneset.

1794-1837 First growth period of the town

In the period after the foundation of the town in 1794, Bergen and

208

Copenhagen were its most important trading partners. However, during the Napoleonic Wars the direct seaway to the continent was closed and it caused trade with Russia to flourish, with official sanction from 1818. After 1814 union with Sweden brought about peace and trade and, from the 1820s, Tromsø entered a period of intense economic growth which lasted until the 1880s. In 1837 Tromsø was a town of export, the headquarters of the local offices of government, with schools and health services in the town.

Morphological changes The period is characterized by the introduction of two development patterns, the wharf and the street grid, city block system. The town plan, provided by a district governor, Sommerfeldt, in 1788, designated an orthogonal city block system based on the street pattern around the cathedral. Within the city boundaries, merchants' houses, docks and warehouses were situated on the first wharf structure in the bay between the church and the customs' house. Houses of craftsmen and merchants were situated in the orthogonal city block pattern behind the wharf. The parsonage and communal storage houses were located outside the city boundaries. The residence of the county prefect, the local storage house and storehouses had been built along Sjøgata (Sea Street). The most important building of this period was a new church, consecrated in 1803. Allemenninger (wide, open squares/streets) were introduced as a principal urban pattern to secure connections with the sea.

In 1837, a lieutenant Due provided a survey map of Tromsø as a basis for further development. According to the map, the orthogonal city block plan of Sommerfeldt, from 1788, was to be extended towards the north. The plan was concerned with Grønnegata, Storgata, Nedre Strandgata, Tollbodnes gate and the road to the farm, Hansjordnes.

1837-1898 Second period of growth

In the 1840s Tromsø developed as a trading town. Two conditions fostered this growth, independence from other towns and the establishment of a new system of commercial credit. Economic services appeared in the town with the opening of a branch of the Norges Bank (Bank of Norway), insurance companies and the postal service. Communications were improved. Development of trade due to the Crimean and French German Wars happened in parallel with the development of industry. International economic cycles caused a trade crisis in 1877 with the lowest point in 1881. This was intensified by exceptionally low profits in the fishing industry. Large fluctuations in international economic conditions and the state of the local economy determined the extent of the crisis locally. After re-evaluation of the loan policy, Tromsø entered a period of a slow growth from the 1890s until World War I.

Morphological changes Physical development was generally characterized by expansion and increasing densities in the wharf and the city block structure, which developed towards the west. In addition, monumental buildings were built as dominant features of the townscape. Plans regulating development were laid out from 1846 and 1875. The most profound changes were caused by a new municipal law, in 1837, giving towns local autonomy. It influenced the change of the town boundaries and altered property rights. Thus, the area around the church, called the 'Vatican', was incorporated into the town. Around 1850, a system of distribution of plots of land was established. Merchants and traders could build houses towards the sea in Sjøgata and Strandgata, north of Strandskillet. Craftsmen were limited to Storgata, Grønnegata and other streets. A religious revival in the 1850s led to the building of several new churches, Tromsøsund Church, the Catholic Church, the Methodist Church and the Cathedral. The school system had been strengthened by the Latin School, an elementary school, the Seminary and the Technical Evening School.

Brick buildings from this period are the Tromsø Bath and the Norges Bank. A museum and hotel were built to commemorate the first centenary of the town in 1894. Other important buildings are, the Town Hall, the prison, the hospital, the fire station and the electric power station. Improvements to existing roads and expansion of the traffic system generally led to displacement or demolition of older buildings. Towards the end of the period low density housing spread towards Hansjordnes Bay, and the city block system extended towards Vestregate which had been constructed on the upper side, towards the Latin School. In the south, the buildings stop at Strandskillet where the museum and hotel were situated as individual buildings.

1898-1918 Development and change

The period of a slow prosperity continued from the 1890s until World

War I. Banks financed motorization and the development of different industries, crafts and trades (Ytreberg, p.56). The Russian Revolution of 1917 broke the traditional trade links. Whale oil, internal trade by steamboats and expansion in agriculture formed the basis of the economy.

Morphological changes The main changes were growth within the city block structure, transformation of the harbour area caused by fires and establishment of the steamboat/northern jetty and housing outside the central area of the city. Dwellings were built in the north and the south, according to the plan of 1875. Summer cottages in the west were transformed into permanent dwellings and a new street network followed the development pattern. New, important buildings were the hospital, 1917, and the shipyard. Three new assembly halls were also built. The network of streets were extended, Kirkegata, Søndre gate and Sjøgata were lengthened and Havnegata (Harbour Street) established. New parks were formed in connection with Skansen (Entrenchment), the hospital and the Gymnastics Club.

1918-1940 The period between the two World Wars

The period of growth during World War I was followed by a period of inflation, bankruptcy, unemployment and political unrest. The loss of Russia and south Europe as trade markets increased the crisis. Nevertheless, this time was characterized by a slow growth. The Arctic fleet was motorized and fish processing and wholesale dealing became important businesses. Tromsø was connected to the national road system and its role as a regional centre strengthened.

Morphological changes Physical development was characterized by the transformation of the harbour and the growth of industries northwards along the coastline. Settlement areas grew in the west. The building law of 1845, based on the grid system of street planning, was replaced by a law in 1924 which introduced zoning and the segregation of functions (Dunin-Woyseth, 1988a). The centre was modernised and its building densities became higher. The fishing fleet was brought up to date, followed by improvements to the harbour. New buildings of importance were a high school, water works and power station.

1945-1958 Rebuilding

After the war, Tromsø was the only town, north of the Polar Circle, which remained constructed of wood. The reconstruction of the North-Troms province took up most of the resources of the region and it was not until the 1950s that conditions for the growth of the town were fulfilled.

Morphological changes During these years attitudes towards the wharf structure and the existing city block structure changed. The quayside became continuous and large single buildings replaced the traditional city blocks, creating high densities in the centre.

Planning activities were very vigorous in the 1950s, and their implementation meant tearing down the existing fabric and replacing it with squares of 3 to 4 storey buildings. The plans of this period are still valid, but have only been partly implemented. The Austad fire of 1948 made building according to plans from 1949-1950 possible. Strandgata was strengthened and Havnegata extended. Single buildings built according to those plans are, Telebygget, NKL/ Domus, the Town Hall, the North Light Observatory and the library. The steamship pier was extended north of the mole in 1958. Felleskjøpet's pier and the NKL-pier were built in wood and the town square and the Nansens plass were given floating stages.

1958-1989 Growth

From being a regional centre for politics, economy and administration, Tromsø became the centre for the whole of north Norway in the 1960s. The role and character of the town were changed by the new bridge, 1960, incorporation of Tromsøysund, 1964, construction of the airport, 1964, and the founding of the university, 1968.

Morphological changes The healthy economic situation caused an explosion of building activity. A continuous transformation of the city block structure and the construction of high, single buildings were characteristic features of this period. Building lines were set back to accommodate traffic. After the fire in 1969, the wharf on the southern side of Torghuken disappeared. Prostneset lost its importance as a central harbour to Breivika, bringing new uses to the harbour area. New housing areas appeared on Tromsø Island, the mainland and Kvaløya. The Sandnessund Bridge was opened in 1973. Concrete buildings were constructed in the centre, the SAS and Grand hotels, the Focus Cinema, some insurance offices and the county administration headquarters. An area development plan from 1975 was used as a guide for urban development, although it was never formally ratified. A tunnel connection between the airport, city centre, university and the mainland has been partly completed. This is expected to have a significant impact on future development when it is finished.

Morphological changes and innovations

1837-1898 The establishment of a new, autonomous municipality, introduced by law in 1837. This changed property rights and land use in the town.

1898-1918 Finance for the motorisation of shipping and industry by new types of banking and insurance institutions. This created demand for new parameters to the harbour leading to its transformation. New trends in housing. A new type of dwelling, the villa, was introduced outside the core of the town. New settlements required new infrastructure. Demands for qualitative public space were realised by the creation of parks.

1918-1940 Cooperative and municipal housing was introduced as a consequence of international trends and political development within the country. New planning ideologies, expressed by the Building Law of 1924, introduced the segregation of functions. This influenced the development of the central and peripheral areas of the town. The automobile appeared, creating a demand for roads and changing the hegemony of sea traffic as the main means of communication. Technological developments resulted in new industries connected with shipping and food processing. The harbour continued to be developed responding to new functional and technical demands from the shipping industry.

1945-1958 New planning and architectural ideas shaped the urban form of the town. The block structure was transformed, land use and floor area ratios changed in the centre. The automobile began to occupy the streets, causing a reduction in the quality of life in the town centre and pushing housing out to other parts of the town. Housing, supported by the state and the municipality, was developed rapidly and extensively. Progress in maritime and shipping industries resulted in the continuous transformation and development of the harbour area.

1958-1989 The motor car, the well-developed road network and the bridges connecting the islands to the mainland encouraged the construction of housing outside the old centre. Technological demands caused the central harbour to move northwards to a new location. Progress in communications, traffic and information technology reduced the isolation of Tromsø. It changed from a single, sea-oriented economy to a mixed structure. Since the 1960s the population has increased radically.

Level 3A/4: Core-centre/homogeneous morphological unit

It was considered that the specific character of the area and the size of the town justify this section as representative of both levels. The core centre is that part of Tromsø Island which was the original town before the incorporation of the settlements on Kvaløy and the mainland. It encompasses the harbour front, the city block structure and the buffer zone in between, consisting of a linear structure of streets running north to south. To the north and south the area is delineated by fire breaks, Skriverplassen-allmenning and Musegata-allmenning. To the west and east are the natural borders of the sea and a mountain ridge (Figure 10.3.). From being a town in its own right this area has turned into a service centre for the region. Initially it contained all the urban functions and although it still retains some of these it has become a specialised area.

Pre-1898

The orthogonal city block structure, or 'Quadrature', the linear structure of streets and the wharf were relatively continuous. The only 'allmenning' (fire break) extending to the water was the market square, which became a natural meeting point. The role of this square was strengthened by the location of the town hall. The linear structure of the streets and wharf was broken at Prostneset, the former property of the church, where the new steamship pier had been built. Important buildings are situated either adjacent to the allmennings or outside the town boundary. Buildings differing from

Figure 10.3. *Development of the core-centre*

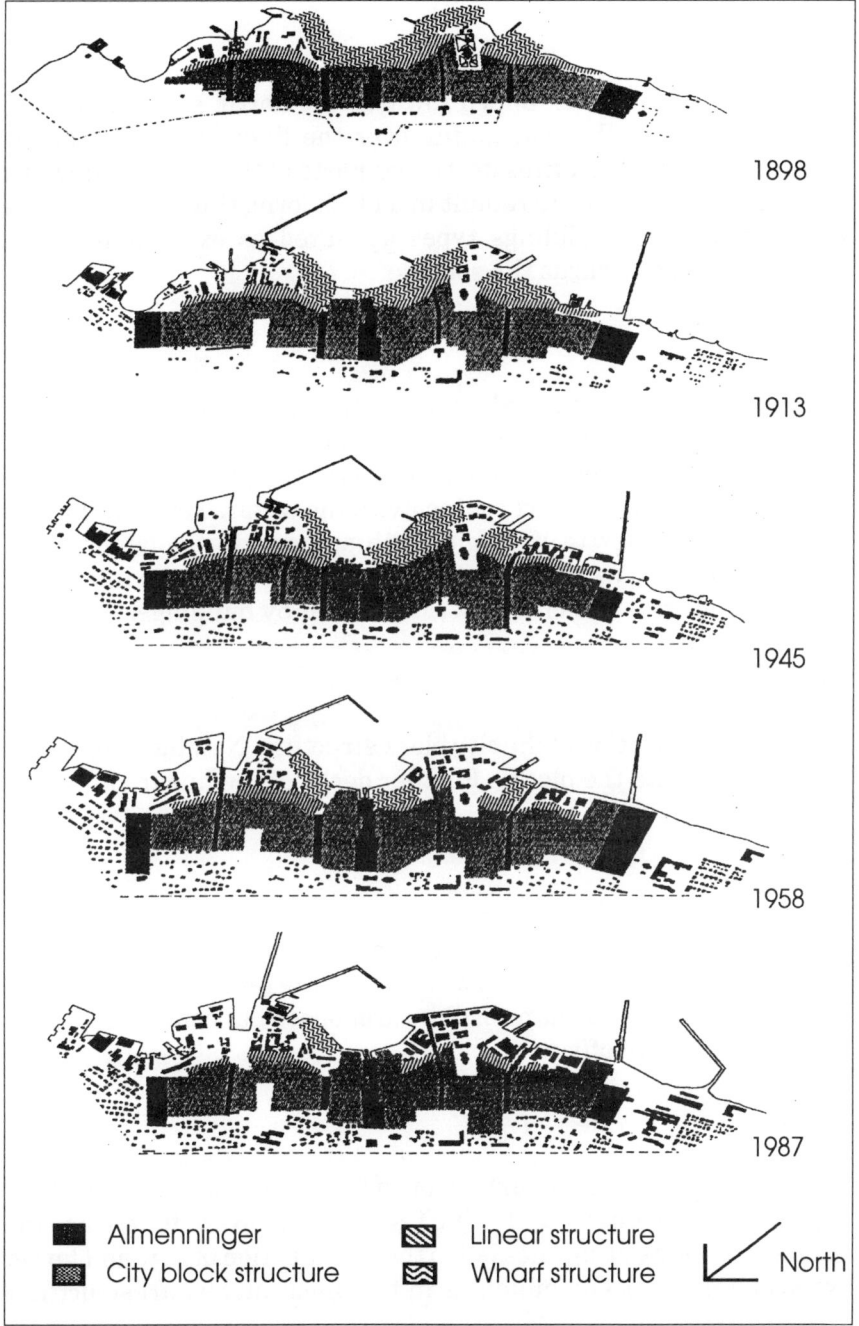

1898

1913

1945

1958

1987

	Almenninger		Linear structure
	City block structure		Wharf structure

North

the usual typology are the brewery, bank and three warehouses on the steamship pier.

1898-1913

The Quadrature was extended towards the north, south and west, and also towards the sea, taking over the linear structure of the transition zone. When fires destroyed parts of the linear and wharf structure, the areas were rebuilt in a block form, thus extending the Quadrature. New buildings types appeared as extensions of the wharf structure towards the south.

1913-1945

The general trend was towards condensation of the city block structure. Extensions to the built- up area do not follow the city block system. A continuous pier along the harbour was developing, separating the old warehouses from direct contact with the water. New warehouses were adjusted to new technologies in shipping and sea related trades. Some allmennings were extended to the sea through the linear and wharf structures. The church was encircled by buildings.

1945-1958

Further condensation of the city block structure. New buildings were incorporated into the blocks. Further decay of the linear and wharf structures.

1958-1989

Storgata began to develop in a linear form within the block structure. Only single buildings of the old wharf structure remain, new buildings imitate the scale and shape of old warehouses but contain new uses such as shops and offices.

General features of development

The general trend was an extension of the city block structure at the expense of the linear and wharf structures. New functions and changes to the existing ones led to the introduction of new and larger types of buildings. Old buildings in the linear and wharf structures were replaced by new ones oriented to the block structure. This was particularly prevalent around the market square. Also, the allmennings were extended from their original locations in the block structure, through the demolished areas of the linear and wharf structures, to the water. The wharf area underwent the greatest transformation. It began around Prostneset, where the new steamship pier was built in 1898. When the area north of the market square was destroyed by fire this created further opportunities for new organisation and types of buildings. The changes were characterised by rebuilding new, larger buildings complying with the technical and functional demands required by the harbour and its associated trades. The new buildings follow the shape of the site, refer to the city block system or only to the pier.

Parts of the linear structure which were still intact generally followed those of the wharf. Since it contained mainly houses and small shops it was not exposed to the drastic changes of use which took place in the harbour and could continue its role as a buffer zone. Also, the building of piers into the sea relieved pressure on this zone. However, the changes were similar to that of the wharf, old buildings were demolished and their replacements tended to extend the Quadrature.

Changes within the blocks were mainly infill, with a tendency towards densification and higher floor area ratios, although there was also some expansion. Allmennings were retained and used as traffic areas or squares. Storgata became the central business district, comprising many small shops with the market square as the focus. The demand and resulting high prices prevented the construction of larger complexes which needed combinations of sites. This has meant that such development has taken place in a belt between the central area and the periphery of the city block structure.

Morphological changes and innovations

The residential function of the block structure has been partly replaced by the tertiary sector. Rich inhabitants have moved to the west, into villas, following new fashions and ideas for housing. New functions in the core centre caused an increase in traffic and a further deterioration of the area. Inhabitants left in the central area moved into state and municipally-supported housing. As the price of land rose, so the density of sites increased.

New ideas in planning and planning law promoted further

segregation of functions. The demands of new functions and technology from shipping and related sea trades transformed the wharf structure. Generally the innovations highlighted in Level 2 are relevant to Level 3A. The 'hard' innovations being those connected with technological development, the 'soft' ones with ideologies, political priorities, institutions and laws.

Level 5: Specific area, the harbour

The harbour and its development has been dealt with earlier, but only in the context of the core centre. At this level an outline is given of the role of the harbour as the nerve centre of Tromsø and its physical development, structurally and in size.

History

1898-1913 New demands connected with developments in maritime trades brought about changes. The construction of the steamship pier on the southern part of Prostneset was accomplished, the wall in 1898, the infilling behind in 1904. Two warehouses were erected on the pier. At the moment of completion the pier was already too small so two traders applied for permission to build a private stone pier. However, this was refused by the municipality since this function was a public responsibility and the pier might hinder further public works. In 1907 the northern mole was completed, protecting the inner harbour. Kirkegata was extended down to the new steamship pier and Sjøgata improved. In 1913 the steamship pier was extended to the north to allow space for local passenger ships. It was decided to expropriate the land further to the north to guarantee space for further expansion, although this was against the interests of the local fishermen, who depended on the old piers.

1913-1945 In the 1920s trawler traffic in the harbour increased. Developments in the production of frozen sea food required direct loading at the pier. The stability of the old wharves was threatened because of the efficiency of the Harbour Commission's new dredger and the weak ground conditions. This meant the introduction of a limit 20 metres seaward of the wharf, which led immediately to the extension of private piers.

A new plan for the city and the harbour was prepared by Professor

Sverre Pederson and approved in 1923 (Ytreberg, 1971, p.274). The building activity and the new way of loading ships heralded the beginning of the continuous harbour front and an extension of the northern mole in 1931 gave still better weather protection. With the completion of the road on the mainland in the 1930s, ferry traffic began. They landed in the southern part of the area rendering it useless for other commercial shipping. Little of the harbour was damaged during the war, but the lack of maintenance left the piers almost completely run down.

1945-1958 Immediately after the war, the last private property was expropriated and the steamship pier extended to the north. Only single storey warehouses were permitted on it. After the extension Prostneset became a fully equipped modern harbour with the local authority taking over responsibility for improving the private piers in the inner portion.

1958-1989 In 1958 it was decided to move the main activities of the harbour to a more suitable location, Breivika, in spite of protests from the local commercial business association. At the same time as the city was extended to become the 'Big Tromsø' in 1964, the harbour area was extended to the length of the island (Ytreberg, 1971, p.684).

General features of development

As bigger boats were used in the fishing industry and the methods of loading altered, the piers were extended further out of the original harbour basin. When the local authority took over the building of piers in 1958 the rights of property owners were set at two metres depth of water. In order to avoid expropriation of land, the authority built a new pier in deep water creating a strip of piers between the old warehouses and the sea. This allowed trucks and other motor equipment to be used on the pier.

As activity in the harbour declined, most of the old warehouses were torn down or given over to new uses. New buildings faced Sjøgata and to some extent the pier. In most cases several plots were joined to provide sites for larger buildings. From an area consisting of warehouses, the harbour became filled with large storage and office buildings belonging to big companies involved in harbour-related activities.

The character of the old harbour was changed when the main

functions moved to Breivika. Once more new functions filled the existing buildings, usually retailing on the ground floor with offices above. When the harbour extended along the coast, Tromsø changed from a port city to a coastal city, with the long coastline giving '...sufficient space for the use of modern systems for loading and unloading merchandise.' (Montanari, 1988, p.178).

Morphological changes and innovations

The technical innovations related to the maritime, economic life of Tromsø, shipping, fishing and sea food processing have already been mentioned in Level 1. At this level they are more clearly 'readable'. As a direct consequence of these innovations the port city of Tromsø has evolved into a coastal city. Old functions have disappeared and new ones are developing.

Level 6/7: Blocks and land parcels

The investigation at this level deals with the Quadrature. It is part of the core centre and its boundaries were described at that level.

History

Tromsø's urban form was shaped in the period between 1800 and 1870 (Bakke, 1983, p. 111). It was a wooden town in the Empire style, with rectangular blocks adapted to the coastline and topography. Since this was a very functional and flexible pattern it was followed until World War II. The Quadrature comprises clearly defined blocks creating long perspective vistas. The difference between public and private space is clear. The grid structure made it possible to build a town, consisting of individual houses and at the same time to achieve a formal unity by creating good public space with streets. This is the urban ideal from the Renaissance which characterizes all towns founded in Scandinavia (Bakke/Hage, 1988, p.71).

In Tromsø, as in other towns, the house walls were shaped deliberately and maintained properly. Gaps between houses were built up by high board fences, creating spatial continuity, so the streets had an unbroken facade. The courtyard in the block was private. Sometimes a grid pattern can be a constraint if it does not adapt to the topography, in Tromsø it compromised with the topography

Figure 10.4. *Development of the urban form*

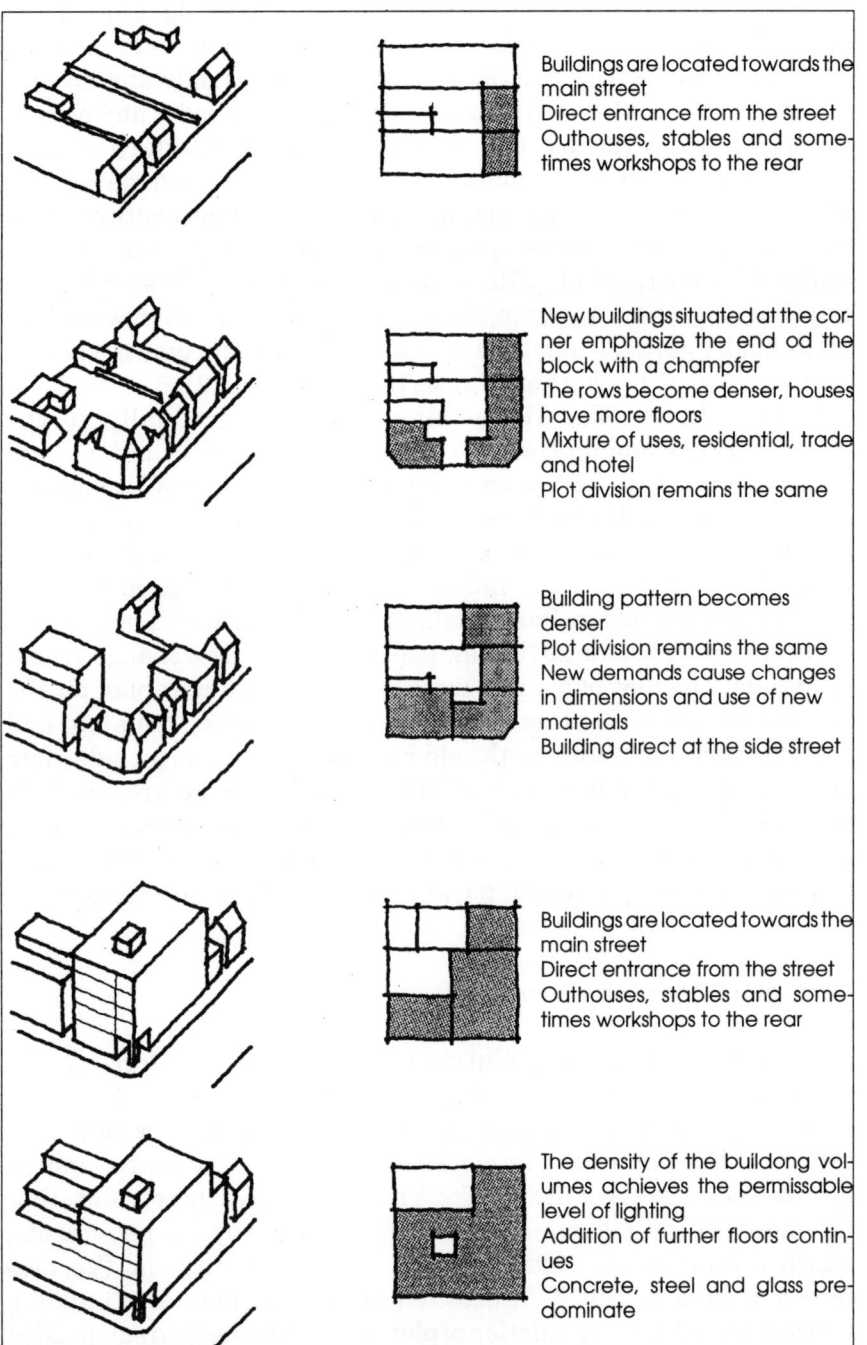

Buildings are located towards the main street
Direct entrance from the street
Outhouses, stables and sometimes workshops to the rear

New buildings situated at the corner emphasize the end od the block with a champfer
The rows become denser, houses have more floors
Mixture of uses, residential, trade and hotel
Plot division remains the same

Building pattern becomes denser
Plot division remains the same
New demands cause changes in dimensions and use of new materials
Building direct at the side street

Buildings are located towards the main street
Direct entrance from the street
Outhouses, stables and sometimes workshops to the rear

The density of the buildong volumes achieves the permissable level of lighting
Addition of further floors continues
Concrete, steel and glass predominate

221

giving the town its specific character.

Within the blocks there were three ways of parcelling the land into plots, the row block with access from the street front and rear lane, the broad block with street access from two sides only and no rear access, the broad block with street access from four sides and no rear access. The plots were built up along one, two or three sides according to the owners' needs or economic situation. The courtyards gave space for many functions: workshops; barns and lofts for agriculture, since this was a normal occupation of citizens until the end of the nineteenth century (Ytreberg, 1971, p.13); stables and wagon sheds; storehouses on pillars (stabbur) and storerooms; servants' quarters in wealthier households; garden; toilet, with the content of courtyards varying from one owner to another (Bakke/Hage, 1988, p.72).

After World War II, housing expanded gradually to the neighbouring communities on Tromsø Island, Kval Island and the mainland. The Empire town was invaded by offices, small industries and businesses. With the absorbtion of the neighbouring municipalities in 1964, the 'Big Tromsø' was founded. The master plan designated areas for further housing, industry, trade and service in the new satellite settlements on Tromsø Island, Kval Island and the mainland.

In the middle of the 1970s new political preferences changed these proposals. The new objectives were expressed in a new plan for the downtown area, dealing with land use, traffic, housing and preservation. From then on the old Empire town was to include more housing at higher densities in the city block structure, initially through offering a higher quality housing milieu. Traffic and parking problems were to be solved by building new traffic connections in long tunnels and constructing car parks in the mountain. This plan for the central area has been adopted.

Development of urban form in the quadrature

An investigation of the evolution of a sample of six blocks in the Quadrature over the last hundred years illustrates six basic areas of change. The example shown in Figure 10.4. illustrates how new functions have appeared in blocks situated near the centre and on broader side streets or fire breaks. These are offices, hotels and other commercial uses. 'Change occurs within plots and by plot increments, which reinforces the initial pattern of land subdivision' (Vernez-Moudon, 1986, p.139). New uses require larger volumes of building which causes the amalgamation of plots and alters the land subdivision

Table 10.1. *Innovation on each level*

Level	Innovation	Soft	Hard
1	Shipping technologies		✳
	Land traffic		✳
	Air traffic		✳
	Information technologies		✳
	Trends to invest in science, teaching and research	✳	
2	Institutional innovations (municipal autonomy, new banking systems, state supported housing cooperative and municipal	✳	
	New ideals for lifestyle priorities, ways of living, attitudes to surroundings	✳	
	New building law, later planning law, segregation of functions	✳	
	Automobile		✳
	New food processing technologies		✳
	New shipping and port technologies		✳
	New construction technologies (bridges)		✳
3/4	Planning ideals, planning law, segregation of functions	✳	
	State and municipally supported cooperative housing	✳	
	New housing ideals, exodus from the central areas	✳	
	Private transport		✳
	New technologies in sea related trades		✳
5	New technologies in sea related trades		✳
	New trends and new demands for a qualitative improvement	✳	
6/7	The same innovations as at level 3 and 4	✳	✳
	Life style preferences, new ideals for ways of living	✳	
	New planning and architectural ideals	✳	

pattern and the character of the city block. In the amalgamated plots the floor area ratios exceed that of the average of the block. Residential use still predominates in side streets and other locations further from the centre. Houses have become larger and their standards and amenities better. Outhouses in the courtyards have been demolished. The structure has proved adaptive to changes. Residential use has survived and new inhabitants are moving in, making the mixed use objective of the plan a reality.

Morphological changes and innovations

Innovations pointed out at Levels 3, 4 and 5 are able to be 'read' more easily at Levels 6 and 7. The city block structure is receptive to innovation. It is able to survive such influences. The influence of 'soft' innovations, changing life styles and priorities of life quality, is evident. The city block area was initially a residential area for the wealthier people of Tromsø. The courtyards illustrated the life style of the residents. When the exodus to the villa residential area took place it signified a new life style and life quality preference. This is equally true for the return to the Quadrature from the middle of the 1970s onwards and the improvement of houses by those who already lived there. Those changing preferences and ideals have a close relation to other 'soft' innovations, those in planning and architecture. The social consensus which was achieved found a means of implementation through planning legislation. Generally, Levels 6 and 7 emphasize the conclusions of Levels 3, 4 and 5, adding a new insight into the 'soft' innovations.

Conclusion

Table 10.1. illustrates a broad spectrum of innovations by grouping them together in levels. Although the picture is not clearly structured it does show that the majority appear at all levels of the town.

References

Bakke, P. Hage, I. (1988), *Empirebyen Tromsø, Arkitekter i Norge Yearbook 1988*, Bonytt-Norsk arkitekturmuseum, Oslo.

Bakke, P. (1985), *Tromsø bys historie, Fortidsminneforeningen Yearbook 1985*, Foreningen til norske Fortidsminnesmerkers Bevaring, Oslo.

Bertheussen, K. (1985), *Eksempel fra Tromsø: Sentrumplan med forhindringer (An example from Tromsø: a centre plan with obstacles)*, Trondheim NTH, Trondheim.

Dunin-Woyseth, H. (1988a), 'Built Form versus Modern Planning Legislation: Genius Loci versus International Influences', *Proceedings of the Third International Planning History Conference, The Century of Modern Urban Planning*, (8-12 Nov. 1988) Tokyo.

Dunin-Woyseth, H. (1988b), 'URBINNO', *Arkitektnytt* 18/1988, Norske Arkitekters Landsforbund NAL, Oslo.

Eilertsen, R. (1984), *Naeringslivet i Tromsø (Trades in Tromsø)*, Tromsø Sparebank, Tromsø.

Eilertsen, R. (1981), *Tromsøboka (The Tromsø book)*, Tromsø Municipality, Tromsø.

Eilertsen, R. (1979), *Byen og menneskene: Tromsøbilleder 1915-1960 (The town and the people: Tromsø pictures 1915-1960)*, Tromsø by museum, Tromsø.

Eilertsen, R. (1978), *Billeder fra gamle Tromsø (Pictures from the old Tromsø)*, Tromsø by museum, Tromsø.

Eriksen, H.E. (1979), 'Troms giennom tidene', in Kristofferson, I. (ed.), *Bygd og by i Norge. Troms*, Gyldendal Norsk Forlag, Oslo.

Flora, P. (1983), *State, Economy and Society in Western Europe 1815-1975 Volume 1, The Growth of Mass Democracies and Welfare States*, Campus Verlag, Frankfurt.

Fulsås, N. (1987), 'Husholdsokonomi og kapitalistisk økonomi Nordland 1850-1950' (Household economy and capitalist economy in Nordland 1859-1950), *Historisk tidsskrift*, B66 No 1, Tromsø.

Hage, I. (1985), *Tromsø-byplan og stilhistorie: 1800-1950*, Fylkeskonservatoren i Troms, Tromsø.

Hagerup, Vegard, Jensen og Larsen, (1981), 'Sverre Pedersen: en pioner i norsk bolig og byplanlegging', *Institutt for By-og regioplanlegging NTH* 1981 No 2, Trondheim.

Kristofferson, I. (1977), *Havn og havneliv i Tromsø (The Harbour and Harbourlife in Tromsø)*, Havnevesenet, Tromsø.

Lorange, E. (forthcoming), *Town Planning in Norway*, Mansell, London.

Moe, K. (1979), 'Industri', in Kristofferson, I.(ed.), *Bygd og by i Norge. Troms*, Gyldendal Norsk Forlag, Oslo.

Montanari, A. (1988), 'A Modern Perspective: The Recent History of Port Cities in Southern Europe, *Mediterranean Historical Review*, 3(1).

Nilsen, H. (1966), *Bergensernes handel på Finnmarken i eldre tid (Bergen merchants trade with Finnmark in history)*, Universitetsforlaget, Oslo.

Prahl Harbitz, G. (1986), 'Fra Longyearbyen til Rosenkrantzgate' (From the Longyear town to Rosenkrantzgate), *Tromsø Sparebank 1836-1986*, Tromsø Sparebank, Tromsø.

Storm Munch, J. (1981, 1985), *Tromsø-byplan og stilhistorie 1800-1900 (Tromsø Urban Plan and History of Styles)*, Tromsø-veiviser, Tromsø.

Troms Arbeidsdirektoratet, (1953), *En statistisk-økonomisk analyse. Kontoret for områdeplanlegging i Troms*, Tiden Norsk Forlag, Oslo.

Tromsø Municipality, *Area Development Plans*. (1979), The Tromsø Centre: Traffic and Parking, subaspect 2. (1983), Beskrivelse av planen for Torgallmenning (Description of a Plan for Torgallmenning). (1983), Utkast til generelt forbedringsprogram for Skansen (Improvement Program for Skansen). (1987), Utvikling av Prostneset og Holmbobryggene (Development of Prostneset and the Holmbo Wharves). (1987), Disposisjonsplan (Disposition Plan).

Vernez-Moudon, A. (1986), *Built for Change. Neighbourhood Architecture in San Francisco*, MIT Press, Cambridge Mass.

Ytreberg, N.E. (1952), *Tromsø som kirkested og by gjennom 700 år (Tromsø as a Church Domain and a Town through 700 years)*, Tromsø kommune, Tromsø.

Ytreberg, N.E. (1971), *Tromsø bys historie (History of Tromsø)* Volumes 1-3, 1946-71, Tell forlag, Tromsø.

Ytreberg, N.E. (1936), *Det gamle Tromsø (The old Tromsø)*, J.W. Cappelens forlag, Oslo.

Part II
INNOVATION AND URBAN
DEVELOPMENT: SELECTED ITEMS

11. INTRODUCTION

In this part of the book we present a number of thematic papers on the development of innovative processes and their impact on land use and urban form. Part One, which addresses innovations in a select urban sample, shows that the urban environment has undergone a number of transformations in adapting to new technologies, new economic processes and the diffusion of new cultures, either separately or, as is often the case, all at once. In some cases the new circumstances were exogenous, in others endogenous. It is very difficult to assess exactly which features are internal or external to the innovation process.

In the effort to come to grips with an extremely complex issue, there is frequent reference: (i) to change and mutation and thus to the conception of new ideas, (ii) to the selection of phenomena of change through their acceptance in society and (iii) to the dissemination of new ideas. In this part most of the emphasis is on the third aspect since it can be observed more easily in urban structures and its impact measured in both space and time. We look at the cultural dissemination of some of the innovative processes which, more than others, have proved to have a significant impact on changes in urban form. In so doing we come back to that basic issue which has to be taken into consideration in every field of research: the relationship between form and information, understood as creative processes that can be convergent or divergent.

Once the methodological basis for analyzing and then classifying innovations were defined, it was proposed to divide innovation processes into two categories, soft innovations and hard innovations as in the chapter on Tromsø. The soft innovations concern the cultural sphere of urban populations: (i) ideologies, (ii) policies, (iii) dynamics of social groups and (iv) economic goals. The hard innovations are mainly the technological innovations in (i) building techniques and materials, (ii) transport infrastructures (roads, railways, bridges, tunnels, ports, airports, etc.), (iii) information infrastructures (communications, computing, etc.), and (iv) technical infrastructures (water, sewerage system, treatment of waste waters, systems for monitoring environmental quality). Dissemination of both soft and hard innovations occurs through other elements, which act as a kind of filter: (i) economic planning, (ii) fiscal incentives, (iii) physical planning, (iv) rules and regulations, (v) capacity of administration to implement decisions.

The thematic approach for assessing the dissemination of innovations and their impact on urban form was based on the following criteria: (i) thematic case studies referred to in Part One of the book; (ii) themes of common interest that had emerged repeatedly during the phases of coordination with the three other WGs of the URBINNO Project; (iii) material available in international literature on innovative processes that had occurred at least from the last century, which could be used to test and consolidate the results of the case studies.

We followed the indications given in the introductory chapter for the space-time variables, namely for the organization of the period to be studied and the definition of the hierarchy of geographic areas.

Hard and soft innovations

Within the framework of the methodological context of the URBINNO Project, and the indications of WG IV, the articles presented in this volume have been ordered according to the following logic: starting from the dissemination of hard innovations and leading on to that of soft innovations. In fact, the division is not so marked: even innovations normally considered to be hard contain some purely soft elements. This pattern of examining first the hard and then the soft components of innovation is evident from the start of Part Two: the first article starts with an examination of innovations in transport systems and then considers a particular area of the city, namely the waterfront, where the territorial impact of important hard innovations - such as those concerning maritime transport - is extended and developed by other more specifically soft innovations such as changes in territorial management systems, or in the attitude of public opinion to the use and management of environmental resources, and the quality of life.

The first paper takes into consideration the impact of a consolidated technological innovation such as transport innovation; the second, the impact of new technologies on space; the third the possible transformation of certain elements of urban structures, such as individual buildings or major areas like city centres, following the introduction of hard and soft innovations. In his article Dunin-Woyseth focuses on innovations that concern life-styles and their impact on urban form. Allpass deals with a closely related theme, namely changes in the relationship between public and private spaces as a means of understanding innovations in the ideology and culture of urban society.

Transport innovation

The study considers the innovations that occurred in the transport sector from the second half of the nineteenth century, with special reference to their impact on the metropolitan (Level 1) and core-continuous built-up areas (Level 2). Transport innovations are divided into two categories: radical innovations, involving a total transformation, and partial innovations leading to a restructuring of existing services. New forms of transport, finance and infrastructures clearly belong to the former; the transformation or the improvement of existing forms of transport to the latter. This situation of soft versus hard, and of radical versus partial is part of the complex interaction between the transport system and the urban structure, which is conditioned by a number of external factors, such as economic trends, population movements, other technological innovations, and events and situations that are unpredictable and imponderable. The author studies the role of transport innovations at each of the urban levels referred to in the Part One. The aim of the empirical study is to answer the following queries: (i) how far does transport innovation influence urban form?; (ii) how does transport innovation arrive?; (iii) how much time is needed to develop a new transport innovation?; (iv) for how long is an innovation awaited? On the basis of the answers to these questions, transport innovations are seen to have a definite but hardly ever decisive impact on urban form. To test this conclusion in methodological terms, the authors refers to a set of empirical studies conducted on some of the case studies in Part One.

New technologies

An in-depth study is conducted on the introduction of new technologies and their impact on land use. Three issues in particular are looked at in depth: the active role played by the introduction of new technologies at the territorial level; evidence of territorial transformations produced by the introduction of new technologies; socio-territorial aspects most congenial to the positive impact of the introduction of new technologies. The following parameters are considered possible indicators with respect to territorial transformation: convergence-divergence and public space-private space. The study refers mainly to Level 4 (districts - homogeneous morphological units) and Level 5 (specific areas).

Blocks, plots and buildings

The purpose of studying the minor components of urban form is to identify what relationship exists between society's demand for change and the financial capacity to satisfy them. Every change, even those involving the very smallest elements constituting urban form, requires corresponding changes in the units adjacent; this is so even where a general overall transformation of a more extended area is not possible. Blocks (Level 6) and plots and buildings (Level 7) represent the smallest available measure for calculating change. Both levels, though small, still contain the space and the capacity required for necessary changes. Once this minimum unit has been established, it is possible to identify the main characteristics: (i) form, (ii) size, (iii) systems of aggregation and (iv) morphological structure. The author shows that there is a constant conflict between the interests of individuals and of society. This conflict is generally solved by introducing changes in the available physical planning tools. However European cities have a very solid structure and do not respond easily

232

to demands for change. As a result, the dissemination of innovation is very slow indeed.

City centres

This paper deals with a more or less homogeneous morphological unit (Level 3a and 4). It can be identified on the basis of its geographical characteristics, the functions it performs and the results of a solid tradition of studies that has developed in recent decades. The authors take into consideration the soft and hard innovations developed in the last century. They look at the changes in administrative structures, in the organization of commercial activities, in urban and physical planning culture and theories with respect to soft innovations. They then consider the technological changes that concern separate components of city centres (construction technology, vertical movement systems, etc.) or the overall area (transport systems and networks, demand for mobility, pollution and environmental conditions) with respect to hard innovations. Cycles of soft and hard innovations alternate, sometimes giving rise to changes in urban form, and sometimes ensuring that urban form remains intact.

Life style

There are several definitions for the term 'life style' depending on the discipline involved and the period. Hence the author has addressed the term, in absolute terms and in relative terms, with respect to its possible use as an indicator of innovative processes capable of having an impact on urban form. The period considered the last century, but in relation to economic changes. Since these economic changes vary in time from one region to the next, rather than referring to periods of absolute value, reference is made to the various phases of urban development that have involved the different urban situations considered. The study focuses essentially on the methodological approach of the relationship between urban form and changes in life style. Thus the reference is to all the levels (Levels 1 - 7).

Public space

Historically European cities were marked by a complex division of public and private space, the former being dependent on the latter for its validity and efficiency. On the basis of this observation, the author defines the value attributed to public space in antiquity and the changes that have occurred in relations with semi-public, semi-private and private spaces, as a result of changes in economic and social conditions. Frequent reference is made to the position of international literature, but the empirical testing of the methodological assumption refers mainly to the situation of Scandinavian cities. Thus in this case the time reference is more precise: pre-industrial cities (seventeenth to eighteenth centuries), the first phase of industrialization (1850-1920), the industrial city (1920-70) and the post-industrial city (from 1970). General indicators are considered for each of these periods: for example, society, family, roads, the facades of buildings, areas of transition. Although the approach is methodological rather than empirical, the paper discusses the impact that innovations in the public space/private space relationship have had on the organization of residential areas (Level 5), blocks (Level 6) and separate buildings (Level 7). Most of the emphasis is on soft innovations, such as the physical planning culture and the influence that this has had on ideological and political attitudes. In a period marked by major ideological reevaluation, the author feels compelled to intervene directly in the critical analysis of the current situation, expressing his hope that steps will be taken to remedy the mistakes of the past and improve the situation in the future.

Waterfronts

By studying the transformation of a specific area - the waterfront (Level 5) - in this last century, the author considers hard and soft innovations and how they have overlapped and had an impact on each other over time. Until quite recently the waterfront area was dominated by the impact of technological innovations on maritime transport. More recently the innovations have been the result of the financial strategy of multinationals, the administrative capacity of local bodies, the demand for environmental quality from citizens of the more advanced industrialized societies. Thus there has been a passage from hard to soft innovations. The following areas are taken into consideration for the period closer to us: the coastal areas of major port cities in the United States, Great Britain, Italy and Japan. All these areas have undergone major transformation during the last century as a result of the simultaneous introduction of innovative processes in all four countries, given that the dissemination of the innovations was driven by changes of a general nature. The technological changes involving ships, systems for unloading and storing goods, the financial strategy of multinationals, the stress on the quality of the environment and, more recently, the decline of property markets following the global economic crisis, have all taken place at the same time and in a similar fashion. All the areas considered are shown to have experienced a similar impact from innovative processes, at times even quite independent of the local circumstances capable of conditioning the impact.

Conclusions

Although conceived within the framework of a joint research programme - which, among other things, was marked by considerable debate on methodological aspects - the seven papers presented in this part of the book are conceptually autonomous. Each author wished to provide his own definition of the term 'innovation', depending on the one most congenial to his or her methodological approach and especially to the theme. Each of the authors was given greater freedom of formal expression. On the basis of the work undertaken thus far, we can draw a number of general conclusions in relation to a list of innovations considered relevant to the transformation of urban form, identify what relationship exists between the various innovations and identify areas of innovation that need to be studied in a further stage of research.

12. THE INFLUENCE OF TRANSPORT INNOVATION ON URBAN FORM

George A. Giannopoulos, Transport and Organization Section, Aristotle's University of Thessaloniki.

Introduction

This paper deals with an investigation into the impact of transport innovation on urban form and is based on the case studies of the URBINNO project Working Group on built-form environment and land use, included in the first part of this volume. These case studies examined the development of urban form and structure in the last 100 years and tried to link the observed changes to the various 'innovations' that occurred throughout the examined time period. The definition of 'innovation' as given by Schubert (1992) encompasses 'new ideas set in motion', or 'new activities or reorganization of activities such as new services, institutions, rules, policies, planning paradigms etc.'. In the field of transportation, 'innovation' can be related to: (i) the introduction of a new mode of (individual or mass) transport; (ii) new transport infrastructure (roads, railways, canals, ports, airports etc.); (iii) the upgrading of performance and service of existing modes of public transport; (iv) changes in the organization and provision of mass transport services and/or traffic arrangements; (v) new means of financing transport infrastructure; (vi) other changes related to the existing system.

An investigation into the influences of transport innovation on urban development and form can be attempted either through a statistical and modelling type of analysis, or through a description and qualitative type of investigation of the experience and evidence that comes through comparative analysis and evaluation of several case studies. The establishment of a definite mathematical relationship between transport innovation and urban form is very difficult because of the extremely complex nature of this relationship and the bulk of detailed data it requires. This was not the objective of the URBINNO case studies. What is feasible and possible within the URBINNO framework is the second type of examination, namely that of comparative analysis of the evidence in the various case studies. This is attempted in the following paper. The analysis is restricted to Levels 1 (metropolitan area) and 2 (core continuous built-up area) because for the other levels the influence of transport has been difficult to substantiate from the case studies at hand.

The paper first sets out a theoretical framework which is based on an examination of the interrelation between transport innovation and urban form. This will help us to understand better the discussion that follows immediately afterwards on each particular case study which focuses on the transport innovation elements and their impact on urban form. Then in the final part some general conclusions are drawn and some recommendations for further study are made.

A theoretical description of the relationship between transport and urban form

In analysing the effects of transport innovation on urban form and development, it is evident that there is an interaction and close relationship between transport and location decisions for individuals and land uses, which (along with other factors) ultimately shape urban form. In order to illustrate this relationship various approaches may be followed. For example before and after studies have been made in order to establish the influence of new transport infrastructure on land prices and land use distribution and densities (Giannopoulos 1979, Giannopoulos Pitsiava-Latinopoulou, 1985). Others have followed a more theoretical and modelling approach (eg, Wegener 1987, Webster and Bly, 1988). From the existing body of research but also as an indirect result of the examination of the URBINNO case studies, a rather widely-accepted definition of the interrelation between transport and the spatial structure has emerged. This relationship is illustrated by Figure 12.1.: (1) the spatial structure of the region determines the distribution of activities in space; (2) activities generate traffic in the transport system; (3) the response of the transport system affects the accessibility of locations; (4) locations with high accessibility attract more development than less accessible ones thus changing the spatial structure.

Figure 12.1. *The interaction process between transport and urban structure*

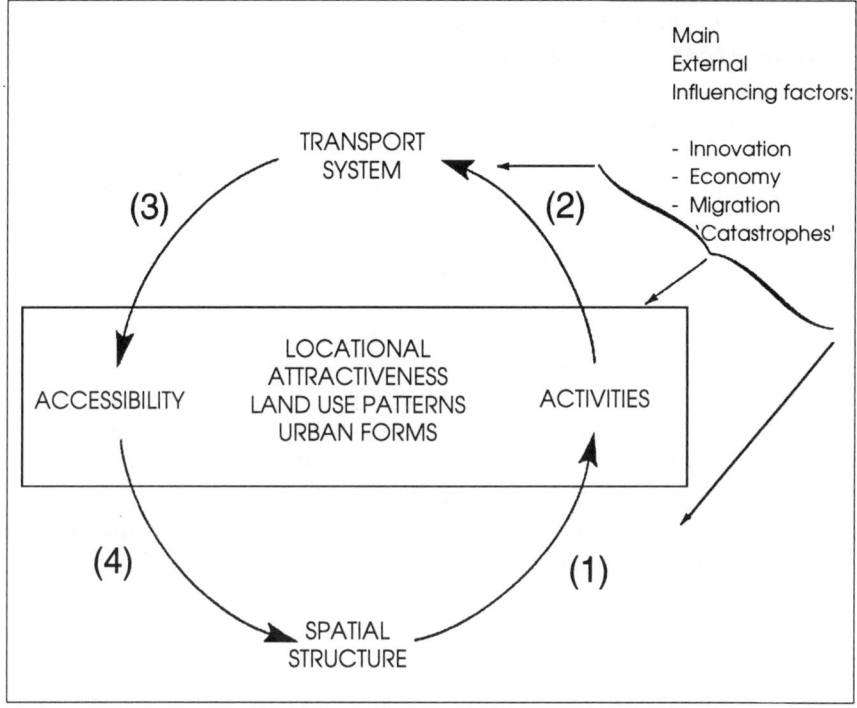

Based on Wegener 1987

Hence a location is characterized by the activities it accommodates and its accessibility. Activities and accessibility together constitute the 'locational attractiveness' of a location which in the long term determines the whole pattern of its land uses and ultimately urban form. The 'locational attractiveness' factor can thus be thought of as the driving force that shapes urban form and change, through the establishment of a temporary equilibrium state (either forced or natural). Observing the urban form at each particular point in time, we are, in fact, observing the result of the temporary equilibrium which prevails at that particular point in time. The 'circle' of interaction shown in Figure 12.1 is usually 'disturbed' by external factors such as the economy, influx of population (through migration), innovation, and unforeseen events which can be termed as 'catastrophes'. In this sense Figure 12.1 represents the interaction between transport/ accessibility and the spatial structure that results in a certain distribution of land uses for a given point in time.

If we add the time scale, we obtain a dynamic picture of successive equilibria that are the result of a process of continual transformation and evolution of the above interaction. To each 'temporary' state of equilibrium in this process, there corresponds a particular form of the urban area. This process is diagrammatically shown in Figure 12.2.

The major external factors that may trigger the dynamic processes of transformation for the new state of equilibrium are: (i) increases in population (through migration or other reasons); (ii) economic activity; (iii) innovation in its general sense (Schubert, 1992, and our definition in the introduction); (iv) the various catastrophes (war, fires etc.).

By accepting the overall dynamic nature of the interaction between transport - accessibility - location shown in Figure 12.2., one can visualize the mechanisms that connect urban form to transport and transport innovation more particularly. The 'static' circle of equilibrium that exists between the transport and the spatial structure at a given area is disturbed at a given point in time by one, or more, of the external factors mentioned above. The influence of these factors is to set in motion an 'instability' phase, shown by the dotted circles in Figure 12.2., which eventually settles to a new equilibrium state in which the new transport structure corresponds to a new spatial structure and urban form. These mechanisms and interactions have been represented with a varying degree of success in the so-called 'Integrated Land Use Transportation Models' that have been developed and which have recently been extensively reviewed through the work of another international study research group for the modelling of Land Use and Transport Interaction - ISGLUTI (Webster and Bly, 1988).

In the following sections we will examine specific examples of this interaction by looking through the processes of urban change in the last 100 years in specific urban areas in Europe which were examined as case studies in the URBINNO project and published in the first part of this volume.

Figure 12.2. *Diagramatic representation of the process of urban change*

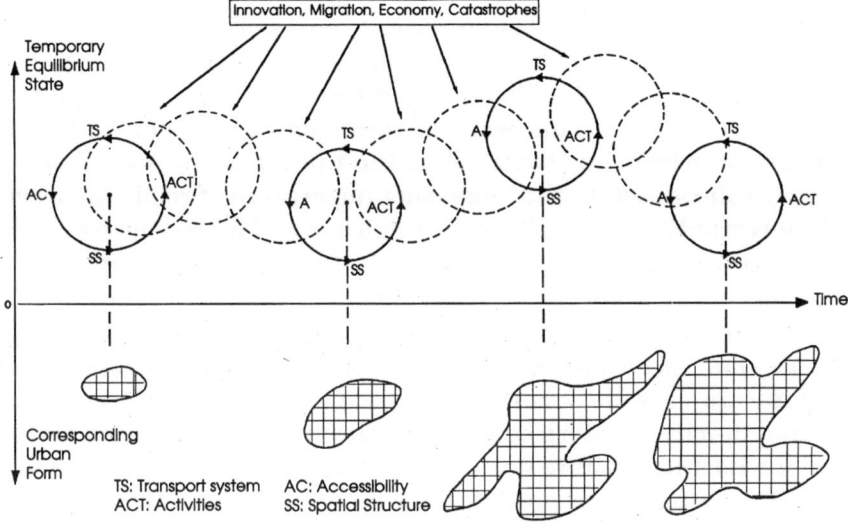

TS: Transport system AC: Accessibility
ACT: Activities SS: Spatial Structure

The evidence from the URBINNO case studies

Merseyside (Liverpool)

The very location of the Merseyside metropolitan area on the north-western seaboard of England is the result of a transport-dominated decision, namely its strategic location as a port. It has been described as 'an urban and industrial island created by the port and superimposed on a landscape of good husbandry' (Smith, 1953). The very plan of Merseyside embodies its port function. The river axis dominates the whole lay-out of the place. Building has proceeded outward from the river in concentric circles to end suddenly against rural country. Over the last century, and more particularly in the last fifty years, a series of 'spokes' of settlements have been developed along the main radial transportation axes from the core continuous built-up area or 'hub' to give Merseyside its contemporary 'hub and spoke' macro-physical structure (Massey, 1983). From the details of the case study for this

241

area it can be said that the description of the whole process of development of the Merseyside area provides a picture of 'waves' of development coinciding with 'waves' of innovation in aspects of the economy, the transport infrastructure, legislation and policy, and the inevitable 'catastrophes'. The embryonic 'spokes' of urban development along the lines of the electrified suburban railway network in the 1920s and 1930s had added a new dimension to the metropolitan physical structure. This was the first major wave of transport innovation which was centered on the railways. The second wave largely coincides with the increasing importance of road transport. Since World War Two railway line mileage on Merseyside has been much reduced reflecting the dominance of road transport. Liverpool's highway network developed gradually in the twentieth century with the modern road system largely growing out of the former network of country lanes which were widened and straightened. Two exceptions to this were the wide circumferential boulevard, Queens Drive, built around the city's eastern boundary in the pre-World War One period and the East Lancashire Road of the 1930s which incorporated a few existing routes. As significant as the physical pattern of road development in this 'wave' has been, the changing patterns of vehicle development and use, from the introduction of tram services in the nineteenth century and buses in the early twentieth century to the boom period of the private car in the post-war years. A third wave, evident in the last decade is beginning to emerge, namely that of the co-existence of road and rail transport within an integrated transport network. The main evidence for this are the new urban rail developments (loop and link system) built to increase efficiency of movement in central Liverpool but also the increasing pressure to restrict car use in the central area of the core. The impact of these new transport-related trends on urban form is yet to be seen.

Thessaloniki

Thessaloniki is an example of a city whose development has been predominately influenced by emigration of population and industrial expansion and to a lesser degree by transport innovation. A big catastrophe, the fire of 1917 which destroyed the largest part of the city, also played an important role. The main road network has remained virtually unchanged all the time periods considered in the study (last 80 years). For the last two decades three notable additions have had serious effects on the city's development. These were: (i) the construction of the inner peripheral road; (ii) the extension of the Egnatia avenue; (iii) the Eastern seaside 'exit' of the city towards the airport and the Chalkidiki peninsulas. Public transport is operated by buses. Its lines largely follow the expansion of the city by extending existing bus lines and occasionally creating new ones. The tramways that existed in the city disappeared just after the war when the wave of traffic increases made their operation with other traffic very cumbersome and expensive.

The transport 'innovations' that can be isolated and connected to the city's growth and development pattern largely fall into two periods quite a long way apart from each other. The first is the pre-1917 period and notably the late 1880s to 1910 when the main transport 'innovative' events were the construction of the main road arteries that still dominate the city's road network, the port and the railway station as well as the construction of all the basic infrastructure of sewage, water etc. The second period begins after 1960 when the three major road infrastructure works mentioned earlier were created. This is the only response so far to the private car revolution, which for Thessaloniki started in the early 1960s and boomed in the 1970s and 1980s completely overtaking the available infrastructure of the city and creating one of the major sources for its problems today. Perhaps the major element of this latest period is the lack of any rigorous planning actions to react to the overwhelming pressure created by the increased demands for mobility in the city and its metropolitan area.

Aachen

Aachen's development is marked by a series of major events in the economy, political organization and change, the two wars, and its geographic position as a border city. Transport innovation played a considerable role too, but secondary (as it seems) to other more decisive events. The first time period of consideration, up to 1900, is marked by 'transport-related' developments which determined the structure of today's inner city. The link between the French 'routes imperiales' and a local street system for the eastern part was closed after 1815. It was a deliberate strategy of local development: a new system of modern boulevards outside the eastern wall connected those (local) streets, and this was the signal for further development in the eastern direction. In the subsequent period, until 1928, the development of transport infrastructure and innovation has also been found fundamental for the division between town and region.

243

The town developed along the main axis to the north-east and the south-east. This dominance of development to the east was influenced by the topography, the community borders, the expansion of the eastern and northern district and by the borders to the Netherlands and Belgium in the West. The period 1929-50 started with worldwide economic problems which left their marks in the statistics of development. The reconstruction after the war saw the rebuilding of the tram network, the outspread of secondary streets to serve new settlements and industrial areas and the widening of the main through-traffic streets from the north to south-east and south, and from the east to the centre. This was the beginning of a profound transformation in history of the town. The scale of streets was changed and gave place to the motor car. Since the 1950s, and following the new German State and economy dynamics, Aachen grew accordingly. During the period 1950-72 the tram network had its first reduction. Some lines were closed. The bus as a more flexible system of public transport increased. The final period considered starts in 1972. In that year most of the community borders of the whole State were changed. As a result Aachen doubled its area by incorporating 6 communities of its surroundings. Special programs were developed by the State. A very important one intended to reduce private traffic. Communities started to change the large scale streets in housing quarters and also partly, in some cases, the dimension of main streets in order to find solutions for a coexistence between inhabitants, local needs and through-traffic.

Tromsø

Tromsø in the northern part of Norway, standing 'isolated' between the Nordland and Finmark Counties, has always heavily relied on transport connections for its development. A line of islands produces a varied coastline toward the sea and the mainland is sculptured with deep fjords. The most direct factors of development and change of the region have been stated as being primarily technological development in transport (shipping) and communications, and the refinement of the main industry of the area which is foodstuff production. Transport and communications have been the decisive factor for development in this isolated area. The coastal liner traffic between South Scandinavia and Troms County has been traditionally the only means of communicating with the outside world and has been in existence for more than 2000 years. Development of land traffic began in Tromsø slowly in 1857. The first state highway reached Tromsø in 1936. Mass public transport was provided by bus services combined with a ferry boat service. The railway has never reached the County. The first scheduled planes reached Tromsø in 1938 at Skattøra. In 1964 a new modern airport was opened at Langnes, becoming a centre for major north-south traffic. The growth of the area has been centered on the port and its surrounding functions and the development of fishing industry at least for the first two periods of growth until the 1950s. During the last 30 years, Tromsø from being the regional center of politics, economy and administration, became the center of the whole of North Norway. Its development followed the main transport innovations that changed the whole role and character of the town.

Lisbon

The first period examined was 1890-1910. This period's development was shaped by new residential areas in Lisbon. The main transport features of this development was the street system which followed the checkboard pattern that marked nearly all urban developments of Lisbon since the sixteenth century (including the Bairro Alto and later the Baixa of Pombal). The next period considered was 1910-26. This period corresponds to the years of the Republican Democratic regime after the fall of Monarchy and was a period of reforms in every branch of Portuguese society and mainly in social affairs and education as well as in consolidation of new urban systems in Lisbon. The public transport system was improved with the completion of the circuits of electric streetcars. The period 1926-34 was marked by the installation of military dictatorship that led to a complete collapse of the economy and the subsequent rise of the Salazar's 'New State'. A major drive for 'modernism' followed which involved construction of a number of new buildings but not major changes in the transport infrastructure. The next period 1934-53 is the period of bold action for the expropriation of large rural areas of the city, and the construction of new residential quarters along the new road axis, the Ave. de Roma, that overlapped the ring rail-road in the direction of Alvalade. In the period 1953-63 the occupation of the whole of today's built-up area was almost concluded and the construction of the University City began. In public transport the opening of the subway system, that followed the traditional radial line of Lisbon's valleys, Ave. da Libertade and Ave. Almirante Reis, repeating the tracks of the electric trams and bus lines, was a great improvement. The final period of consideration is

1963-80. This period was marked by the change of the political scene to a democratic regime. In transportation there was a shift towards private car transport and several viaducts to favour traffic flow were built. Perhaps the single most important transport innovation in this period was the construction of the 25th of April Bridge over the river Tagus profoundly altering the structural dynamics of production of urban land in the region, 'transferring' the greatest growth rhythms to the South Bank after 1960.

Kecskemét

Kecskemét is a medium-sized Hungarian town of 100,000 inhabitants, which provides a good example of how infrastructure for a rather neglected mode of transport - walking - can change the face of a small city. Kecskemét is situated in the centre of the Great Hungarian Plain between the Danube and Tisza Rivers, along highway E5 that links Northern Europe with the Balkans. It is first noted that it was railway constructed in middle of the nineteenth century that brought about an upswing of corn growing and thus the economic upheaval of the area. The original idea of prominence through the creation of major road infrastructure resulted in the planning (since 1893) and gradually implementing of the main road across the town to create Rakoczi Street by street widening and straightening. The 42 mt. wide, 500 mt. long promenade was completed in the years 1950-60. By the end of the 1970s it was decided to divert transit traffic from the centre. The idea of a pedestrian town centre was born and implementation of a project for pedestrianisation began by closing off the main square to traffic. The by-street that connected the main square and a small near-by square was converted into a 'pedestrian street'. The street which had been closed off by a small works gained new life as a shopping centre without traffic at the end of the 1970s and has remained since then the central element of social and commercial life in the town.

Athens

Perhaps the single most important factor that influenced development of Athens up to date, was its rapid population expansion after the second world war. Transport innovation always followed pressing needs for the movement of persons and freight of the ever-increasing population (Kokkossis and Shubert, 1988). The gradual outward spread of Athens and the evolution of urban expansion in the plain was marked by transport 'innovation' as follows. By 1900 the major core of the built-up city of Athens was still tightly developed around Acropolis. Since the establishment of the Greek State the major urban expansion was to the southwest around the port of Piraeus, to the south along coast of Faliron (summer resort for Athenians) and to the north (suburb of Kifissia) and the old city of Aharnai. In 1920 all horse-drawn carriages for urban transport were replaced by fixed rail tramways. This was also the period of the introduction of bus service in the area. In 1942 a special organization was created to organize the bus service in Athens. In 1930 the urban area expanded dramatically, linking Athens and Piraeus to create almost one urban area and started spilling to the south towards Glyfada and the north towards Kifissia. In 1950 the urban area expanded to fill the open spaces between settlements in the previous period solidifying a continuous built-up area from Piraeus and its western worker suburbs to Athens and the eastern refuge settlement areas. This development took place along the line of the railway connecting the port of Pireaus to Athens. From 1953 to 1961 the tramways were replaced by trolley buses and buses and by 1960 the first traffic signals and one-way streets appeared. The development of an international airport on the coast to Sounion changed rapidly the dynamics in that area as southern suburbs spread further along the coast. The new road artery of the Vouliagmenis avenue in 1967 intensified further these patterns of outward spread of the city while similar phenomena occurred in the other direction (Mesogion Ave. in the north-east, Kavalas Ave. to the west and the new National Road 1 to the north). In short, transport infrastructure in the case of Athens has been the consolidating 'agent' of development already existing there, and not as (is usually the case) its initiator.

Bari

In the eighteenth century, the area of Bari (Apulia) had very bad roads. Nevertheless, it had a considerable amount of harbours and therefore traded by sea with Venice, Trieste and other Mediterrannean ports, much more than by land with the rest of the Kingdom. In the 1960s all the roads of the network around Bari were linked together, with crossings at different levels, and with a 16km by-pass road that has freed the city from excessive traffic. The latter (perhaps the most important public work of that decade) is also linked by the two

247

junctions which lead to the highway tolls. The A16 highway reached Bari in 1965, from Naples, the southern city where the 'Autostrada del Sole' ends. A few years later the A16 reached Taranto. It connects Bari and all of Apulia to the national highway system, and in Canosa di Puglia it splits up into the 'Adriatica' highway which goes to Pescara, Ancona and Bologna. Some years before the Unity of Italy, a new harbour started to be constructed as the existing one was totally inadequate. This bold initiative, which took up the old idea (that apparently dates back to Emperor Frederick II) of a westerly harbour, in the bay between the city and the S. Cataldo penisula, is proof of the importance maritime transportation has always had for Bari. As years elapsed, the harbour was enlarged, connected with the railroad (1916), equipped with an oil wharf (1937) and cereal siloses (1960). More recently (1980) it was equipped with a new passenger station at the service of the many tourists which take ferries to the other shore of the Adriatic Sea, Greece and the Near East. Only after National Unity, did the national railway system reach Apulia; it reached Bari in 1864, coming from Foggia, Pescara, Ancona. In 1869 Bari became the first railway junction of the Apulia region. In 1882, besides the Foggia-Lecce line and the connecting line to Taranto, an internal one to Barletta (via Bitonto, Terlizzi, Ruvo, Corato and Andria) was built; it was made up of a steam narrow gauge tramway, almost always along the edge of the road. In spite of the very simple technical characteristics, this 'economical railway' (that is how it was called at the time), built with foreign funds, achieved the goal of connecting Bari to centers other than those along the littoral Foggia-Barletta-Bari line. In 1915, out of the twenty towns of the Bari metropolitan area only two (Bitritto and Cellamare) did not have a railway station; whereas Modugno and Grumo Appula had two (National Railways and Calabro-Lucana). The Bari junction embraces the four lines (FS, FSE, FCL, FBN). Regular flights with Rome became operational at the beginning of the fifties. Until 1975 planes used the runway built before World War I, in the military airport of Palese, an otward-lying littoral, west of the city. At the end of the sixties, the construction of the civil runway began, located not far from the military one. It became operational in 1975. Concerning the overall regional organisation as seen in conjuction to the above changes it should first be noted that the city has at present a greater industrial importance and plays a more qualified tertiary role. Industries are mainly located in the western part of the area, site of the new airport and the future railway cargo area now under construction. This area is already equipped with many transportation infrastructures and is located next to the highway and it will soon be better connected with the harbour.

Transport innovation

Types of transport innovation

Transport innovations, of one or other of the six types defined in the introduction of this paper, can be identified in almost all case studies at some point in time. The most easily identified innovations are those relating to the 'introduction of a new mode of transport' and 'new transport infrastructure'. For these types of innovations there are distinct references in the case studies and, therefore, some specific conclusions can be drawn which are presented in the following sections (Giannopoulos and Curdes, 1990). As regards the other types of innovations, e.g. upgrading performance or service levels of existing modes, or changes in the organization, financing, etc., there is evidence that these too have played a traceable role in shaping the transport and spatial equilibrium process shown in Figure 12.2. However, their influence, compared to that of the first two types of innovation, is relatively less pronounced. While keeping to the basic typology of 'innovation' that was put forward in the introduction (and which was followed in presenting the results of the examination of each case study) one can further distinguish two classifications of transport innovations. Those that are found in all areas at related (though not always coinciding) time periods and which can thus be called 'universal', and those that can be found in almost all urban areas but at different time periods covering almost the entire 100-year period of our consideration. These latter types of innovation can be classified as 'local' in the sense that their timing, and extent, are 'site-specific' and depend largely on the local (sociopolitical or economic) conditions prevailing in the area at the time of their introduction. Following the above further classification and the case study examinations one can say that the first two types of transport innovation, i.e. the introduction of a new mode, and new transport infrastructure, are fairly 'universal' types of innovation while the subsequent four (see introduction) are more 'local'. In the above sense it is tempting to relate the universal types of innovation to the so-called 'hard' (i.e. of technical nature) innovations mentioned by

249

Hornberger (1980) and the 'local' ones to the 'soft' innovations mentioned by the same author.

Timing and time of occurrence of transport innovations

One point to be made first concerns the differences observed in the timing of transport innovation in the sense of the time periods of their introduction, maturity and fading out as factors affecting urban form. Even for so-called 'universal' innovations, the timings for the above 3 stages, although falling in the same broad time periods for all case studies, do not coincide in each urban area. That is to say, although these types of innovation (by definition) have a universal existence at broadly the same time periods, their specific introduction and development at different places occurs at different times and at different rates. Figure 12.3 shows the generally established time periods of occurrence of so-called universal innovations. However, by looking at the specific timing of each type of innovation in each urban area examined, we obtain the results of Figure 12.4. and 12.5. From the comparison and analysis of the results in these figures some interesting conclusions can be drawn that give some answers to the questions put forward in the introduction.

a. When does an innovation 'arise'?

As expected, innovations arise earlier in regions and countries which have achieved earlier a higher level of development as opposed to regions which have become developed later and less intensively. There are exceptions to this rule. For example, while - following our expectations - Aachen is significant for the beginning of the implementation of most innovations, the horse tram appears earlier in Athens. Apart from a few exceptions, it becomes evident that a North-South divergence of implementation does not really exist. For example, take the implementation of the railway. Except for Liverpool it appears in Aachen and Thessaloniki almost simultaneously and only a little time later in Kecskemét. Again the times of the introduction of the tram do not differ essentially in Lisbon, Liverpool and Aachen, and Lisbon and Liverpool opened their airports at the same time (both become continuously extended). More significant are the differences concerning ring-roads and motorways. Here are differences which span almost half a century. For example the development of the ring-road in Liverpool began in 1903 and was finalized in 1933.

250

In Lisbon, the development of the ring-road began in about 1948 and will take until the 1990s to be completed. Regarding time, there were no essential time-lags between north and south Europe (except Liverpool) during the middle of the last century; during our century, however, differences increase concerning those innovations which are related to individual traffic.

b.   How long does construction and development of the network last?

This aspect is also illustrated in Figures 12.4. and 12.5. which show how much time it takes for a town to realize the advantages and the full effects of the junctions of the new network. Railway lines were, in most cases, developed by building at first one line which was followed, in later times, by further lines. Thus, the indicated time-period is the time from the beginning until the completion of (most) services. In Bari, it took 49 years, in Lisbon 38 and in Liverpool only 10. The very short period of time referring to Thessaloniki is only the year of opening. Extremely long times of development and construction become obvious regarding the ring-roads (Figure 12.5.). Ring-roads are a necessity for the outer connections of quarters within radial-concentric networks of about 3-6 km in diameter. Lack of planning and the number of necessary procedures for land expropriation may explain these periods.

Figure 12.3. *Time periods of 'universal' transport innovations*

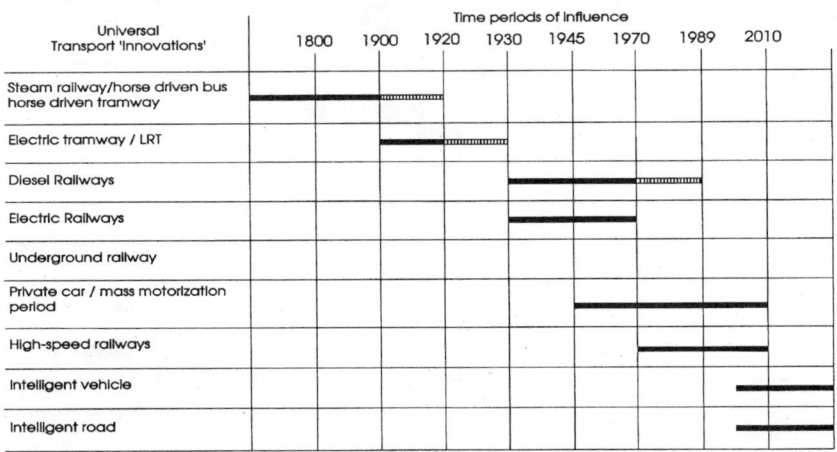

Universal Transport 'Innovations'	Time periods of Influence							
	1800	1900	1920	1930	1945	1970	1989	2010
Steam railway/horse driven bus horse driven tramway								
Electric tramway / LRT								
Diesel Railways								
Electric Railways								
Underground railway								
Private car / mass motorization period								
High-speed railways								
Intelligent vehicle								
Intelligent road								

Figure 12.4. *Timing and time span of infrastructural innovations in public transport*

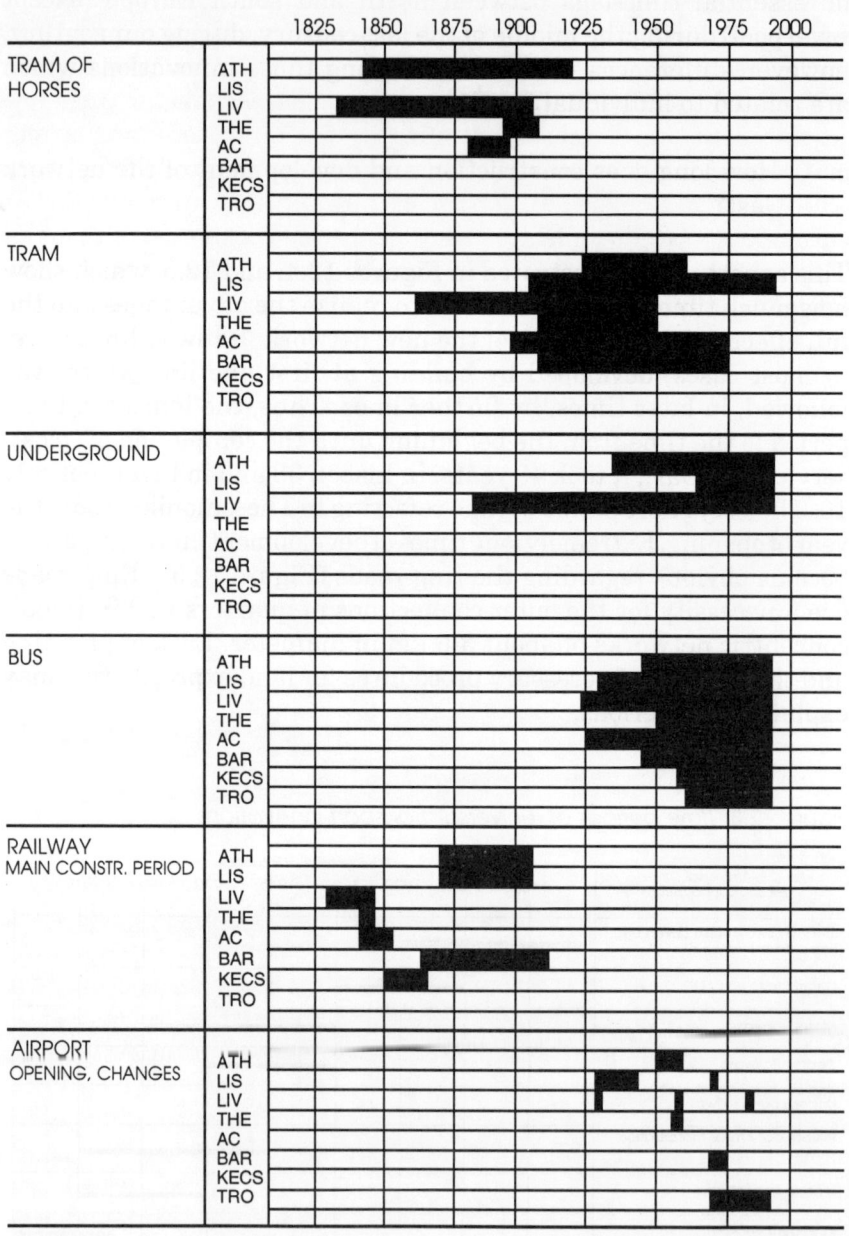

Based on the case studies of Athens, Lisbon, Liverpool, Thessaloniki, Aachen, Bari, Kecskemét and Tromsø (Curdes, 1989).

c.   How long were the 'innovations' in demand?

In other words what was the life span of innovations no longer in existence? These are dimensions of time which are very short within the life of a town. Urban railway lines within the physical structure of the urban areas examined show long time periods for their full development (e.g. 60 years) that should not be interpreted as the time that is necessary for these effects to take place. Evidence from other studies (Giannopoulos, 1979; Giannopoulos and Pitsiava-Latinopoulou,1985) indicates that the time for the effects is much shorter and in the order of 5 to 15 years. Table 12.1 shows the average life span of transport innovations that no longer exist in the respective urban areas.

Table 12.1. *Life span of innovations which do not exist today (years)*

	ATHEN	LISBON	LIVERPOOL	THESSALONIKI	AACHEN	AVERAGE
HORSE BUS			40			40
HORSE TRAM	85		38	14	14	38
ELECTRIC TRAM	41	82*	(50)	43	80	59

* Continuing to present day. Source: Curdes, 1989

Effects of transport innovation on urban form

From the 'innovations' regarding new transport the one with the most devastating impact on urban form was the advent of the motor car and the mass motorization that started after the war and continues up to date. The old traffic and circulation system that was based on the small and narrow streets winding around squares and pedestrian areas in the central cores of the old towns and cities gave way to large streets and parking areas and helped transform radically the old pattern of urban movement from a type based almost solely on mass transport modes to another that depends mostly on the private car. It consequently gave rise to the well known urban sprawl which characteristically followed the pattern of major transport axes (see Tromsø and Liverpool). This became evident when we compared the

Figure 12.5. *Timing and time span of infrastructural innovations in private transport*

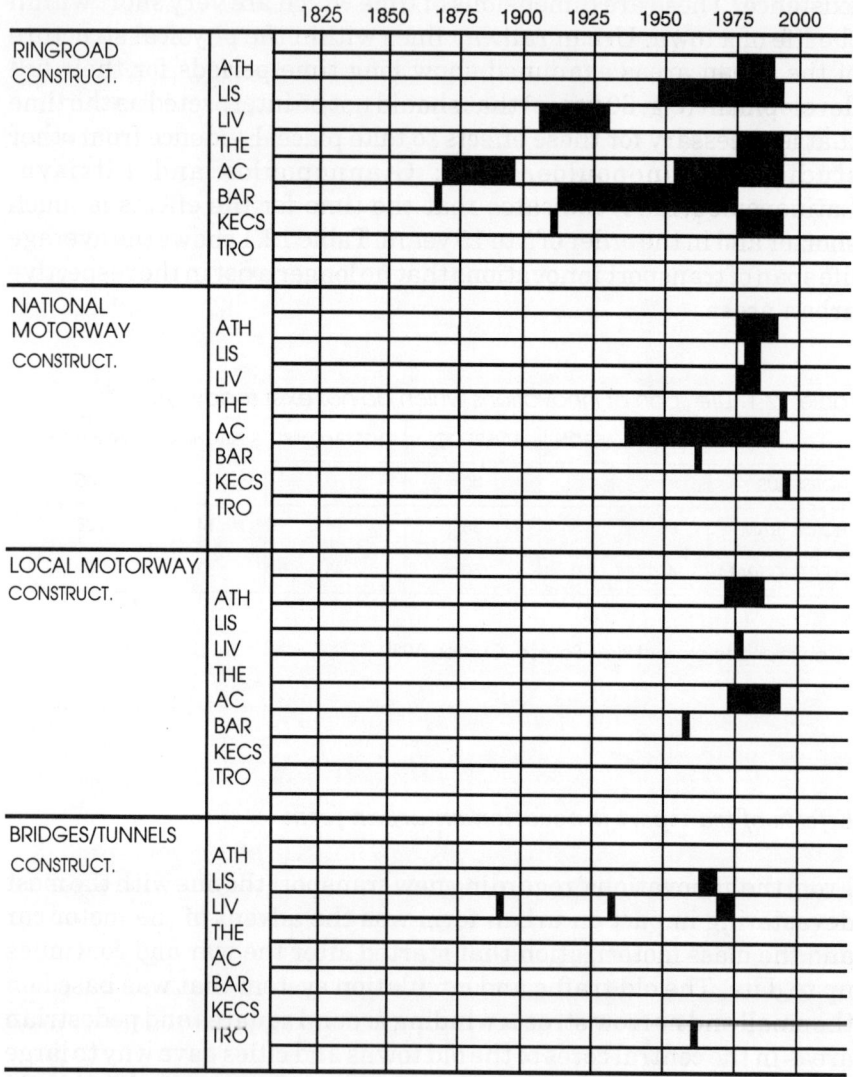

Based on the case studies of Athens, Lisbon, Liverpool, Thessaloniki, Aachen, Bari, Kecskemét and Tromsø (Curdes, 1989).

254

road network in the areas examined, in comparable time periods. The development of these networks through time, shows a remarkable resemblance to the forms of overall development in the area.

However of great importance was the advent of innovations regarding other modes of transport. A detailed study, in 1909, of the large electric tramway network of Aachen concluded that it had helped create both special suburbs and a general widening of the city (McKay, 1976). And since the tramways extended far into the countryside and served also as light railways, they permitted peasants in the area to give up farming for the mine and factory while continuing to live in their houses and villages. In Thessaloniki the introduction of the horse-drawn tramways in 1893 and the electric ones in 1907 had a major role in the expansion of the city to the east where the main residential quarters of the city are today. Even walking, as a mode of transport for which one has not to plan new infrastructure, has influenced the urban structure considerably. The 'universal' innovation of major pedestrianizations in the 1960s and 1970s with implemented plans for a horizontal segregation changed many urban centre in almost all case studies. A most notable 'local' type of innovation in this sense is the policy in a very few urban areas (Liverpool from the case studies examined) to implement a vertical segregation through its 'walkways' in the city centre. The above points about 'walking' and pedestrianizations can also be thought of as innovations of the second type mentioned in the introduction, i.e. as new infrastructure. However, there are other more notable effects on urban form concerning new infrastructure. The construction of new roads, road or rail bridges over natural barriers or other infrastructure (e.g. extension of metro lines) is usually followed by a 'ribbon' type development along or around them. The examples for the cases of Tromsø (new bridge and road), Lisbon (new bridge), Liverpool (extension of rail network and new motorways), Aachen and others of the case studies examined, are characteristic (see discussion and drawings given in the corresponding case study chapters). Other examples of the 'new infrastructure' type of innovation are the construction of the inner ring road in Thessaloniki which almost doubled land prices in its vicinity and considerably changed the distribution of land uses (Giannopoulos and Pitsiava-Latinopoulou, 1985) and the new port at Breivika, in Tromsø, which diminished the importance and central function of the port at Prostueset, a fact that brought about a total reorganization of land uses and values in the area.

With regard to the other types of innovation, i.e. policies and organizational structures (upgrading of performance, changes in organisational structures, new financing etc.) significant effects can also be found. In almost all urban areas examined, (especially after the Second World War) comprehensive land use/transportation studies and extensive traffic management schemes (one way streets, parking restrictions etc.) have been carried out. Although from the material supplied by the particular case studies there is little connection to be established between these types of innovation and urban form, there is considerable consensus among the researchers of the URBINNO project that these innovations helped mostly for residential uses outside the city. In an opposite sense, new policies concerning these innovations (e.g. provision of parking space) are recently being used as instruments for bringing back residential uses to the city centres. Other examples of the impact of these types of innovation can be found. As Horne and Maund (1975) comment, 'the introduction of 1 penny fares in 1899 was one of the most important events in Liverpool's transport history. It resulted in an enormous increase in traffic, because it enabled many working people to become regular tram passengers. The ultimate social effect was very far reaching as the spread of suburbs which had already started with the development of railways and other transport media in the nineteenth century was accelerated and many more workers were able to seek living accommodation outside the traditional industrial areas'. An important point to be made concerns the differences in the extent and importance of the effects that similar innovations had on urban form. For example the creation of new transport infrastructure with the inner ring road in Thessaloniki did not have the same impact as the new bridge and road connecting the island of Tromsø to its adjacent land in the mainland. In the second case the impact in the sense of the new development that followed was much more pronounced and rapid.

Conclusions

In conclusion, it can be said that all transport innovations influenced the form and extent of the development of the urban areas examined, but none of them has determined the form completely. A visible outcome of our work is, therefore, that transport-related innovation is one of a number of influencing factors that affect urban form, without however being the only one. Towns are socio-physical systems with a high resistance against short-time transformations. Any urban morphology has its own logic, pattern and scale. Technical innovations have short life cycles compared to the longer life-span of urban morphology. Therefore, it can be said that innovation and urban development build the synthesis between new needs and technologies in a form which is comparable to the specific (and different!) urban scale. In this understanding, the most important innovation is an innovation of contextual creativity (Curdes, 1989).

The following innovations seem to have left significant marks in the form of towns: (i) transport - the railway (barriers, lines, front and back areas), the motorcar (street widening, ring roads, suburbanization, decentralization, motorways, parking places, garages, transformation of streets and crossings), the airplane (airports, new extensions towards the airport, industries around airports); (ii) town planning - morphological theories (Blocks, rows, ribbons), the concept of the garden city, the Charter of Athens (1933), concepts of parks and green belts, the concept of flowing space, concepts of decentralized centres; (iii) housing - the ideal of the individual family house.

None of these innovations formed a town in total. There is always a mixture of influences from different fields like economy, social behaviour, planning ideals, geometry of the network, the role of a town. Innovations are only part of this process. However, from the towns in different stages of development that were examined, it can be seen that the influence of innovations on urban form is high in an early stage of development and lower in dense and upbuilt towns. Innovations for upbuilt towns have a chance only if they are more or less compatible with the usual structure and scale of towns.

References

Curdes G., et al, (1989), *The Influence of Innovations of Urban Development and Urban Form: A Comparison of the Development Stages of Athens, Lisbon, Liverpool, Thessaloniki, Aachen, Bari, Kecskemét, Tromso*, Institute for Town and Country Planning, Aachen University of Technology, unpublished.

Giannopoulos, G.A. (1979), *Transport and Regional Development: The Case of Countries with Less Advanced Economies*, ECMT 9th International Conference on Transport Economics, Istanbul.

Giannopoulos, G.A., Curdes, G. (1990), 'Innovations in urban Transport and the influence on the urban form', *Transport Reviews*. No.4.

Giannopoulos, G.A., Pitsiava-Latinopoulou M. (1985), 'Some Findings on the Interaction between Transport and Land Use in Greece', *Transportation Planning and Technology*, No.1.

Hornberger, K. (1980), *Interdependenzen Zwischen Stadtgestalt und Baugesetz*, Zurich.

Horne, I.B., Maund, J.B. (1975), *Liverpool Transport: Vol.1: 1830-1900*, The Light Railway Transport League, London.

Kokkossis H., Schubert S. (1988), *URBINNO Group 4: Athens Case Study*, paper presented at the URBINNO Meeting, Aachen March 1988.

Massey, P.W. (1983), 'Conserving and Renewing the Built Environment', in Gould W.T. and A.G. Hodgkiss (eds.),*The Resources of Merseyside*, p. 131-46, Liverpool University Press, Liverpool.

McKay, J.P. (1976), *Tramways and Trolleys: The Rise of Urban Mass Transport in Europe*, Princeton University Press.

Schubert, U. (1992), 'Conceptual Ideas on Innovation', in Drewett, R. and Montanari, A. (eds.), *Innovation and Urban Development: the Role of Social and Technological Change*, Avebury, Aldershot, forthcoming.

Smith, W. (1953), 'Merseyside and the Merseyside District', in Smith, W.A. (ed.), *Scientific Survey of Merseyside*, Liverpool University Press, Liverpool.

Webster, V., Bly, P.H. (eds.) (1988), *Models of Land Use and Transport Interaction (ISGLUTI)*, Gower, Aldeshot.

Wegener, M. (1987), *Transport and Location in Intepreted Spatial Models*, Gower, Aldershot.

Whitehand, J.W.R (1987), *The Changing Faces of Cities: A Study of Development Cycles and Urban Form*, Blackwell, Oxford.

13. SPATIAL EFFECTS OF 'NEW-TECHNOLOGIES': AN APPROACH TO PROCESS ANALYSIS

Andrea Haase, Lehrstuhl und Institut für Städtebau und Landesplanung, RWTH, Aachen.

Introduction

The choice of subject was determined by the intention to examine the effect of very recent innovations, thereby attempting to fill the gap between the historically-oriented approach of case study analysis and current questions arising from multi-disciplinary research into urban transformation. Therefore, the aspect of 'new' technologies is only a secondary one, selected to represent the dominating innovation of the present day. It also seemed appropriate as a basis for reflecting in principle on the complexity of innovation-processes and their interrelated factors, since this cannot be examined only from data-analysis. The approach of this paper is different from other studies because there are few empirical data available from literature and none from the case-studies. The 'complexity' can be initially defined by underlining the fact that new technologies in history have always been a result as well as a condition of socio-spatial transformation. However, their importance as a motor for urban structure transformation cannot be acknowledged except for the periods of main production innovations (machines) and of transport innovations (railway, individual traffic). The role of 'new' technologies, information and communication technologies[1] in urban transformation is still dependent upon the results of current and future processes of diffusion and of implementation. Thus, the paper focusses on the question of how this complexity can be investigated by multidisciplinary research. The paper offers suggestions for deriving goals for the next stage of the URBINNO project, concentrating mainly on process-analysis and its methodology.

The stage of multidisciplinary urban research

Questions of research arose, when uncertainty about future fields of investment and about future development created the need for research in order to guarantee economic progress and to provide a basic scale for a prophylactic compensation of environmental and living conditions. The discipline of spatial analysis and spatial planning still has difficulty in finding its place among other disciplines in urban research, though there are attempts to widen the range of contacts.

History of innovation survey

During the 1920s, primarily economic approaches dealt with the theme of 'innovation diffusion' (Burns, Kuznet, Schumpeter). Urban analysis concentrated on goals of planning within which the discussion of environmental quality only had an important role as far as the actual potentials for urban growth were concerned. During this period, an immediate interrelation between spatial analysis and the discipline of economy was not evident, though, indirectly, the discipline of spatial planning was used to contribute to the spatial organization of society and economy by implementing the idea of function separation. During the 1950s, 'innovation diffusion' research focussed attention on sociological and geographical aspects with a specific interest in potential application for specific products, while economic and morphological aspects of implementation-effects were missing. Though urban theory still was bound to the ideas of the 1920s, the discipline of town planning did not deal with 'environmental quality' apart from concern with urban growth and the modernization of old urban areas by a network of public and private infrastructure. During the 1960s, 'innovation diffusion' research was mainly influenced by the sector of economics referring to basic processes of decision making by potential innovation adopters of industry. Links with research about changes in urban structure quality were not known. This can be related to the stage of urban development which at this time had set itself the task of realizing economic progress. During the 1970s and 80s, research evinced a lack of investigation concerning environmental changes during the process of diffusion and to changes in the innovation itself during that process. Within this context the theories of Schumpeter gained new importance, referring to 'clusters of related innovations with the potential to affect... even the economy as a whole'. Present urban research gives importance to multidisciplinary links between different subthemes as well as to different spatial levels of investigation within an international network of dependences, selectively represented by the following terms: 'global cities' (Esser and Hirsch, 1987), 'changing hierarchies' (Läpple, 1987), European towns stages of 'disurbanization' (Friedrichs, 1977; Kujath, 1986 and 1987; Läpple, 1987), 'hyperindustrialization' (Esser and Hirsch, 1987), functional and social segregation of inner town areas (Esser and Hirsch, 1987), segregation and differentiation of regions and towns within prosperous as well as in declining areas and the thesis of an increasing internationalization of processes (Esser and Hirsch, 1987), 'footloose' offers of employment with advantages for the choice of residential locations (Läpple, 1987).

These very general issues are neither definitely proved by sufficient empirical work nor are they based on a theory which is related to perspectives of future development. But they are illustrative enough to raise the leading question: which of these phenomena and their outcomes are influenced mainly by innovations and especially by 'new' technologies? This question is raised to emphasize the process character of urban transformation and the difficulty in identifying single influences within the network of factors taking an active part in the process. The following section regards approaches to the field of 'spatial effects', questions their potential function for a process analysis and gives attention to new technologies as an example of innovation, as far as basic material is available.

Approaches to 'spatial effects'

Existing investigations into the innovative effects of new technologies on space mainly concern the macro and intermediate levels of investigation (Henckel, 1985; Henckel et al., 1986; Henckel, 1987; Stöhr, 1986, p.8, 11-13). The 'micro-level surveys' available from literature mainly concern the history of high-technology centre development, dealt with in a prevalently descriptive way. Conclusions referring to spatial effects are rare. However, existing surveys at the intermediate or macro-level draw some links between the institutions of technology centres (science parks, spatially isolated from the 'context' of a town and originating from external investments) and the structural response of the surrounding region by regarding primarily the spatial dimension. They do not deal with the specific effects of innovation diffusion on the socio-spatial structure. Fundamental to this kind of evaluation is an understanding of the 'pervasive character of new technologies' (Stöhr, 1986, p.8, 11-13) which implies at least the principle of an interdependence between existing conditions and resulting effects. Referring to the rule that spatial effects depend on the appearance of a 'critical mass' (Schubert, 1992) of single and different innovation implementation steps in the transformation process, the macro-level analyses and their issues (e.g. the distinction of structure-types), (Henckel, 1985; Henckel et al., 1986; Henckel, 1987) are to be regarded as the respective contexts within which the development of a 'critical mass' should be investigated. One elaboration is mentioned in the notes as an approach to micro-level investigation; however, it mainly regards 'company history' and thus remains in the field of economics (v. Geenhuizen, Nijkamp, Townroe, 1989). In order to gain socio-spatial knowledge about locations and about the kind of process and the factors involved, further attention has to be given to micro-level investigation within a major and multidisciplinarily-oriented level of survey. This hinges on a very comprehensive understanding of change. In theory, there is the term of 'endogeneous change'[2], derived from economic discussion about innovation diffusion, but also implying an approach to spatial aspects at different levels. The term implies an understanding of spatial surroundings as one influencing factor among others and, thus, highlights the close interrelation between existing conditions and resulting effects. Therefore, it seems appropriate to link up these two levels to a multidisciplinary scale. This approach promises to be usable for an investigation into the effects of new technologies which until now have been difficult to define, because it acknowledges the existence of a 'hidden' change which still has not reached the stage of implementation of an obvious 'critical mass'[3].

Basic features

The idea of applying economic categories to aspects of urban and socio-spatial research is used as an example: 'consumer' and 'producer profiles'[4] as well as 'demand profiles' (firms) and 'supply profiles' (socio-spatial potential) are categories related to industrial location theory (Davelaar and Nijkamp, 1986). The respective interdependence implied in each of the pairs of terms links them together in a process of complementary development which continuously establishes new patterns of relationship. The difficulty with the 'endogeneous change'-analysis in dealing with a variety of local specific characteristics could be resolved by a system of parameters which would guarantee, as the main part of a desired major scale, the comparability of the internal logic of the investigated processes from an external point of view. The major scale could be concentrated on the complementarity of urban structure and its transformation. An appropriate major scale for this complementarity has to be determined in principle and under mulitidisciplinary conditions, as there are always different 'profiles' within urban structure. The complementarity of the use of private and public space and its physical appearance seems to be wide-ranging enough to serve as a methodological basis and as a socio-spatial framework which encompasses every context of time and space, every spatial level of investigation and different factors in the process. The reasons for choosing the complementarity of 'private and public space and its use' as a major scale are:

i. The relationship between private and public space has been relevant since the beginning of urban history and was fixed mainly by Roman law. The distribution of space and its use, influenced by the right to own private property, has led to the distinction between 'private' and 'common' areas ('*privare*', Latin = deprive). Thus, distinctions between private, common and public have developed.

ii. Debate about private and public space became a factor in urban research history around 1970, when Richard Sennett introduced a combination of sociological and spatial aspects of investigation. He focussed on the influences of social consciousness on space.

iii. One dominating social force has always been the collective belief in social and technological progress which influenced private and public space concerning form, use and function. Current belief is strongly affected by new technologies and is still based on the belief in saving time by cable-communication. On the other hand, the complementary effect of this is a redundance of time which affects urban space through rates of unemployment, leisure activities at local to international levels, the development of new images of location advantages for the land market (concerning all sectors of housing and industrial location choice), developing social values for collectively accepted symbols of socio-spatial identity (change of view from the local factory to the main regional, national or international centre of service-agglomeration).

iv. Changes in the interrelation between public and private space also have an international level. The question thus arises as to whether local specific influences will have any importance at all in the future and, regarding old industrialized declining areas, which processes of international migration and adjustment will determine future spatial development (European city-centres are already hardly distinguishable from each other as a result of the standardization of the organisation, the lay out and furniture of urban public space).

v. The question of government influences has changed to that of economic influences, which are going to become internationally more important than any national government. This role will be reflected by urban space.

The proposed major scale 'public and private space' will have to give specific attention to the view that users of urban space are mainly categorized as 'consumers'. Therefore the setting might be related to the respective profiles of location advantages or disadvantages, distinguished by the categories of 'consumers' privacy' and 'consumers' publicness'. The 'consumer's' side is taken here as a primary characteristic and is based on the thesis that urban structure reflects the successive implementation of political and economic power into space which therefore has become a 'consumable item'. The relevance of this thesis, especially for the present, has to be discussed firstly in relation to the socially-protected ideal of 'liberty' to use and to form the spatial organization of society, which is said to have destructive effects on spatial quality. The thesis also has to be discussed within the context of the political efforts of self-help groups and associations, developed in the 1920s and revived during the 1970s, which are an expression of the goal to give users the right to 'produce' their built-form environment through their own initiative and, in this way, to give them the possibility of escaping from the social and economic restrictions of 'consumers'. But, nevertheless, the above-mentioned thesis shall be followed, based on a historical review and on the thesis that existing 'liberty' is really only one side of a profile which forces people to fulfill the 'consumer's' role. In former times, during feudalism and especially during absolutism, but also in the later periods between the seventeenth and the nineteenth centuries, dominating forces were defined by the respective superior political authority and by state control. Only since the nineteenth century has there been a change towards the increasing influence of the economy which has become more and more self dynamic, though still subject to government control (for example, town enlargement planning in the Ruhr-area was mainly economically reasoned but initiated and controlled by the Royal Prussian Government, (Kastorff-Viehmann, 1984). In Germany, French influences in the areas west of the Rhine brought a different understanding of development control, reserved to the decentralized authority of local communities and their specific economic needs. The 'freedom' of local communities to decide about the aims of local development (eg. in the FRG) and the 'freedom' of the legally-guaranteed right to use and to form private space individually has become a force which essentially reflects the socially-created and individually-interpreted aims of taking part in the general process of social and technological progress, guided by international economic forces. Thus, the 'freedom' of the individual (community or single person) is normally limited by standards which are collectively accepted, individually consumed and reflected in the artifact of built-form environment, and which either conform to well-known building patterns or may lead to new and innovative ones. Examples of the socio-spatial view on the current process of urban transformation under the influence of new technologies will illustrate the preceding reflections, with discussion of the following questions:

i. What relevance have 'new technologies' for the current process of urban transformation within local, regional and international contexts?

ii. Are there any 'new' effects on urban structure to be found and if so what kind of transformation do they contribute to?

iii. What kind of socio-spatial conditions are appropriate for conceiving new technologies within the urban structure?

Examples of questions and answers

Our starting point here is the supposition that new technologies are only one category of affecting forces. They have been selected from the range of different influences which work in parallel with each other, depending on the stage of urban development within a specific spatial context. An indepth analysis of different spatial situations would be necessary to give sample answers to general questions. Where such analyses are not available, answers will be given hypothetically by confronting subjectively-selected empirical knowledge with theoretical conclusions from macro-level analyses. As there is no 'finished chapter of history' to be considered, questions and answers are expected to change with the influence of future development.

1. What relevance have 'new technologies' for the current process of urban transformation within local, regional and international contexts?

From a macro-level analysis the conclusion is that 'information-technology does not create spatial trends of a specific kind.' (Warner, 1972, p.5)

This answer is restricted by being general and including local, regional and national as well as international points of view. It takes account only of the supposed active role of new technology and does not say anything about the effects of new technologies reacting with other influences; for example social values, their implementation within the economic system, mediated by the built-form environment and its symbolic character. Thus, examining the relationship of new technologies with existing conditions and with their transformative forces could provide a different point of view and could lead to a different answer.

Considering the main consequences of new information technology, we can see on one hand the socio-spatial problems or 'negative' effects on specific areas, caused by changing the structure and number of jobs, and by separating people into groups of those who still have to save time and those who have more than enough time because they are jobless. The differentiation and social segregation among small spatial units which results from these effects can be seen as an influencing factor itself. Whereas, on the other hand, the new 'qualities' of urban areas, architecture and civic design, bring about a specific new kind of urban transformation, centring on aesthetics and on the concept of taking part in technological and economic progress thereby forming a new 'urban pattern' which tends to be regarded as an example to follow. This is particularly obvious and concentrated within metropoles such as Frankfurt, London and New York. The social consciousness which is fundamental to the implementation of this spatial trend of urban transformation is undoubtedly of a very privileged kind, but only appears a 'free' choice because the privilege is forced by economic needs of competition. The symbolic reflection of this competition in space is fundamental to the purely functional and, afterwards, spatial consequences of diffusion, influencing the local to international socio-spatial structure. The local specific implementation of the new patterns of 'social urban life' into space, its rapidity of transportation and its dimension of effects, however, depend primarily on locally different conditions, in order to become adopted for example available space, waiting for transformation on one hand, the necessary economic power and the mentality to be open to change on the other hand. Nevertheless, considering the current process of internationalization, we may suppose that the effects will become reconciled in the long term, despite appearances. This means that the influence of new technologies is fairly fundamental, because the latter tend to provoke 'socio-spatial answers' which may change the traditional distribution of location advantages rapidly by highlighting them, by neglecting them or by creating new ones. In consequence, the existing system of spatial and economic relations may change and be replaced by a different one. It has often been proved in history that original location disadvantages have turned into advantages (cycles of land use in agricultural and urban areas). The following hypothesis starts from the idea that the appearance of an 'insufficiency' is causually necessary for innovation while the theory of a 'critical mass' only takes account of the factual implementation of an innovation but does not regard its inception.

Hypothesis:
An 'inadequacy' will arise when new technologies have influenced those locations which offered all the advantages, demanded for the 'privilege' of new technology implementation. This inadequacy will apparently concern potential areas for investment. Consequently, it will contribute to questioning the logic of investments.

The result from this basic question will either be to change the existing location disadvantages in areas which are still not effected by new technologies (eg. through measurements of environmental improvement which are actually occuring in the Ruhr-area, FRG, for the International Building Exhibition 'Emscher-Park'), or to change the understanding of disadvantages and transform such location disadvantages into advantages for new social and economic aims (which are expressed as a critical counter position towards the potential effects of the above-mentioned IBA 'Emscher-Park'). Since November 1989, the question of a real 'inadequacy' has been postponed in Germany because of the huge amount of new locations in the East, waiting for investment. Therefore, the attempts to turn round the conditions and the image of disadvantaged locations have lost their obvious urgency. But the discussion about an evaluation of location potentials has become more important in principle.

2. Are there any 'new' effects on urban structure to be found and if so what kind of transformation do they contribute to?

From a historical point of view, 'new' effects on urban structure concern the idea and implementation of 'industry parks'; the 'new' elements of urban structure are given by the combination of industrial land use and of a representative green park. The image-factor of the green space reflects the ideal of labour combined with leisure and evokes the idea that the originally luxury good of leisure is given as a benefit to this kind of labour. Thus, two 'messages' are transported through the environment and the lay out of this kind of industry, apart from the representation value for the respective firm itself: firstly, the social image of leisure is highlighted, secondly, the type of industry and technological progress appear to carry the responsibility for saving time for a better life in terms of aesthetics and leisure activities. Of course, the idea of 'industry parks' also reflects increasing public awareness about environment protection as well as being a consequence of a marketing-strategy. The example may be related to the contexts of 'private/ public space' and 'internationalization'.

Hypothesis:
The impact of transferring international standards to urban space on the relationship between private and public space may be summarised in the following way: the very specific local complementarity of private and public space will change through the influence of 'public' standards onto private space. The social attention and the financial support which are given to the improvement of this kind of private space and its spatial lay-out (eg. by adding public green to private green parks in order to stabilize the image of the area preventatively) is tending to create a new complementarity of private and public space. This is caused by an enforced gentrification which nowadays mainly concerns the environment of new buildings located

in inner town areas of mostly former industrial use. Following the rules of location competition, this gentrification is meant to improve location advantages in order to attract new investors, even in areas which already have a good image. If this includes housing investment, it is hoped that new inhabitants will be attracted. Existing users will either try to take part in the new development or, if necessary, they will migrate to other unaffected areas which they can afford. These consequences are, in general, well known since urban area improvement has become part of urban transformation, which is not specific to a certain time period in urban history. But the recent effects of location competition on the complementarity of private and public space suggest that the type of public space within which the consumer does not have to spend money or can behave in a non-conformist way will decrease continuously. This may cause different and unpredictable changes of paradigm[5]. Private space is still threatened by standardization influences which may appear to be attractive even for those who might not be able to afford either rent or land prices. Thus, the question of environmental 'quality' becomes one where 'image' has to be confronted with 'use' for specific groups of 'users'. This means that international standards have to be confronted with local specific socio-cultural and socio-economic conditions of life. These reflections should not be misunderstood as an overestimation of the socio-spatial effects of new technologies caused by gentrification. However, they serve to underline the fact that environmental improvement, originally promoted by 'green power groups', has become an item of mainly economic interest and, in this way, while resolving economic or environmental problems, may cause even more serious social problems, unless new problems can be prevented through appropriate measures. Apart from effects on the labour market, effects on the quality of public and private space could be guided. One preventative compensation would be achieved by maintaining decentralized identities of public space through a mixture of housing, shops and work places which would guarantee the coexistence of various 'privacies', possibly open for neighbourhood contacts, but not based on standardized rules of social 'progress'. If changes of paradigm will not take account of this necessity, public space will not provide any 'niches' for specific groups. If changes of paradigm do take account of it, there is still the risk that a 'revival' of the values of local identity may change the objective of guaranteeing social 'stability' for those who had been excluded from 'progress' to that of creating 'privileged areas with local specific identity'. This could be brought about through reviving factual environmental qualities as well as by highlighting old industrial areas in order to rediscover the lost identity of the origins of the present. This way, such areas would be brought within the influence of the market once again.

3. What kind of socio-spatial conditions are appropriate for conceiving new technologies within urban structure?

The new tertiary/industrial use of space favours locations within central or suburban areas. These locations are characterized by a high degree of privacy within the surrounding area (of residential or of tertiary use). This privacy must be seen as the sum of different single privacies, separated from each other by a lack of real public space (public space in tertiary sector areas suffers from the dominance of the short-term function of consumption which is not sufficient for linking up different privacies to each other; the same is true of suburban public space which even lacks consumption activities). Locations which are characterized by a different quality of public space and public life (eg.: residential quarters without the image of a 'respectable adress') will be primarily unattractive to such a use unless investment interests alter their idea of 'image'.

Conclusions

Having considered some aspects of the stages of theory and empirical work, I will now discuss the links between past and present, referring to the idea of space as a 'consumable item'. At the present time, the rapidity of transformation seems to be a main characteristic of current processes which are essential for the structure of decision making. History cannot be explored sufficiently for us to be able to distinguish details of the climate and outcome of social consciousness. But we can suppose that our consciousness might not always be able or quick enough to question the climate of any future outcome and, at the same time, to guide technology in a way which could use the potential of technology in order to improve the environment independently of the needs of the economy. As history has shown that most of the former stages of social and technological progress have caused a subsequent lack of (environmental) qualities, research has

272

the task of showing effects which may not be positive. One specific task is to show the degree of impact caused by a widespread process of internationalization. This branch of urbanization research has already begun to become established, focussing on the distinction between 'convergency/divergency' (Hamm, 1988, p.43) in international urban development. Reference to the issue over the interdependence between public and private space promises to provide a duality of settings which offer enough variables to determine the course of investigation. The following hypotheses for both the major settings take account of all the spatial levels implied (micro to macro spatial), adopting the principle of questioning 'shortcomings' in order to define the change of paradigms and thus to approach future demands and their implementation in a qualitative manner. The historical change of paradigms does not give any hints as to there being any linear logic to the process but only the very vague one that a change of paradigms happens after the existing system has shown an essential inadequacy which disturbs the system (of minor relevance) or the organization of society as a whole, and creates the need for improvement; or when an invention illustrates deficiencies by offering new or different possibilities.

The hypotheses are also based on the awareness that the tradition of culture(s) has historically often been broken by migration and by various economical and technological influences and that the aim of preserving and renewing regional and national identities of culture as a possible future value must necessarily take this fact into consideration.

Hypotheses concerning 'convergency/divergency': (i) local identity in terms of large and small spatial units is at risk from international economic influences and population migration (migration of residences following the migration of work places and temporarily recurrent mobility because of international tourism); (ii) the introduction of new technology has evoked a new paradigm for certain groups of society. The paradigm has changed from simply saving time to demanding a high quality of environment for spending the gained time within; (iii) the internationalization of the land market primarily destroys the socio-spatial identity of inner-town and low-income quarters and secondarily creates external demand, resulting in gentrification.

Hypotheses concerning 'private/ public space': (i) public space in inner-town areas which had offered a variety of qualities to respond to the needs of different private demands will be replaced by a kind of public space which is dominated by high-income private demands without any qualities for local social life (eg. among the inhabitants of a quarter); (ii) public space which has changed in this way will be subject to future change when the new inhabitants realize the loneliness of privacy and demand 'real public space', offering opportunities for interaction between different social realities; (iii) public space will have to be based on a different system of decentralized organisation of supply (modification of the hierarchy of centres, mixture of use) to guarantee 'real public space'; (iv) public space will also have to provide space for different kinds of private life (that which still casually exists either in metropoles or in workers' settlements); (v) private space will have to be bound closer to public spaces in order to provide easier links between them; (vi) private space will have to lose its separate, private character in order to become part of socio-spatial units; (vii) socio-spatial units will have to provide space for private life in order to guarantee freedom alongside public life. This should not not only concern individual privacy but also collective privacy; (viii) the amount of private space should be reduced for the benefit of public space; if private space continues to dominate, the owners of private space will have to provide 'private-public' space which does not exclude public users by law or spatial measures in order to maintain a socio-spatial transparancy of large and small spatial units.

The question arises as to for how long and with what consequences the three traditional goals of capitalist societies and of urban life, 'open competition, community and innovation' (Warner, 1972), will have any relevance for social welfare and economic success. This question can only be dealt with multidisciplinarily. Any investigation of indicators and factors concerning the transformation process would necessarily be bound to indepth (micro-spatial) analyses regarding quantitative and qualitative aspects. Even the quantitative category of a 'critical mass' is, related to the factors it is caused by, always an expression of qualitative conditions preparing a change. The link between quantity and quality is given by social, cultural and economic influences, implemented and unimplemented values, and by their effects on socio-spatial reality. The movement and the dynamic between influences and effects is an important part of the transformation-process which cannot be said to be consequent to its earlier stages. In principle, there cannot even be any definite distinctions between influence and effect, because the process is flowing, with material and immaterial factors of influence and effect overlaying each other. Temporally these factors can be distinguished by the fact that they work in parallel to each other. In the dimension of space and social reality they have to be regarded as supporting, excluding or multiplying each other. This means that even a 'critical mass' of phenomena need not be significant only for the one aspect of change that has become visible, because there may be forces which already work contrarily but which cannot be seized by quantitative categories.

The post-War II period shows a variety of paradigms overlaying each other. Most of them have provoked opposition, and though critical voices in some cases have prepared another stage of paradigm, the original aim was never realized in the way it had been meant initially. But the aim had been used as a 'vehicle' for realizing another aim which had in one way become slightly similar and in another way essentially different to the original one (eg. improvement of inner town areas instead of continuing renewal-activities). The following overview illustrates the change of 'identity' which affects locations over time, distinguished according to two spatial levels (Table 13.1.).

Any attempt to derive a certain relevance for the future from the past should be restricted to the knowledge that innovations can, but need not, be based on the social awareness of an 'inadequacy' which can be derived by investigating the present. There is no certainty about this because inventions may cause 'needs' which cannot be predicted. And even if an 'inadequacy' can be predicted, attempts to compensate for it may not fulfill the needs which may follow from it.

Notes

1. The term 'new production and storage technologies' embraces a broad spectrum ranging from new single techniques to new computer-aided concepts of production. The utilization of computer-aided design or computer-aided quality control is included as well as industry robots or flexible production cells, integrated and automatisized production-installations and automatic highshelf

Table 13.1. *Identity-value of locations of public and private space*

Time	Small-spatial units	Large-spatial units
Medieval	Street, market-places, quarters nearby towngates;	Local main market-place, city-hall, church;
1800-1900	Quarters, belonging to social or religious groups, nearby factories;	Local central area of trade, service, culture, station, local green parks;
1900-1920	Quarters, distinguished by use and social groups;	
1920-1950		
1950-1960	Loss of the importance of small-spatial units, replaced by the increasing importance of individually owned or rented standards, oriented on the separation between working and living-place;	Regional city-centres
1960-1970	First revival of quarters, related to improved environment and housing standards versus 'milieu';	Regional to national centres, influenced by preparing the new administrative structure of incorporating little com-communities (FRG), public buildings of social infra structure;
1975-1980	Street/quarter, restricted to areas which still offer environmental qualities of supplyment and green areas;	Sub-centres, main local city-centres, national and international centres;
1980-1985	Street/quarter, with location advantages of environment (individually usable green space, public parks);	
1985-1989	Street/quarters of a good 'image', mostly located in the extension-areas of the period 1890-1920.	

Source: the overview refers to Dunin-Woyseth, 1989.

storages. The main areas of utilization are those of industrial production, and services such as trade and transport or product oriented services.

Within the sector of production, the technologies are utilized in all fields from development and planning to finishing and assembly and storage with a different mass and a different dynamic of diffusion (Henckel et al. 1986, p. 51, transl. by A.H.)

2. Endogeneous Change. The basic idea is that one sector reacts to change in another sector so that sequential development occurs. These sectors need not be economic but could include public administration, political, religious or strictly social and cultural. A good deal of spatial feedback in the urban system occurs via the political process. Changes in spatial arrangement will usually result but not necessarily. This concept will be called the 'going concern': there is a monumentum generated from past actions which provoke future actions so that the system appears to have a life of its own. This mechanism is intuitively appealing but all but impossible to handle analytically (Curry, 1986).

3. Innovation at the micro-level occurs at all times. For the outside observer, seen from a macro system point of view they may appear to be random. It is only those innovations that are adopted or imitated by a suffiently large number of urban actors that have an impact on urban change.- Critical mass can be reached by several processes. At the one end of the scale there is innovation by a single decision unit which becomes binding for all actors in the urban system. At the other end an innovation has to be picked up by a large number of other small decision units to become socially effective (Schubert, 1988).

4. The process of innovation - conceived as a creation of technology associated with the transformation of environment- is fuelled by a process of learning which sets in as the result of an innovative choice and the carrying out of innovative processes of production. Learning ... concerns both the production and the consumption side of the economy. It takes the form of the acquisition of higher ... skills and ...of a change in the consumers' preference system' (Amendola and Gaffard, 1986). 'Each moment in a diffusion process is simultaneously a moment in one or more structuring processes. - The structuring of space is inseparable from social structuring processes. In the becoming of regions, the social becomes the spatial and the spatial becomes the social through the interpenetration of structuring processes of different temporal

depth and geographical extent (Pred, 1986).

5. If conflicts and contradictions in social relations lead to patterns of behaviour which do not correspond to the normal conditions of their reproduction, strategies of acting break the dominating system and its regulations of social relations within which the incompatibility of reproduction and regulation then changes the form of social practice and, consequently, the configuration of social relationships. Prigge, 1987, transl. by A. H.

References

Amendola, M., Gaffard, J.-L. (1986), 'Innovation as Creation of Technology, A Sequential Model', paper presented at the *Conference on Innovation Diffusion*, 18-22 March, Venice

Curry, L. (1986), 'Endogeneous Geographical Change in the Economy', paper presented at the *Conference on Innovation Diffusion*, 18-22 March, Venice

Davelaar E.J., Nijkamp, P. (1986), 'Spatial Dispersion of Technological Innovation: The Incubator Hypothesis', paper presented at the *Conference on Innovation Diffusion*, 18-22 March, Venice

Dunin-Woyseth, H. (1989), 'Genius Loci versus International Influences' in *Planning History, Bulletin of the Planning History Group*, vol. 11, no. 2

Esser, J., Hirsch, J. (1987), 'Stadtsoziologie und Gesellschaftstheorie. Von der Fordismuskrise zur 'postfordistischen' Regional- und Stadtstruktur' in Prigge, W. (ed.), *Die Materialität des Städtischen*, Birkhäuser, Basel

Ewers, H.-J., Goddard, J.B., Matzerath, H. (eds) (1986), The Future of the Metropolis. erlin - London - Paris - New York. Economic Aspects, de Gruyter, Berlin/ New York

Freeman, C., Perez, C. (1986), 'The Diffusion of Technical Innovation and Changes of Techno-Economic Paradigm', paper presented at the Conference on Innovation Diffusion, 18-22 March, Venice

Frick, D. (ed.) 1986), The Quality of Urban Life. Social, Psychological and Physical Conditions. de Gruyter, Berlin/New York

v. Geenhuizen, M., Nijkamp, P., Townroe, P. (1989), 'Company Life History Analysis and Technogenesis', paper presented at the *URBINNO-Conference*, 24-28 July, Chapel Hill

Hamm, B. (1988), 'Sunbelt and Frostbelt - Ein Testfall für die Konvergenztheorie?' in *Soziale Probleme von Industriestädten*,

Dokumentation des vierten polnischtschechoslowakisch - deutschen Symposiums zur Stadt - und Regionalsoziologie in Wisla vom 25.-30. November 1985, German Commission for UNESCO and the Academy of the Chamber of Architects, North-Rhine/ Westphalia, vol. 25

Henckel, D. (1987), *Räumliche Wirkungen neuer Technologien,* lecture presented at the Rheinisch-Westphälische Technische Hochschule Aachen, Abteilung Architektur, 2nd of July, Aachen

Henckel, D. et al. (1986), *Produktionstechnologien und Raumentwicklung,* Deutsches Institut für Urbanistik, Deutscher Gemeindeverlag W. Kohlhammer, Stuttgart, vol. 76

Henckel, D. (1985), *Stadtstrukturelle Wirkungen: Zukünftige Anforderungen an die Planung*, Institut für Städtebau und Wohnungswesen der Deutschen Akademie für Städtebau und Landesplanung, München, Manuskriptreihe No. 2.10

Kastorff-Viehmann, R. (1984), 'Frühe Stadtbaupläne in Duisburg und Ruhrort. Der Weg zur öffentlich-rechtlichen Planung im Ruhrgebiet' in Fehl, G., Rodriguez-Lores, J. (eds), *Stadterweiterungen 1800-1875*, Christians, Hamburg

Kujath, H.J. (1987), 'Wandel der Stadt als Raum der Konsumption' in Prigge, W. (ed.), *Die Materialität des Städtischen*, Birkhäuser, Basel

Kujath, H.J. (1986), *Regeneration der Stadt, Okonomie und Politik des Wandels im Wohnungsbestand*, Christians, Hamburg

Läpple, D. (1987), 'Zur Diskussion über 'Lange Wellen', 'Raumzyklen' und gesellschaftliche Restrukturierung' in Prigge, W. (ed.), *Die Materialität des Städtischen*, Birkhäuser, Basel

Pred, A. (1986), 'Power, Practice, Knowledge. Structuring Processes and Innovation Diffusion, paper presented at the *Conference on Innovation Diffusion*, 18-22 March, Venice

Prigge, W. (1987), 'Raum und Ort, Kontinuität und Brüche der Materialität des Städtischen' in Prigge, W. (ed.), *Die Materialität des Städtischen*, Birkhäuser, Basel

Schubert, U. (1992), 'Conceptual ideas on Innovation', in Drewett, R. and Montanari, A. (eds.), *Innovation and Urban Development: The role of social and technological change,* Avesbury, Aldershot, forthcoming.

Stöhr, W.B. (1986), 'The Spatial Dimension of Technology Policy: A Framework for Evaluating the Systemic Effects of Technological Innovation', *Interdisziplinäres Institut für Raumordnung, Wirtschaftsuniversität Wien*, Discussion 33

Warner Jr., S.B. (1972), 'Tradition as Determinant' in *The Urban Wilderness - A History of the American City,* Harper & Row, London/ New York

14. SPATIAL ORGANIZATION OF TOWNS AT THE LEVEL OF THE SMALLEST URBAN UNIT: PLOTS AND BUILDINGS

Gerhard Curdes, Lehrstuhl und Institut für Städtebau und Landesplanung, RWTH Aachen, Germany.

Introduction

Plots and buildings are the smallest units from which the urban structure is built. They are individual and flexible elements. Their usefulness depends on their shape, situation, degree of openness and legal obligations. Since the demands on the size of plots and buildings can change with time, especially if the type of use alters, or a peripheral area becomes more central due to urban expansion, then the geometric form and shape of plots and the size and form of buildings have a decisive meaning for their usefulness in a changing context. The fundamental problem of every aggregation of plots is their geometry. This determines whether single elements are able to be linked in larger units and whether these can be further combined into larger parcels or quarters. In this paper the subsequent level of unit is termed 'block'.

While in earlier phases of urban development and in the establishment of a town there is freedom of choice in determining these characteristics, they are very difficult to alter in later stages. The most commonly observed changes are: extension of the building into the unused portions of the plot, increasing the number of storeys, linking a number of plots into large unit for common use. What is much more difficult is the alteration of the outer boundary of the block, since this usually requires the alteration of adjacent blocks. In most cases this only happens on particular occassions, for example after catastrophies or for extraordinary functional bottle-necks. The probability of change decreases in proportion to the increasing size of the aggregation with its associated costs. Consequently there is an inbuilt inertia within established structures against fundamental changes. Therefore, what tends to happen instead of fundamental change is a temporary balance between the demands of society on the spatial structure and the expenditure necessary to transform in this direction. The user adjusts to the conditions which the structural characteristics allow and changes only what is possible and necessary. The general rule has been a compromise between the actual requirements and the total costs of realisation within the existing structure. When structures are able to be renewed to a useful level in spite of such compromises then they clearly possess characteristics which are suitable in the longer term. Consequently they are stable elements of the town because through their favourable form of arrangement, plot or building size and their internal flexibility they are always able to accommodate new requirements. On the other hand plots and blocks which are too small or too large or have unsuitable geometric shapes will not hold out because they do not have a similar combination of characteristics.

Fundamental forms and sizes of plots

Theoretically there can be as many forms of plots as there are geometrical configurations. In practice a few major forms have evolved which reflect experience and expediency. The following order illustrates the frequency with which these forms are to be found: (a) rectangular-deep (b) narrow-deep (c) square (d) parallelogram-deep (e) rectangular-wide (f) triangular (g) many-sided in different combinations (h) irregular (Figure 14.1.). The straight line always emerges as the commonest form of separation, because no unused space is allowed to remain in plot development, but also because it is technical easy to measure and unambiguous. The form and size of plot differ according to historical period and culture. In the towns of the middle ages in Mid-Europe the small, sometimes very small, deep

Figure 14.1. *Form of parcels*

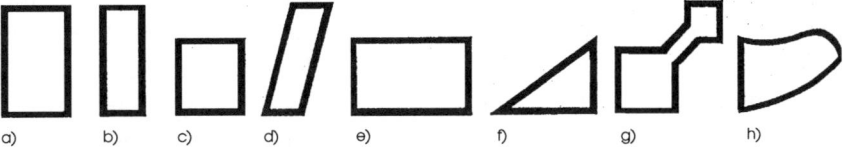

a)   b)   c)   d)   e)   f)   g)   h)

Figure 14.2. *Form of blocks and land division*

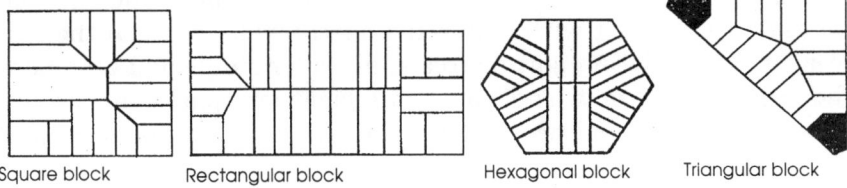

Square block   Rectangular block   Hexagonal block   Triangular block

Figure 14.3. *Regular nineteenth century blocks*

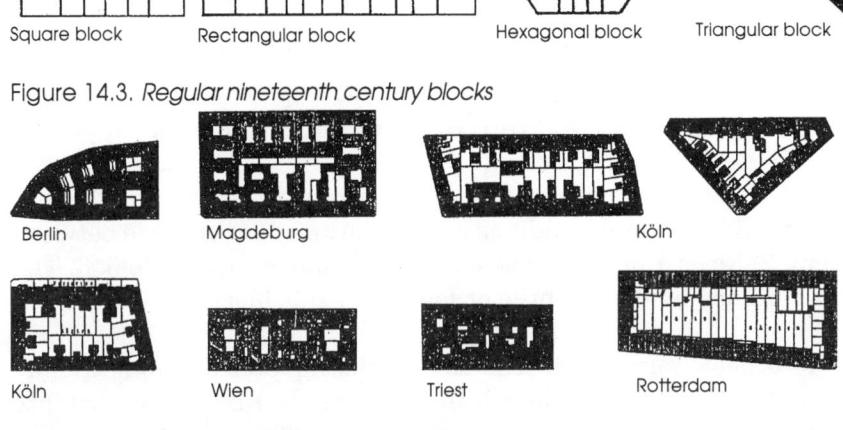

Berlin   Magdeburg   Köln

Köln   Wien   Triest   Rotterdam

Figure 14.4. *Irregular blocks (Aachen)*

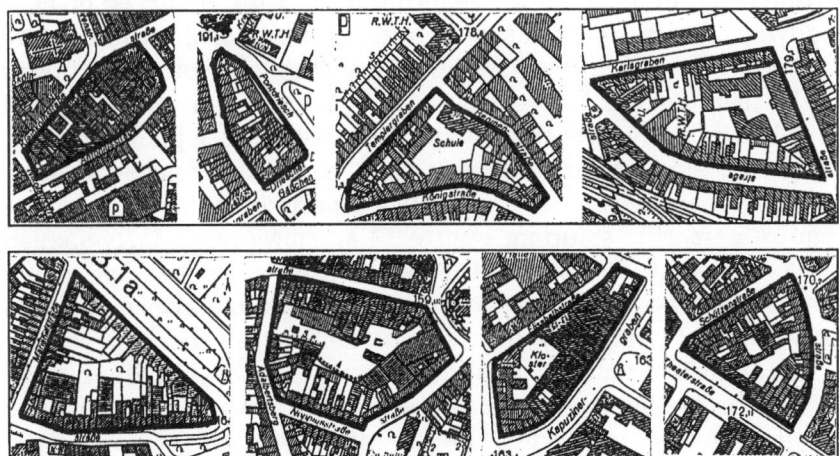

plot was used (Figure 14.1.b). In towns with half-timber construction the widths varied from 5 to 10m. Southern countries with a tradition of stone and arch construction had larger house types and broader plots. Still larger plots are required with the Mediterranean courtyard development. Deep rectangles and squares occur in most towns. Small deep plots result from a shortage of building land and when as many buildings as possible must be located on a given line, for example, markets and harbours. The Netherlands, England and the north German coast have developed especially small and rational housetypes, in part with only a five metre wide front. In the last century in the USA eight metres sufficed for single and tenement houses while 32 m. was deemed suffucient for apartment houses. Generally the size and form of plot was determined by the predominant house type.

Basic forms of aggregation of plots and buildings

Three basic forms of spatial aggregation can be distinguished: 1. Groups of individual buildings with large amounts of space between them. 2. Rows alongside a street or road (one or both sides). 3. The flat arrangement with three or four open side blocks).

Two basic forms of blocks can be distinguished: geometrically regular and irregular. The regular type can be further divided into three additional categories: rectangular, oblique and polygonal. The various irregular forms are generally the result of curved or bent roadways. Figure 14.2. illustrates different forms of block and Figure 14.3. the typical built form from various European countries during the last century. The blocks in planned towns are easily distinguished from blocks in towns with irregular morphologies. In planned towns the result is a standardisation of plots and building types while in irregular morphologies there are always particular situations which require a specific solution. Figure 14.4. shows irregularly-shaped blocks from the Middle Ages in Aachen. An overview over many centuries of town planning history shows that unusual forms such as hexagonal or triangular blocks have not proved themselves. For instance the triangular blocks which were part of the circular central area of Circleville, Ohio, were redeveloped in rectangular form after only forty six years (Carter, 1980). This shows that the form of the block is not only determined from the inside, but just as strongly from the outside through the geometry of the streets and the context of the blocks in the urban use structure.

The interrelationship between blocks and street networks

The street development system is the fundamental grid for spatial organisation. Topologically it is a question of a linear connection system, to which the total use structure of a towns organisation is related. Uses which are not connected are in the end not possible. On closer examination this linear system separates into many dimensions: linear, network formation, network geometry, and network hierarchy. The universal dimension of the linear structure is the assignment of points, plots and blocks to a connecting line. Complete elements which are adjacent to this line are connected to each other. Through one series of movements along this line all these elements can be reached. The linear accumulation is a system of rational addition of uses and their simultaneous connection. This addition can be continuous or discontinuous. These linear characteristics are independent of the straightness or crookedness of the linear form. Lines can be part of a larger linear system and through this the number of accessible uses is increased. The meaning of the individual sections also increases because they become part of an extended spatial system. In addition any movement along a line of accumulated uses will distribute information along the length in as far as this is perceptible through specific characteristic external properties or directed offers of information.

A further important property of linear spatial development is the formation of the network. Through crossings the network involves many linear elements. These connecting points are the crossing points, the junctions. The situation, form and capacity of the junctions defines to a great extent, together with the capacity of the lines, the properties of the network. Additionally, through the network development, middle zones and junctions arise as areas of especially favourable access from the settled area. Through these hierarchies of parts of the network, junctions and site locations come into being. After the network has reached a minimum size, particular linear connections also take on hierarchical characteristics. There are those elements which, because of their position in the geometry of the network, link to many other linear elements thereby producing a connecting function to important parts of the inner or outer areas. These are the through streets, main streets, radials or arterials and the inner area links. With time they build a graded hierarchy of linear systems from the functions of the region, city and quarters. The degree of hierarchy is an indicator of the level of differentiation and maturity of network systems. When networks have been built and consolidated through use on the edges and in enclosed spaces they are near to ultra stable. It is only possible for society to alter the basic characteristics of such systems in special cases and at tremendous organisational and financial expense. This applies equally to networks of spatial structures. Through this blocks have a double function. They are the aggregate of plots and buildings and at the same time the smallest parts of the network. They are affected by two principles of formation: adaptation to the requirements originating from inside and the influence of the outer, surrounding network.

Morphological structures

The building morphology of the city is formed from the type of network, the shape of blocks and buildings as well as through the integration of other space-determining elements, for example, railway-lines, rivers, canals, specific large buildings, inner area, open space and areas with special structures such as for industry. The morphology of the city consists of the horizontal form and order of these elements and their vertical dimension. The characteristics of the smaller units determine to a meaningful extent the characteristics of their aggregation, and vice-versa, the large-scale morphological order influences the functions and qualities of the single elements. Accordingly, the amount of resistance which built structures have against changes means that changes within plots and blocks on the individual decentralised level (incremental) are still relatively easy. These are generally the adaptation of structures. Adaptations often take place in the following hierarchical order and sometimes more than one of these steps take place at the same time. (Table 14.1.).

It can therefore be ascertained that there is a type of mechanism of adaptation of building structures for new requirements, which normally use the parts of the structure with the least resistance. In this way unbuilt areas or those with modest values serve as a buffer. The town and the elements are scanned for reserves and their potential resistance, weak spots and potential areas for completion are sought and in this way the old structure is transformed. The morphology of towns can also be understood as a network, an aggregation of elements which are connected to each other through interdependent relationships. Four levels of elements operate together: buildings, blocks/quarters, town, region. Every grouping on these levels has relative autonomy, so that the levels are dialectically connected in such a way that every part possesses elements of the level below and is itself an element of the levels above. Change within one level is carried out in such a way that the flexibility of the existing structure is fully used. (Curdes et al., 1989, p.143).

Table 14.1. *Hierarchy of morphological transformations*

---

1. Change of uses on sites and in buildings.
2. Reorganisation within the building.
3. Extension into the unbuilt areas of plots and blocks, densification.
4. Increase in the number of storeys.
5. Linking of plots.
6. Alteration of the whole or relevant part of a block.
7. Changes to the size of blocks through alterations to the street network.
8. Alterations to a large area consisting of a number of blocks.
9. Changes to a whole quarter or part of the town.

The transformation of morphology is an objective process which can be confirmed through structural comparisons of maps, but in comparisons over time the causes and steps of change tend to be lost. Simply the results, hardly the processes can be proved. The condition of the morphological structure at a particular point in time, as it is recorded on maps, is the product of autonomous and controlled exertions of influence and stands in contrast to the strategies of individual participants who must bring their requirements into unison with the given circumstances. Here an important new phenomenon appears which can explain changes. The phenomenon is related to the aged process of building structures. With increasing age the relationship between the costs of alteration and conservation in comparison to new buildings becomes unfavourable. When the transformation of a structure can no longer satisfy the basic necessities of a function, then a new construction on the same site, a separation of sites, or a move to a new site becomes necessary. Because the interruption of business while a replacement building is built on the same site is not possible, many institutions use this time for constructing a completely new facility on another site. This provides the possibilitiy of realising constructional and plant innovations at the same time. Consequently new buildings become a means for carrying out a combination of innovations in a single step. The new site is often sought in the weaker zones of the morphological structure or on the edge of the town. In this way the fringe belt or irregular edges of the settlement have an important function in the completion of the structure. (Curdes, 1990).

Changes to morphology between two given points of time are shown in part by such institutional processes of expansion and spatial formation through innovation. An examination of important institutional transfers opens a window to the causes of changes in the body of the town and in the individual spatial elements.

Aesthetic and functional concepts for the organisation of space

The spatial arrangement of uses follows on the one side the preveiously described forces of an inherent structural and behavioural logic. However these do not necessarily generate a result which corresponds with the formal aesthetic norms. To accept the arbitrary result of a public and private debate about power as aesthetic messages of societal reality, stands in opposition to leading ideas from almost all times in town planning history, which require such processes of aesthetic control. These ideas bring together, for particular periods of socio-cultural development, the ideal forms of an epoch in aesthetic norms, which are incorporated as formal guiding instruments for the influence of development and transformation processes. They influence the shape of buildings, the form of the network, the location

Table 14.2. Innovations in basic urban arrangements and in the concepts for basic urban elements

Time	Innovation	Location
1100-1500	Medieval irregular towns	Middle Europe
1200-1400	Medieval regular towns	France, south west Germany, Baltic Sea, east of Elbe
1500-1700	Renaissance town concepts	Italy, France, Germany, USA
1600-1900	Baroque town concepts	Rome, Paris
1800-1830	Classical grid/block reverting to renaissance principles	Krefeld, Prussia
1800-1880	Geometric town design	Middle Europe
1850-1900	Haussmann: axis concept, circus, triangle, boulevard, point de vue	Paris
1857	Ring concept	Vienna, Cologne
1889-1930	Sitte, Henrici, Unwin: artistic movement	Austria, Germany, UK
1898-1903	Howard, Parker and Unwin: Garden city	Letchworth
1902-1970	Garden city movement	Worldwide
1900-1930	Modern blocks	Netherl. Germ.
1920-1930	Corbusier, Taut, May, Gropius: Rationalism and 'Neues Bauen'	France, Germany
1930-1945	Fascist neo-classicism	Italy, Germany
1945-1975	Flowing space and free	
1975-to date	Reurbanisation: reverting to blocksystems	Europe
1975-to date	Postmodernism	Worldwide
1985-to date	Deconstructivism	Western world

of uses and specific buildings. Table 14.2. attempts to illustrate the ideal forms from important stages in European town-planning history. Each new ideal of form and function can be interpreted as a town-planning innovation.

Spatial arrangement of buildings as a conflict between aesthetic concepts and private interests

Over five thousand years of town-planning history the same phenomenom always shows itself: layouts controlled by private interests have more tendencies to irregular, unhomogeneous arrangements. Coincidence, individual requirements and possibilities determine location and form. The aesthetic norms of developments controlled by society are clearly more homogeneous, often with regular, geometric, order. While irregular arrangements do not always originate without aesthetic planning control, regular arrangements take this for granted.

The estimation of the irregular order was low during most periods. In ancient times the planned arrangements of towns dominated. In medieval times two contrary aesthetic concepts existed alongside each other. Whereas the *Gründungsstadte* were built following geometrical concepts, the form of the irregular medieval town, and the classical village, was characterised by narrow, winding streets, irregular building blocks and an apparently random structure. Renaissance and Baroque with their clear geometrical structures turned away from the medieval aesthetic, which was reevaluated during the Romantic period. The restructuring of Rome by Sixtus V is the first major attempt of the new period to guide 'chaotic growth' aesthetically using principles of geometric order. The second large project of this type was the breaking through of the axes in Paris by Haussmann. The nineteenth century with its street corridors following planned building lines took on a total geometric aesthetic form, which has only recently been reevaluated. This was replaced through the onset of the artistic approach to town planning before the turn of the century and through the idea of the Garden City and through the still favoured concept of the *'Gegliederten und aufgelockerten Stadt'*. Nevertheless, the background to these models always contained the idea of a far-reaching transformation of the whole urban structure through one single aesthetic ideal. This has changed within the last twenty years. Today these qualities of individual arrangement of form are regarded as historically important contributions to the character of towns, their identity and their value for tourism. While the importance of the individual is increasing, as the basic social innovation since the Renaissance, the reduction of the potential for economic and aesthetic self realisation of the individual becomes more and more difficult to be implemented. This can also be related to the interests of enterprise and administration. Already in the 1920s the form of functional buildings such as schools, offices and theatres had freed themselves from the context of the surrounding morphological order.

The abandonment of the building line, as a functional and aesthetic means of controlling the development of streets in most middle European countries, led to an increasingly arbitrarry variety of building arrangements. Consequently there has been an increasing loss of form within the newer areas of town expansion, industrial estates and the transformation areas of the city. Therefore, in many German, French and Italian towns, a morphological structure development becomes more amd more heterogeneous, a 'patchwork' or 'Collage-City', more a product of chaos-theory than the classical idea of the unity of urban structure. This is obvious through the tendency of larger, functional buildings and complex projects of urban renewal to ignore the context of the surrounding structure. Here there is a relationship with the previously mentioned tendency towards individual autonomy, which includes land values and rents, and the separation of the planning process from the local context through the standardisation of solutions and investors in attempts to save time and costs. Changing architectural fashions are also part of the process, especially the ideas which place the value of the single object above that of the context. The result in many towns is a 'very unpleasant stage of the fabric..' (Rowe and Koetter, 1984) which has contributed to the crisis of the 'Modeme'. In the end, whether towns adjust to transformation tendencies and how far they consider the morphology of historical parts of the town as a scale for change, depends on a willingness to guide the morphology. In almost all countries it can be seen that adjacent to historically valuable areas this type of guidance has been introduced, therefore the means for this aesthetic control are available.

Within this crisis, two different planning opinions confront each other (and can also be linked up to each other). The supporters of 'comprehensive solutions' and the supporters of the 'combination of contradictions' or 'colection of fragments'; monists and pluralists. As a way to resolve this dilemma, Rowe and Koetter suggest a synthesis of both opinions through the selection of ambiguous solutions, which accommodate the undeniable rationality of stylized, location-independent solutions and nevertheless respond to the conditions of the actual context.

Figure 14.5. *Free arrangements of buildings*

1951 Helsinki-Tapiola

1957 Bremen-Neue Vahr

1961 Frankfurt-Nordweststadt

1979 Nürnberg-Langwasser

On closer investigation the conflict emerges between the aesthetic ideal of definite (commonly more regular) order and deviations from these as a conflict of power between the individual and society, between private interests and public concerns. In the 1920s the innovation of flowing space and free arrangement of buildings accommodated those interests which wanted to break up the existing relationships in homogeneous city structures. Planning control was increasingly criticised legally as authoritarian and against the freedom of property rights. While up until the 1930s the arrangement of buildings was extensively defined through the street network and with it the model of the continuous city continued to have an effect in planning with only architectural language changing, the removal of buildings from the street line coincidently made the functional order and form of the buildings easier (Figure 14.5.). This prepared the way from the nineteenth century city to Collage City. Innovation in town planning directed itself principally towards traffic-regulated linkages and a course control over the separation of uses. The form of the town reflects only the consistent and increasing accidental nature of intervention and investment interests. By purchasing whole blocks powerful investors take advantage of the partial autonomy at this level. Architects use large-scale projects as a means to demonstrate architectural innovations. Recently the Deconstructivist Movement delivered the theory of the fragmented form of the city and the dismantling of homogeneous environs.

Conclusions

The dense fabric of European towns strongly resists changes, especially those which run counter to the logic of the existing structure, by confronting them with individual, economic and legal contradictions of time and culture. Giving respect to a qualitative and homogeneous context demands a greater investment of time and planning and tends to lead to one-off solutions. The small-scale fabric of European towns appears to be contradictory to the requirements of multi-national companies and the increases in scale within public and private units of management. Conformity with the context not only influences the scale of buildings, but can also restrict the freedom to adopt organisational and architectural innovations. This results in a tendency towards an exodus from the narrow network of the morphological structure into peripheral locations where a greater degree of self control over the built form is possible.

The local planning culture determines how those conflicts should be resolved. In general it can be argued that remarkable differences tend to decrease the value of the existing structure whereas a successful synthesis can be productive. However, it is also true that differences can give specific accents to an environment with little or no character, or can introduce a basic change of structural scale in an area where there is none or the existing one is not to be maintained.

References

Carter, H., (1980) *Einführung in die Stadtgeographie*, Berlin-Stuttgart.

Curdes, G., Haase, A., Pasternack, S. (1988) Die Entwicklung der morphologischen Struktur und der Einfluß von Innovationen: Fallbeispiel Aachen. *Institut für Städtebau und Landesplanung der RWTH Aachen Schriftenreihe Städtebauliche Arbeitsberichte*, Bd.3/5,

Curdes, G. (1989) *The Influence of Innovations on Urban Development and Urban Form. A Comparison of Development Stages of Athens, Lisbon, Rome, Liverpool, Thessaloniki, Aachen, Bari, Kecskemet, Tromsö.* Assistance: Haase A., Haneda F., Pasternak S., Schwan C.. Institute of Town and Countryplanning, Aachen University of Technology, 1989.

Curdes, G. (1990) Development Logic of Spatial Systems of Towns and the Influence of Innovations on the Spatial Structure of European Towns. 38th. European Congress of the Regional Science Association, Istanbul 8/90.

Rowe, C., and Koetter, F., (1984), *Collage City*, Basel, Boston, Berlin.

15. INNOVATION AND THE EUROPEAN CITY CENTRE

Bill Chandler and David Massey, Department of Civic Design, University of Liverpool, England.

Introduction

The experience of European city centres over the last one hundred years has indicated diversity rather than uniformity in regard to the nature and timing of innovative events and processes and also in regard to the scale, type and rate of building responses. Although innovations have been international and covered long time scales in their occurrence, it is clear that individual historical, geographical and cultural circumstances have conditioned and shaped the forms urban development has taken. This contribution explores some of the general themes and particularities relating to city centres which have emerged from the URBINNO Working Group on land use, built form and the physical environment. Given the enormous diversity of European urban traditions and experiences, our analysis has indicated that inter-urban and international comparisons made in strictly morphological terms within closely set time periods result in high order generalisations which serve only limited descriptive and explanatory purposes. They may contribute background information but do not necessarily add to our understanding of the phenomena recorded. It is argued here that a thematic approach (that is taking a particular innovative event or field and exploring its incidence and morphological consequences in case study city centres) can be a more useful alternative method. This should make possible comparative analyses which are based on local individualities and empirical evidence, yet which are open and responsive to the more general European context of innovations and innovation processes.

Defining the boundaries of the city centre

In all instances the general location of that assembly of functions which can be deemed characteristic of the central areas of cities is not difficult to identify. Broadly speaking the prime functions cover civic and administrative matters (local, regional and national government and public authorities); cultural and institutional functions (religious, educational, health service, arts and culture); commercial functions (markets, finance, professions, administration, retailing and wholesaling); residential (for different income and status groups, and, those who are homeless); and, industry (artisan-craft work, services and manufacturing). What is more problematical is the definition of the precise physical extent of this range of interconnected functions in terms of patterns of land use, buildings and infrastructure in particular city centres at particular times. Measures and techniques for defining city centres on a consistent basis of greater or lesser complexity have been provided by social scientists, particularly urban geographers, and there have been applications of North American concepts, for example, the central business district, in European contexts. This concept is considered to be too restrictive in European circumstances and urban concept analysts need to have regard to a wider range of functions and to conceive of local boundaries in relation to degrees of intensity within the range of functions and to specific historic and cultural factors. The approach thus adopted was to look for comparative indicators in terms of, for instance, relative degrees of intensity within the range of central activities and hence the density of physical development or the distribution of land values. These rarely provide precise boundaries in terms of streets or other morphological features, rather they draw attention to the need to relate general features to specific local morphological or property market circumstances over periods of time and suggest that the results will not necessarily be precise or fixed. In general, however, it is possible to identify a city centre land-use and built-form environmental area, which at the time of analysis houses relatively-stable, regionally-oriented functions, and, adjacent areas which exhibit transitional characteristics. The latter will house associated fringe activities of fluctuating fortunes which may depend on sub-markets created by proximity to the enduring prime functions. They may also be located there through historic circumstances (e.g. defence purposes) and have more recently become mere residuals of once predominant functions and built elements.

Selecting innovations

Innovations can be identified as affecting either individual central area functions, or related clusters of functions, or, all functions simultaneously. Of particular interest to town planners are those innovations which affect the linking elements in city centres, for example: technological advances in transport systems for the movement of people and goods; the facilitating of building development through institutional, statutory and administrative changes; and, the application of new architectural design ideas and planning theories. It is possible that some of these innovations will be more or less universally and contemporaneously accepted in widespread localities; other will remain regional in provenance ie. in place and time; but most will be taken up through a more progressive diffusion process at different times in different localities. For the purposes of the URBINN0 study attention has been focused on an examination of a number of strategic innovations and their influence upon central area land uses. The selection made thus illustrates the influence of innovatory approaches outside the discipline of specific place and time constraints. The themes identified have regard to administrative and statutory changes, innovation in commercial and trading practices, and finally, the application of new architectural design concepts and planning theories.

Developing innovatory themes in city centre development

The European-wide acceptance of innovatory ideas and resultant system changes is not and has not been instantaneous. Although broad parallels can be drawn for the initiation and diffusion of specific innovations between localities over basic periods (such as, pre-world war one; inter-war; and, post-world war two), cultural and local system differences were observed to account for more specific differences in the date of implementation and extension. Different stages in the European experience of urbanisation, as explored in the earlier CURB project, have conditioned both the need for and the capacity to initiate and absorb new influences, and, since the 1940s political system differences between western and eastern European nations have determined different approaches to the redevelopment and conservation of city centre structures. Although in specific places the presence of international building styles and forms may present an unexpected degree of similarity in morphological terms, innovatory 'collective' (or wholly public sector) measures have produced different results from those found where mixed economies or free market initiatives are operational. This is evident at a micro level (within otherwise comparable city centres) and at the macro level (where national conditions have dictated the channelling of investment to the centres of metropolitan areas rather than to ring centres). Such differences make it even more necessary to define categories of innovation which can subsume locally observed variations under thematic groupings of issues and events that are sufficiently broad in their scope to encompass most situations. In addition it has become clear that many seemingly independent variations are in fact closely related, either through their nature or through their sequence. Indeed it could be that it is more important to analyse sets of local relationships and their consequences rather than to detail individual events. It is possible for instance, to show that technical innovations in particular building types in particular land use zones have had an influence which extended beyond that of the individual sector to affect the functions and form of the central area system as a whole. These phenomena may progress to the extent that overall system changes become innovatory in themselves, for example, through the invention and application of new planning and design concepts. Care is needed, however, to distinguish first between the factors affecting physical development and organisation/management systems, and secondly, to draw out the fundamental relationships between these factors. In many European countries since 1945 innovations in administration and management techniques have made possible widely spread comprehensive urban renewal and redevelopment of a scale and complexity previously unobtainable or limited to individual cities. Innovatory statutory and financial measures were necessary to permit these developments to take place and latterly to protect areas where extensive change was not deemed desirable. Thus the new possibilities for urban change arising out of war-damage opportunities and post-war economic growth and decline cycles spawned a series of innovatory urban design and town planning concepts, the implementation of which was dependent upon prior administration/institutional changes and new commercial practices. At the same time the need to regain a lost identity has lead to a precise reconstruction of pre-war physical elements (buildings, streets, squares and parks) in other centres. Each of the categories of innovation previously stipulated is now developed through the selection of particular events to illustrate both cause and effect relationships within that element, its immediate sectoral context and the wider ramifications for the overall central area system.

Administrative changes

Innovatory measures in 'administrative' terms will mean more than merely the evolution of organisational structures in response to generally improved methods of management. The criterion adopted here is that an administrative or institutional change is deemed innovatory when it is specifically designed to make possible necessary and desired physical change which could not otherwise be achieved. In regard to city centres the measure could, for instance, be selectively directed on political grounds to enable local authorities to employ public funds and land ownership roles in new interventionist roles previously considered unthinkable, or at freeing private sector developers from close public scrutiny and control. Such changes could make possible partnerships between the public and private sectors leading to the setting of joint aims and of the amalgamation of financial resources, project management skills and land transfers which are an essential precursor to rational proposals for, and the implementation of large complex and integrated city centre redevelopments.

In many countries the obligations of their early town-planning legislation were administered through existing departmental structures, particular responsibilities being allocated to the technical sections such as the city engineers, the architects or the surveyors departments. This practice sufficed when the requirements were perceived as straightforward, technical matters related to measures for public health and safety, with less concern for aesthetic matters or public participation in issues of social significance. Change is evident in the post-world war two context when the comprehensive nature of wartime destruction and the wider social and economic goals of new and revised legislation called for new attitudes in regulating and/or promoting central area development. The increasing management complexities lead, in some countries and in some cities, to the establishment of separate Town Planning Departments with new strategic and tactical roles. The setting up of corporate management systems allowed more effective coordination and working relationships to be effected between all the municipal departments concerned with the control of development and the provision of essential services. In some countries or regions, where lower priority was accorded to the control of the urbanisation process, the lack of an effective administrative infrastructure has inhibited the application of innovations commonly observed elsewhere. In others the ability of elected local representatives to control all development in the central areas of their cities has been reduced, through the transfer of powers over substantial areas of land to other bodies with more limited terms of reference. An example from England is the setting up of urban development corporations under the Local Government, Land and Planning Act 1980, with centrally-nominated memberships tasked solely with the restoration and physical development of derelict areas, some of which were of such a magnitude as to defy the efforts of the normal processes of urban management over previous decades. Such urban management innovations could be a reflection of both the effects of delayed building and infrastructure maintenance during the war and post-war recovery years and of the rapid rate of property obsolescence experienced since that time due to innovations in commercial and trading practices.

Commercial and trading practices

The raison d'etre for the modern city centre is the provision, at a point of maximum accessibility, of premises and facilities for governmental and commercial organisations. The purpose of these central agencies is to supply services and goods for extensive hinterlands, and it follows that innovatory changes in their trading practices or service delivery systems will be reflected in the nature and scale of their accommodation demands and needs. Such requirements will usually be allocated locally through a real estate market for existing land and buildings and proposed redevelopment, or an administratively determined allocation reflecting the priorities of the public authorities; or, sometimes a combination of the two. The growing requirements by national and local governments for office space in city centres have been met by public and by private sector building agencies. Where the former have taken the lead the result has frequently been the establishment of separate civic office enclaves; expressing in architectural design and style formal perceptions of the dignity of public service and high office. Where the latter has been the case the various service departments have become scattered throughout the city centre in speculatively-provided accommodation. Where historic area cultural values have been significant and/or where congestion and rents become perceptibly too great, efforts have been made to decentralise these functions to a 'second' city centre, to the suburbs or even beyond.

Locational imperatives for financial houses, trading and associated commercial firms create a recognition of 'prime' sites particularly for headquarters units and major retail outlets in city centre land and building markets. The traditional physical result is a typical dense, compact mass of high-value buildings and sites at the core of the city centre. Negative aspects due to congestion effects may then also be observed, leading to an outward movement of secretariat and routine functional units to peripheral sites, perhaps grouping again around subsidiary transport nodes as urban sub-centres or 'business parks' or in smaller regional towns. Where the office task is so specialised (eg. information processing) that it can stand alone (but linked via modern communications media to central policy and decisionmaking units), a yet greater degree of dispersal even to countryside locations is evidently possible. Thus pressure in the central area office zone may be relieved, but the operation must be large enough to justify the financial investment and the allocation of human and building resources to new projects quite distant from the centre of operations. Innovatory approaches to both retail and wholesale trading practices and operations have lead in the post-second war era to new forms of shopping outlets and storage depots outside city centres with consequent effects upon the traditional shopping centre. These have been most dramatic in North American city centres, but for cultural, economic and morphological reasons European city centres have been less deeply influenced. In part public planning and infrastructure investment policies have underpinned the city centre as a dominant retail feature. Commercial innovation has meant greater organisational rationalisation with company acquisitions and consolidations in the drive for a greater market share, closer investigation of market potentials and consumer behaviour, and, deeper analysis of city centre locational advantages. The result has been extensive modifications to, or the total rebuilding of, individual retail stores and large sectors of central areas, including new means of mass transit access to the centre and pedestrian/vehicle segregation of substantial areas.

The scale of the threatened clearances in historic centres stimulated the establishment and spread of 'protection societies'; citizens and special interest groups concerned with the loss of particular environmental heritages and valued historic buildings. Legislation has been enacted in many countries to govern what may or may not be done in building terms in established 'conservation areas' and innovatory urban management approaches have been used to find appropriate functions to complement new architectural design approaches in the search for appropriate building form and style in such visually sensitive areas. Commercial rationalisation in stock buying, storage and distribution practices has produced new patterns in warehouse building locations and forms. Transport distribution networks have been extended on a city and regionwide basis, some to encompass direct links with manufacturing sources. The outcome has been a general outward movement of earlier wholesaling warehouses from city centres and their peripheries which has provided space for the expansion of more relevant functions and a reduction in heavy-goods vehicle movements with advantage to the central area traffic network. This is seen as a good example of the inter-related nature of innovation effects, where an operational change in one element feeds through to influence others and indeed the entire city centre. Another example is the change widely evident in the practices of the organisations which together comprise the property development industry. Innovations in methods of market analysis, in land acquisition and exchange procedures, in joint-financing methods (involving new sources as well as new arrangements with traditional sources), in the formation of different forms of property holding/managing/developing and marketing companies, have together opened up new possibilities in the scale and complexity of redevelopment that could be implemented as a routine, but on a comprehensive and planned basis.

Technological advances

The period 1880 to 1990 has produced a multiplicity of technological innovations which qualify for consideration. But for the sake of brevity only examples which have had an influence beyond those of individual structures are discussed. The choice therefore focused upon evidence of particular innovations which had wider ramifications for the integrated systems of related activities and linkages found in city centres. With this as the primary objective and to emphasise the inter-related nature of urban development explanations of general effects were sought through particular sources. For example, accommodation pressures inherent in the concentration and growth of central activities resulted in continuing increases in the densities of building development. Horizontal expansion was effectively blocked by the physical limits of surrounding buildings and the innovatory measures which enabled the accommodation demands to be met were those which encouraged buildings to extend vertically. Of particular interest in this category are factors which relate to building construction methods and mechanical movement systems, the impacts upon the central area transport networks, and, the environmental issues stemming from tall, closely-packed buildings.

Construction technology

Masonry towers are found in early history, but the invention of cast iron, steel and ferto-concrete frame techniques provided the essential freedom for building designers to produce very tall buildings on existing site sub-divisions in forms best suited to contemporary needs. Both their spatial and economic advantages led to rapid widespread dissemination of these techniques. Later innovations in advanced construction methods, such as prefabricated industrialised systems, allowed building times to be shortened and thereby reduced substantially the disruptive effects of construction operations upon other city-centre functions. Facade treatments also changed markedly as the new systems imposed their own disciplines upon architectural aesthetics and choices. The universal nature of the application of this new technology carried with it common problems observed in the central area townscape of so many European cities. The visual blending of high and low buildings has been haphazard at best and the imposition of a common international stamp in building form and facade design was seen by many as the antithesis of responsive, sensitive indigenous or regional architecture. Recent years have produced reactive fashions in design with some neomodern examples producing eminently successful urban design solutions, but where nothing but the elevational style has been emulated, conveying no more than a spurious visual variety.

Vertical movement systems

Innovation in ways and means of moving people and goods in a vertical dimension had to occur in concert with advances in building technology before their joint advantages could be exploited. The extensive use of elevators (lifts) and escalators provided a mechanical means of conveying people and goods effortlessly within the tallest buildings, whether office blocks, department stores or housing tenements. Elevators provided the means for direct, close inter-office accessibility and of retaining close contact between office staff and clients in a limited physical area, i.e. in the same building or in adjacent ones. Escalators allowed the mass movement of customers between floor levels, which permitted multi-level shopping and a compaction of the sales areas to within easy walking distance both internally and externally, both above and below ground. The obvious result of such compaction is that more people and more goods are enabled to move and be moved within a limited compass and greater congestion effects become apparent in the city transport systems and networks.

Mobility requirements, transport systems and networks

In considering innovations in transport systems and networks care must be taken not to lose sight of the symbiotic nature of traffic routes and the land-use and buildings which flank them and are at their origins and destinations. The nature and intensity of the latter may affect route design possibilities for the former and in turn the characteristics of an established route may restrict the land and building use alternatives. In most European city centres the basic networks of streets can be traced back to historic rights-of-way determined by field boundaries and land-ownership patterns enshrined in law. Usually adjustments have been made through limited street widening and junction realignments to cater for growing traffic movements, but innovatory ideas, new legislative powers and determination of high order are necessary to alter major parts of the transport systems and their networks in the face of still increasing activity and building densities in central areas.

Mass transport systems impinge directly upon central areas in two ways; in the nature and extent of their routes and in the terminals and service depots they may require. For example, the wide spread construction of surface railway lines in the nineteenth century created impenetrable barriers to other forms of movement and central terminal stations established nodes of development which competed with older historic cores and commercial areas, often creating a land use and morphological duality. Most such projects were completed by the late 1880s, but a similar innovative response in the largest cities had to await the invention of practical systems of electrical traction. In passing beneath the centre the underground 'metro' systems which then became possible linked regional and local suburban routes as well as providing for circulation within the central area itself. The station points on the surface which served the centre allowed a wider range of non-central locational choices for developments of a kind best served by that mode of transport, say, long-distance commuters, resulting in deeper changes in metropolitan spatial form as a whole. The other significant late nineteenth/early twentieth century form of fixed route mass transport was the local tramway system in its various modes (originally horse drawn, later electrified), followed later by the trolley bus. Based upon such transport systems, growth of suburban extensions in a radial fashion increased the demands upon the city centre. In city centres consequent increases in commercial opportunity and employment potential were more marked than was abrupt physical change. Route limitations had consequences for these transport modes due to a combination of factors. Tramways in central areas were forced to occupy the middle of chosen major roads within an already established highway network. Logical and cost-effective as the system may have been on its own, it unfortunately inhibited the free passage of newer modes of personal and goods transport which by the turn of the century were set upon an exponential growth curve, ie. the private car and goods vehicle.

The problems lay in passenger safety and congestion effects. Coupled with the lack of flexibility in suburban route choices, these factors lead to the demise of tram systems in some European cities by the 1960s although metro systems were often introduced or extended. The key to new up-dated systems employing tracked vehicles has been the adoption of special routes whether at surface level, above or below the ground. In the 1950s and 1960s, using innovatory techniques such as origin and destination surveys and desire line predictions based on tripgeneration factors, analysis by new breeds of transport engineers introduced new concepts in alternative patterns of road networks. Where local circumstances permitted selective demolitions and where the political will and financial resources could be found, the imposition of 'ring and radial' road systems served to divert unwanted traffic from city centres while still allowing direct access where necessary. Visible in alternative forms as urban motorway 'boxes', their principle also permitted the creation of safer 'environmental areas' where movement is restricted solely or predominantly to pedestrian traffic.

Seen as a solution to the ever-growing traffic volumes and the inability of the older road system to cope, the building of urban motorways in the 1960s and 1970s has had a material effect upon many city centres. It can be argued that in many instances the cure has been worse than the illness. The massive physical and visual scale of the civil engineering constructions, such as grade- separated junctions and interchanges, is most intrusive. The heavy concentrations of traffic on such hierarchical systems can cause an aggravated and unacceptable build-up of pollution in particular sectors of the city (noise and atmospheric pollution). The large overhead traffic lane direction boards which are a necessary adjunct to the system do not fit comfortably into the urban scene, obscuring many fine buildings and negating attempts at sensitive, humane landscaping. In addition highway traffic management systems are also becoming more common. Generally they attempt to regulate the flows through computer controlled synchronised traffic-lights and permit time-phased access only to certain areas by certain modes of transport. Controlling access in this way does rely upon, for example, changing practices in the distribution and reception of goods and it therefore stresses the need for co-ordinated management action if central functions are not to be impaired. It is a practice which holds out most hope for historic city centres. Each of these approaches faced a common problem, ie., what to do with parked vehicles. Whether provided free or charged for, the concentration of parking spaces in large, off-street, multi-storey car parks has produced a new building form. The nature and scale of their architectural treatment usually echoes the least-cost engineering solution and is difficult to harmonise with adjacent building facades. That multi-storey car parks are essential to the contemporary city centre, however, seems to be beyond dispute given public responses to car use opportunities. 'Park and ride' from outer parking areas has been successful where rigorously enforced, but evidence exists to suggest that personal and collective attitudes to this approach are ambivalent at best.

Environmental issues

Throughout the industrialised regions of Europe pollution from factory flues and household chimneys became an endemic, city-wide problem in the nineteenth century. The solution was sought and found in the early environmental health legislation. These innovatory statutory measures to control emissions have been developed in Western Europe to include the growing menace of ever-increasing road traffic fumes. Although not yet universally adopted, for example in Eastern Europe cities where politicaineconomic factors for so long dominated decision making, the link between adverse micro-climatic conditions and urban functions is now well understood. An issue which has largely been ignored throughout Europe is that of micro-climatic conditions brought about simply by the existence of extensive dense urban structures, especially in the central areas. Close groupings of massive structures had predictable environmental effects in terms of the micro-climates created and the conditions inside and outside of the buildings were inevitably adverse in terms of human comfort. Internal environments suffered from the impedance of natural lighting and ventilation. Technological responses were to install advanced systems of artificial lighting, ventilation and air-conditioning. Once these were established as the norm, it was but a short step to develop the 'burolandschart' model for office space, leading to deep rather than narrow office structures and eliminating the hollow cores of traditional street blocks.

The exclusion of direct sunlight for many hours of the day coupled with induced higher and more turbulent wind velocities at street level were not conducive to human comfort in northern climes. When coupled with low temperatures, rain or snow the conditions could become most unpleasant and indeed unsafe. In southern climes historic models for dealing with the effects of excessive sunlight and hot dry winds were well understood and arcades were a frequent urban form. To cater for improvements in the pattern of pedestrian movements within and between street blocks it has been necessary to reinvent climate-controlled arcades and covered atria; sometimes used according to the dictates of fashion, but fundamentally as a sensible response to micro-climatic conditions. In regard to daylight and sunlight penetration, concern for conditions on adjacent sites as a result of surrounding high buildings was expressed through the form of codes of practice. Advisory recommendations for daylight and sunlight control criteria were produced and in some instances codified into legislation. If used willingly or enforced they caused fundamental changes in building forms. Tall, slender, freestanding towers or narrow slab blocks set back upon a podium construction of two to four floors extending over the bulk of the site, became a viable architectural form, in that such an approach could meet both lighting and density controls. Sporadic applications of this design concept had unfortunate results in urban-design terms when valued, traditionally-uniform street facades hitherto imposing in their continuity were rudely disrupted.

Mechanical means for redressing adverse internal environmental conditions have the unfortunate characteristic of being prodigal in the use of energy. Since the oil crisis of 1973 architects, planners and legislators have addressed the problem with varying degrees of success. Building regulations have been tightened, for example in regard to the thermal insulation values of external walls and this has altered the 'solid to void' ratio of fenestration patterns for many buildings. Completely glazed curtain wall' structures will be expensive if triple glazing is required to meet the standards demanded. But beyond the cladding systems, should the energy issue return in greater force (or when it does) consideration will have to be given to factors which are more fundamental in principle. They could be more far-reaching in terms of architectural form e.g., that a cube provides the greatest usable volume of space per unit surface area and that artificial lighting and air-conditioning as a remedy for poor urban design will not be acceptable.

Urban design concepts and planning theories

Nostalgia for the urban street scene of the past seems to be a universal public reaction to many post-world war two city centre developments. On the basis of the visual evidence this is understandable, but what is not made equally clear is that the municipality and the individual entrepreneur of the nineteenth century, have also been supplanted. Large corporate institutions more concerned with property portfolios and potential capital growth in the asset values of their holdings now dominate the property development industry. Decision-making procedures have changed markedly, especially in client/architect relationships where development projects have become larger. Complex 'fast-track' projects depend upon the prior assembly of sophisticated professional skills project management, financing and cost accountancy, legal, engineering (construction, mechanical, electronic), architectural, quantity surveying and marketing) all of which have to be integrated through joint-working practices. Aesthetic parameters rate as only one group of concerns among many, and institutional and company board decisions reflect their own priorities.

It was not until the challenges of post-world war two reconstruction had to be faced that planning and urban design efforts showed concern about much more than ad hoc street widenings, monumental beautification schemes or simple controls through building regulations and zoning measures. Public and private sector development roles and objectives were well defined and the existing basic legislation reflected the accepted practices. Where central area redevelopment took place infrastructure changes were the responsibility of public sector agencies, but most building work resulted from individual initiatives in a piecemeal fashion. Straightforward controls upon building frontages set common building lines and heights and it was left to the detailed architectural design of building elevations to achieve visually coherent streetscapes, most ofien successfully, sometimes not. Co-ordination was not seen as a necessary objective other than in special set-piece examples of 'civic design'. The development of innovatory planning approaches during and after world war two had much more drastic consequences, largely due to the assumption that war-damage replacement could not be achieved effectively through piecemeal practices. Untouched surrounding

309

areas were deemed to be obsolete, functionally, structurally and economically and if remedies were to be sought then it seemed logical that the cleared and obsolete areas should be considered jointly. In Great Britain in particular this elevated the planning profession to a central role and the powers of compulsory property acquisition and far-reaching controls upon development and land-use seemed to call out for greater innovation in the search for comprehensive solutions. However, inadvertent side effects such as planning blight lead to area dereliction, and the often sterile, monotonous single-use building complexes in segregated land-use zones were castigated by professionals and laymen alike. Despite the idealism of the period, a more powerful factor was the fixation with real estate property to the exclusion of social factors and therefore, an inability to appreciate wider concerns or more appropriate roles for professional planners in city centres. But growing public and community involvement in the planning process, especially where enshrined in law, has been evident in the more sensitive attitudes and responses to conservation. There has been a more ready acceptance by planning officers of piecemeal redevelopment, though now within a more structured and coordinated framework for public and private development initiatives. A richer mix of activities is sought, but not through negative controls only as the assembly of promotional packages has become accepted as a normal task.

Implicit in the planning objectives now favoured for many European city centres must be the acceptance of form and functional variety, but clearly there is continuing uncertainly as to the limits which should be set in visual terms. Recent innovations which affect the townscape stem from new design requirements, for example, for the many new and old buildings which are to accommodate computer-based information technology systems. New forms of data collection, storage and processing have additional, special space requirements and need complex flexible power supply and servicing provisions which can affect the height and design of individual buildings, e.g., through the insertion of service-only floors in new buildings or raised sub-floors in old buildings, for easy access to service conduits for alterations or maintenance purposes. When considered together with the energy-saving issues discussed earlier it can be appreciated that any form of planning or design controls that prevent change of this kind occurring (either through omission or commission) are bound to fail when commonsense prevails. Such changes have to be separated form the architecturally fashionable and idiosyncratic. 'High Technology' finds architectural expression in dramatic, exciting, stimulating, novel building designs or facile, inconsequential, inappropriate, self-seeking, ugly structures; but all are part of valid design experiments. Serious problems arise in relation to the settings of such buildings rather than their intrinsic value. Another form of functional rationalism has seen the exposure of previously hidden ducts, utility runs and piped services upon the outer face of buildings. The pragmatic grounds argued are similarly for ease of access, but more ofien the theoretical base is part of a fundamental design philosophy which seeks to express all the structural elements of a building as an overriding aesthetic statement. The final form, with exposed and vividly coloured services and structural members, produces a stylistic idiom which can be in stark contrast to more traditional structures which express a different design approach. Both can be appreciated as valid architectural statements, but as yet unresolved is their integration in a unified design composition as part of an urban design policy for the city centre. However that objective is seen by others as being a fallacy in itself - urban designers being regarded as merely self-indulgent in wishing to blend architectural styles in as harmonious a composition as possible.

Conclusion

An effort has been made to relate innovatory causation to direct morphological effects in regard to changes in physical characteristics of city centres in Europe over the last century. While on occasion it has been possible to attribute a particular effect to a singular innovation, it became increasingly clear that, in the main, this is not sufficient to trigger sudden major urban changes. It is the co-incidence of innovation in a number of associated fields, plus the cross-fertilisation of ideas and the driving force of a more universal pioneering spirit which has been essential. The latent value of an early-stage innovation, in for example a technical field, may only be realised later, when new ideas, say in regard to public authorities, company groupings and financial resources, are available to provide the essential resources for the exploitation and implementation of morphological change. Another, perhaps ironic, example can be taken from attempts to prevent disruptive morphological change. Innovatory thinking and measures in the fields of public administration and legislation have been necessary in order to preserve the physical status quo. For instance, proposals for change have been denied where the visual and environmental value of historic components of cities have been deemed by the public authorities to provide a more worthy return than the opportunity costs implicit in radical renewal programmes. In essence it is the need to cater for interdependent decision-making practices and procedures in a range of commercial, professional and political disciplines which constitutes the activity within and determines the form of the fabric of city centres, and this belies the notion that morphological change has its genesis in fashionable whimsy.

References

Hammerson, I. and T. Hall, *Growth and Transformation of the Modern City,* Swedish Council for Building Research, Stockholm, 1979.

Holliday, J. (ed), *City Centre Redevelopment: A Study of British City Centre Planing and Case Studies of Five English Town centres,* Charles knight, London, 1973.

Ministry of Town and Country Planning, *Advisory Handbook on the Redevelopment of Central Areas,* HMSO, London, 1947.

Simpson, B.J., *City Centre Planning and Public Transport,* Van Nostrand, Wokingham, 1987.

Swedish Planning of Town Centres, The Swedish Institute, Stockholm, c. 1966, *Traffic in Towns: A Study of the l'ong Term Problems in Traffic in Urban Areas* (the 'Buchanan Report'), HMSO, London, 1963.

Universiteit can Amsterdam, *Urban Core and Inner City: Proceedings of the International Study Week,* Amsterdam, 11-17 September 1966, Brill, Leiden, 1967.

Whitehand, J.W.R., *The Changing Face of Cities: A Study of Development Cycles and Urban Form,* Blackwell, Oxford, 1987.

16. CHANGING LIFE STYLES VERSUS URBAN BUILT FORM

Halina Dunin-Woyseth, Oslo School of Architecture, Norway.

Introduction

On interdisciplinary issues

This paper investigates the issue of life style and its relation to urban built form over the past hundred years. Life style itself is a highly complex notion. It creates a sphere which is highly influenced by different disciplines. The spatial aspect of the topic makes the problem even more complex. Consequently, the investigation attempts an interdisciplinary approach. There is a broad consensus in the academic world as to the importance of cross-disciplinary research. But: 'if interdisciplinary research is so good, why is there so little of it?' (Fischoff, 1981, p.579). Among the many reasons, one is especially important: no one is trained to do it. Another one is that scientists are anxious not to lose quality control and conceptual clarity as one leaves traditional fields. A final problem is that there are no convincing models of how interdisciplinary research might be conducted (Fischoff, 1981, p.580).

Method

The approach applied in this study is based upon several steps: (i) a *status questionis* with regard to the nature of life style has been carried out; (ii) in order to investigate and define life style as an innovation two spheres of life style are delineated: the human being himself, through a short presentation of some elements of theory of motivation; and the human living environment with its changes over time, through a historical sketch, outlining the sociopolitical, demographic and economic backgrounds which have formed life styles; (iii) the proposed definition of life style is at the same time an instrument and a guide for indicating changing life styles; (iv) finally, the interdependence between changing life styles and urban form is pointed out and some conclusions are drawn.

The method is partly based upon a Japanese study (Kogane, 1982).

Life style: status questionis

The issue of 'life style' has been discussed by philosophers, psychologists, economists and other scientists for a long time. 'In the Marxian order, life style is a phenomenon determined primarily by an individual's objective position in the production process, that is, in the structure that loosely shapes values and attitudes and determines critical life experiences' (Sobel, 1981, p.7). Max Weber is more clearly concerned with life style. In his famous essay, 'Class, Status and Power', he considers the sociological meaning of these three words. He points out their sociological importance and then brings in the term *'status honour'*. 'In contrast to classes, status groups are normally communities. They are, however, often of an amorphous kind' (Weber, 1954, p.68). The status situation is not determined by the economic situation, but by a specific social estimation of 'honour'. '...*status honour* is mainly expressed by the fact that above all else a specific style of life can be expected from those who wish to belong to the circle' (Weber, 1954, p.69). Veblen's interest in life style is also derivative, a by-product which does not receive more investigation. His argument is based on the hypothesis that esteem is primarily orginated from superior proficiency. In early societies this proficiency was expressed by aggressive activity such as war. In industrial societies, prowess needs alternative measures of display. Ownership and the possession of wealth is the primary confirmation of success and the main grounds of esteem. But in order to bring esteem to its owner, ownership must be translated into conspicuous symbols (Veblen, 1966, p.40). Weber and Veblen both regard life style as observable manifestation of the structural source of prestige (Sobel, 1981, p.10). The more recent approach to life style is circumvention. Tallman and Morgner attempt to feine it in the following way: 'The lack of constituent meaning for a concept such as 'life-style' makes any set of indications vulnerable to the criticism that they are not approriate measures and do not tap 'significant' aspects of the phenomenon. We view life-style as a broad rubric under which a number of behavioural activities and orientations can be included, each of which requires a distinctive investment of the individual's resources of time, energy and money' (Tallman and Morgner, 1979, p.337).

Towards a definition of life style as an innovation

On human needs / demands

The starting point to the reasoning here is the individual with all his/her psychological characteristics, archetypical human needs (Figure 16.1.). Human needs are the basis of the classical theory of motivation of Maslow. He points out the so-called physiological drives as basic needs (Maslow, 1970, p.35). A person suffering from want of food, safety and esteem would probably yearn for food stronger than for anything else. But the moment the physiological needs are satisfied, new and higher needs arise. And when these, in turn, are fulfilled, again new and still higher needs emerge and so on (Maslow, 1970, p.38). Physiological needs being satisfied, a new set of needs emerges. These are the safety needs (security, stability, protection, freedom from fear, need for structure, order, law, limits ....etc.) and, among the higher needs, the need for a sense of belonging. Not much scientific information is given on these needs, but they are a frequent theme in literature, particularly in the newer sociological literature. But the need for neighbourhood, for one's territory, for one's own 'kind' is still underestimated (Maslow 1970, p.43-44). All people have a need for a stable high evaluation of themselves, first for self-esteem, and second for the esteem of other people. Satisfaction of the self-esteem need gives, in turn, a feeling of self-confidence, worth and adequacy. On Maslow's ladder of human needs, the next level belongs to the need of self-actualization, self-fulfillment, the desire to become realized in what one is potentially (Maslow, 1970, p.51). A still higher one is the desire to know and understand. Finally, the highest ones are aesthetic needs. At the same time, this need seems sometimes to be the basic one: some people crave actively for beauty, and their craving can be satisfied only by beauty (Maslow, 1967, p.314). Evidence of the need for beauty is found in every culture and in every period, back to the cavemen. One is aware of the fact that the conative and cognitive needs overlap. Such needs as the need for system, for symmetry, for structure, can be ascribed to conative, cognitive or even aesthetic needs. The general framework of the theory of motivation is created by three components: incentives, expectancies and behavioural tendencies (Veroff, 1980, p.5).

Figure 16.1. *Changing life styles versus urban built form*

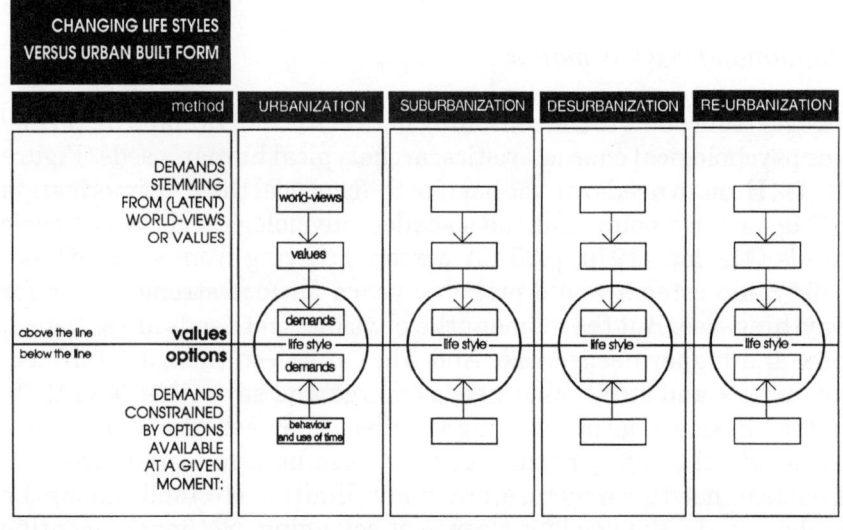

Historical sketch

a. Political development

According to Stein Rokkan's Macro-Model, there have been four main processes of political development in Europe: state formation, nation building, mass democracies, welfare states (Flora, 1983, p.16-26).

The concept of social welfare has a long history of some hundred years. Poor relief had been an early form of social security since the sixteenth century. No more profound changes occurred until the end of the nineteenth century, when social insurance was introduced. First, it was applied to industrial accidents, sickness and old-age and later to unemployment. The difference between the old poor relief system and the new institutional welfare system lay in the fact that whereas the first was applied to a limited range of citizens, the latter covered all those making up the economically active population. The modern welfare state appeared just after the Second World War when social security was extended from covering the dependent labour force to the whole population (Royle, 1983, p.25). Its extent became similar to that of the franchise (Flora, 1983, p.25). Other government activities, often parallelly introduced, were the implementation of housing and employment policies and the development of public and health systems. Compulsory education brought together a new spectrum of citizens' rights: from the right to have the basic needs of economic welfare and security guaranteed to the right to bettering oneself in society. And a still higher degree of satisfying human needs was achieved when access to secondary and higher education was provided in the majority of European countries. Opportunites have been given for individual growth, for the realization of talents, wishes and hopes together with secured social conditions (Dahrendorf, 1979, p.30).

b. Social development

Social development has changed greatly the structure of society during the past hundred years. Using the occupational structure as the criterion for division into social classes, the following social strata may be found: the middle-class of professional and managerial people and higher technicians; the heterogeneous lower-middle class of non-manual employees, small proprietors and the 'blue-collar elite'; and the working class. Above them stood the small group comprising the 'governing classes' (Halsey, 1981, p.24). At the beginning of the nineteenth century, over three-quarters of the employed and self-employed population were occupied in manual work. By the mid-century, the proportion had fallen below two-thirds and since then to roughly a half. If the previous occupational structure could be compared to a pyramid, the trend is towards status symbolized diagrammatically by an electric light bulb (Halsey, 1981, p.25; Mendras, 1984, p.246).

c. Demographic development

Demographic development is characterized by two main phases in the European population process: (i) a 'first demographic transition' which in central Europe was carried out in the 1930s, with a low mortality rate and fertility reduced to replacement level. 'A normal nuclear family' pattern was established, with more than 90% of the adult population being married; (ii) a 'second demographic transition' which in central and western Europe began after the mid-sixties, with fertility below the replacement level, a decline in marriage and increasing divorce and cohabition (Strohmeier, 1992).

d. Economic development

The keyword of economic development is the standard of living, and for life styles consumption is a visible expression of standard of living. At the turn of the century, the standard of living of working classes improved but people still worked 56 hours a week and they earned little by current standards (Dahlström, 1959, p.121). But technological innovation in production dramatically increased per-capita productivity some twenty years later. The advent of mass production changed the character of work. It became more efficient but also more fragmented and tedious (Sobel, 1981, p.34). Between the First World War and the end of the fifties, consumption doubled (Dahlström, 1959, p.120). In Sweden hours worked per week dropped to 48 as a result of a law of 1919. In the 1930s paid vacations were introduced in many European countries. People earned more, had more time for leisure, and their consumption increased dramatically. Higher production efficiency caused by continual technological progress gave rise to mass production and a still higher rate of consumption. The constant increase in consumption has been a *conditio sine qua non* of the further economic progress of society (Sobel, 1981, p.35; Mendras, 1984, p.246).

Some recent definitions of life style

Four recent definitions of life style seem to be appropriate for further reasoning: (i) 'life style are all the observable characteristics of a person, eg. his manner of dress, way of speaking, domestic habits .... which serve to indicate his value system and attitiudes towards himself and aspects of his environment' (Bullock, 1977, p.349); (ii) life style is a 'distinctive and hence recognizable mode of living, attitudes, values and behavioural orientations' (Sobel, 1981, p.28); (iii) 'life style is a variable composed of labour-force participation, household structure and marital status' (Strohmeier, 1992); (iv) 'life style is a person's pattern of choices' (Earl, 1983, p.129).

A definition of life style as innovation

Life style is a multi-level and cross-disciplinary notion. It is based, on the one hand, on the 'permanent data' of human needs which express the archetype of human nature, and on the other, on 'variables', the sociopolitical, demographic and economic aspects of life style. The historical sketch which outlines the 'variables' in the sequences of ongoing events, is pointing towards a contemporary reality which is distinctly different from that of previous decades. Life style is expressed in different areas of one's life: income-related activites, family-related activites, recreation and relaxation, religious and organizational participation, socializing activites (Chapin and Logan, 1970, p.315). The three main spheres are therefore: labour, dwellings and leisure. Labour and the size of real income being the given point of departure, dwelling and leisure remain a subject of choice, providing that the income gives a certain scope of options. Life style influences the individual in his/her choice of place, residence and leisure. The choice is conditioned by the breadwinner's biographic careers, the kind of labour and the stage of his/her life cycle. The choice of living place also has other dimensions; it is a manifestation of belonging to a certain group of people and is a conspicuous expression of one's qualitative preferences. The choice of living place and its character affect more than one group of human needs. Certainly, having one's own home gives one a feeling of security, of 'belonging' to a certain, very special place and a certain neighbourhood. But it can also improve the self-esteem of the owner, and induce the esteem of the other people, as the home can express and symbolize one's status in a consumption-oriented society. Furthermore, the home can satisfy one's aesthetic needs, one's need for beauty. The notion of dwelling is thus related to almost the whole spectrum of human needs. Bearing in mind all four recent definitions of life style, one should emphasize the last one which refers to life style as a person's pattern of choices. Choice means freedom to choose. Life style is the characteristic of a person, a group of persons, a society, who 'can afford' free choices. On the following pages life style will be addressed as: distinctive, observable, resultant from demands stemming from people's (latent) world-views and values and demands which are constrained by the options available at a given moment. Life style is thus a resultant of people's values and options. Life-style innovation lies in the growing degree of freedom given to individuals to organize their lives and to choose their life styles (Strohmeier, 1992).

Spatio-temporal salience of life styles

According to the urban life-cycle concept, the definition of successive development stages is based on socio-economic development, its main features being the changes in the structure of the economy and of income level. These stages are: (i) the conversion from a largely agrarian to an industrial society; (ii) the conversion from a mainly industrial economy to a tertiary economy; (iii) the growth of the tertiary sector to maturity (v. d. Berg et al., 1982, p.24).

These changes have been followed by urban change in a sequence of phases: 'urbanization, suburbanization, desurbanization, and (hypothetical) reurbanization' (Strohmeier, 1992). These four phases are the point of departure in the search for the spatio-temporal salience of life styles.

Urbanization

Urbanization, following industrialization, began at different moments in western and central Europe: in the eighteenth century (England), in the mid-nineteenth century (Germany), and even at the end of the nineteenth century (Norway).

But the moment of the beginning of the next phase, suburbanization, happened more simultaneously in European countries. There were different reasons for the phenomenon, but one of them were the upswings and downswings of the economic cycle which affected the whole of Europe (Stromeier, 1992). Another reason was the growing degree of population mobility with the railway and horse-driven bus and tramway until 1875, and the railway and electric tramway until 1918 (Lichtenberg, 1986, p.21). In the last decades of the nineteenth century, there were no innovative systems of religious, political or scientific thought comparable to those of the Renaissance or the

Figure 16.2. *Changing life styles versus urban built form: stage of urbanization*

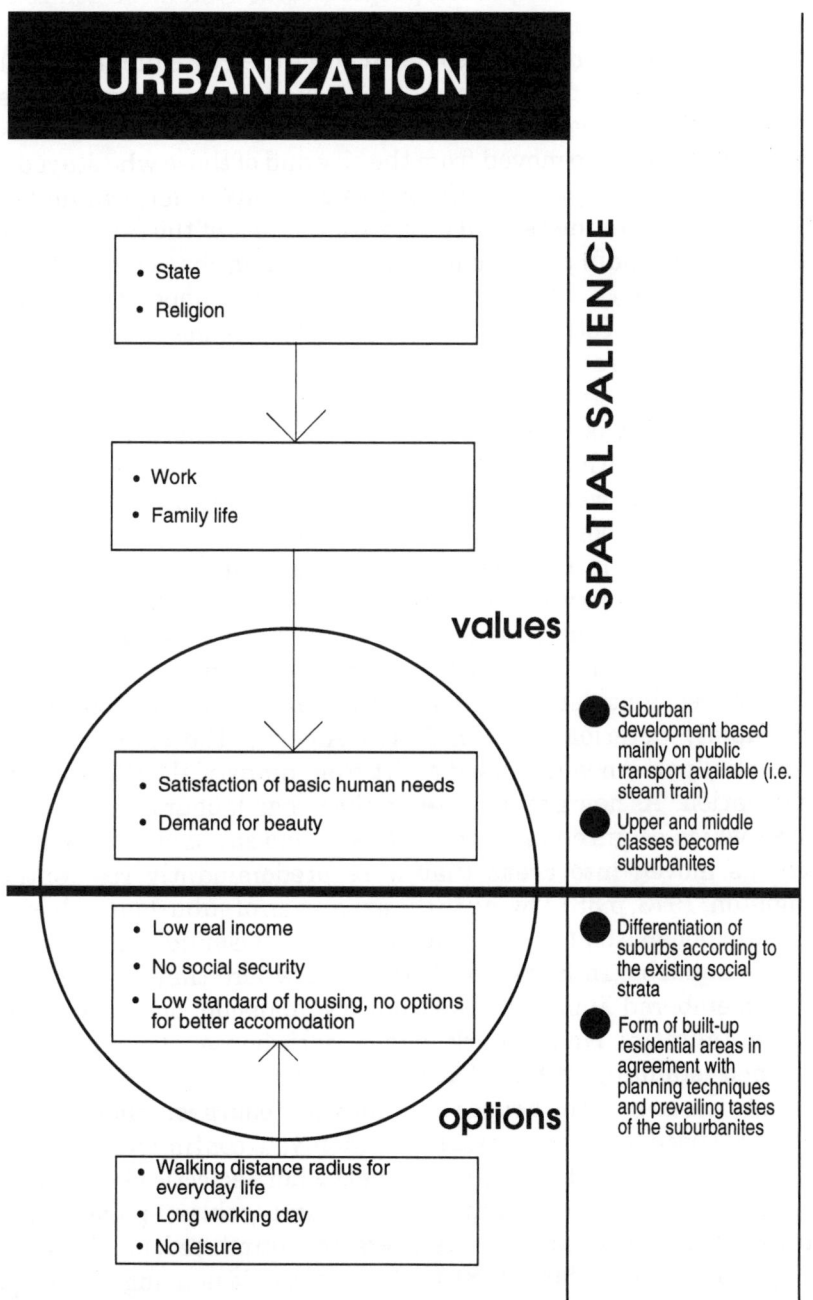

Enlightenment etc. which could provide a foundation for new values (Kogane, 1982, p.21). Nevertheless, the state and religion made an indirect - if not direct - impact on people's value systems. Until 1918, the degree of affluence of the majority of the population was not sufficient to enable ordinary people to choose their own life style while there were still the upper class and a part of the growing middle class who enjoyed privileges. What, then, was the life style like of those wealthier people who moved from the city and of those who stayed in the inner city? The *pater familias* was a breadwinner, commuting from the suburb to the central city, with the rest of the family living in idyllic surroundings and avoiding the seeming chaos and disorder of the central city. The family's everyday life depended on the institution of domestic service (Figure 16.2.). The majority of people still had limited prospects for betterment, their basic demands being insufficiently satisfied (Dahrendorf, 1979, p.30). Their residential neighbourhood was characterized by the filth and noise of the industrial city. With 56 working hours a week and low real income, their life styles were rather an *a priori*-determined life line. Industrialization began to change the residential pattern of the city rather early in the nineteenth century, although in ways which differed from country to country (Dunin-Woyseth, 1988, p.15-26). The factory replaced the residential workshop as the *locus* of manufacturing (Singleton, 1973, p.38). The result was that the wealthier residents left the previously heterogenous neighbourhood, now rapidly deteriorating. In North America the change of the residential pattern was caused apart from industrialization also by immigration. 'As new groups entered the urban labour market, older groups which acquired industrial skills or had saved enough of their earnings moved into areas that were predominantly residential' (Singleton, 1973, p.38). Owing to the expansion of suburban railways, still too expensive for the less affluent classes, the middle class moved into the suburbs. In England, it was here where they re-created a half-remembered image of the country mansion or the country cottage. Imitating Nash's Park village, each house was, if possible, 'detached', standing within its own plot even if the plots were narrow. Closer to the centre, the houses built for petty bourgeois shopkeepers and clerks used to be, for reasons of economy, in continuous terraces. The architectural style was almost everywhere eclectic Gothic, with high-pitched roofs, frequent dormers and gables, polychromic brickwork and stuccoed, decorated by factory-produced, neo-Ruskian foliage (Risebero, 1985b, p.188). In Germany, villa housing of the type in the villa district of Grunewald, remained a retreat for the rich. Low density suburbanization was not a widespread phenomenon until the turn of the century (Sutcliffe, 1981, p.3).

Suburbanization

In the First Modern Age, the world view was influenced by the belief in progress. The development of mass production techniques brought previously unaffordable goods within the financial range of large circles of the population. 'Status and conspicuousness, hitherto reserved for the few, now become possible in a modified form for many. Houses became the object of displays of new bourgeois affluence, not on a grand scale but at least sufficiently to show that one was no worse off than one's neighbours' (Relph, 1989, p.90). With mass-produced goods and with substantial increases in disposable income, such a phenonemon became manifest on an unprecented scale. The middle class grew, becoming the dominant part of society. The suburb became its abode. In the 1930s, its increasing expansion influenced the change of the character of the cities.

Contemporary papers wrote about the new, prevailing life style: 'Investigations demonstrate that a family consisting of parents and two children is a standard family. The husband is a breadwinner who also manages the economy of the household. The wife is living at home. The opinion on child upbringing has changed. Childhood becomes a status as an independent period in one's life. Residential standards improve, houses become bigger, more spacious and they have more light. The standard of living rises and the new 'nuclear family' acquires a new consumption pattern. More efficient production gives ordinary people the right to have paid vacations. New behaviour patterns develop, with weekend leisure in the country and the family being mobile owing to the private car. Political rights are won, the battle for roles of both sexes belongs to the future' (Århundrets Krønike, 1989, p.498). The 'nuclear family' is thus in focus. For most people, the family was a standard social unit, 90% of the adult population being married (Strohmeier, 1992). Social security began to encompass the whole labour force. Society was homogeneous and life careers of the great majority were similar. Basic human needs seemed to be covered. What were the values which attracted people to suburbs? What were the values which people no longer gave priority to in leaving the cities and moving to the suburbs? Life in

323

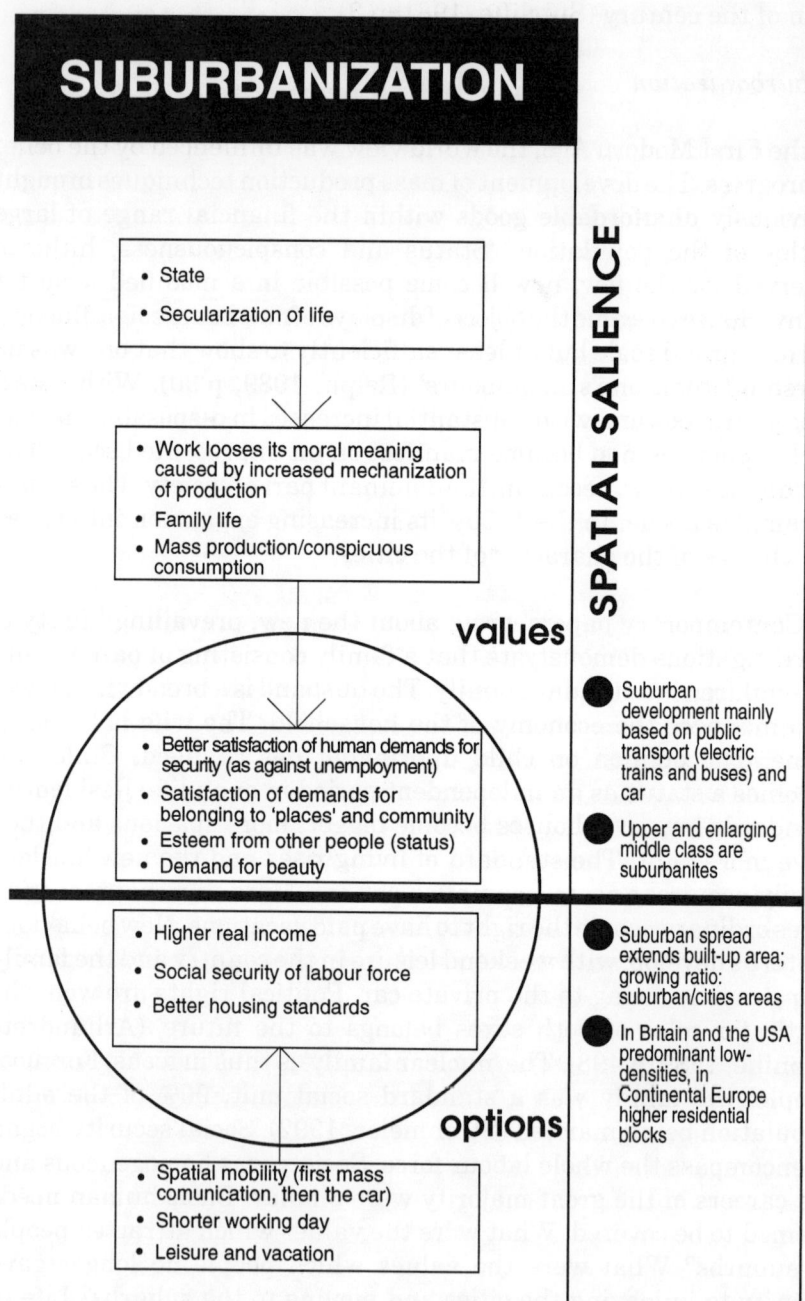

cities was associated with (Figure 16.3.): (i) high density, coupled with great social diversity of the population; (ii) variety, contrast and chance encounter; (iii) a complex network of kinship, uniting generations; (iv) cultural freedom, wide choice in friends, shopping, all within easy reach.

The new suburbs were characterized by: (i) homogenity and conformity; (ii) because of low densities of population, reduced human interaction; (iii) impoverishment and standardization in the range of shopping goods and therefore in the quality of material life.

In three ways, the city and the suburbs embodied antithetical values: (i) mixture versus homogenity; (ii) concentration versus dispersion; (iii) specialization versus the middle range (Hall, 1968, p.111-112).

The phenomenon of growing segregation of the city was the object of investigations and theories in different disciplines. Three models of city structure are to be mentioned in this context: (i) the classical explanation is that of the sociologist Burgess, whose 'circle hypothesis' suggests that special features of different areas in the city are a function of their distance to the city centre (Jansen, 1959, p.213); (ii) the economist Hoyt's 'sector hypothesis' puts forward that special features of different areas in the city are not only a function of the distance, but also of the direction and location in the city. He points out that a place's inherent qualities induce high land values (Scargill, 1979, p.42); (iii) Harris and Ullman's cellular diagram of city structure explains the tendency of urban functions to cluster and segregate, especially under conditions of decentralization. Functions concentrate in order to achieve economies of scale (Scargill, 1979, p.42).

Hoyt's theory is especially interesting as it touches on the relationship between quality and quantity in the context of city models. As an economist he suggests rental value as a surrogate for housing quality. Residential land uses were inclined to be arranged in sectoral fashion, radiating outwards from the city centre along transport routes. High class residential areas developed in a particular direction because of physical and social attractions; middle-range housing bordered the expensive sector to take advantage of the latter's reputation; whilst the cheap housing was 'banished' to the least advantageous sectors. The economic status of the different sectors was sustained as they grew outwards (Jansen, 1959, p.212-213). There were two phases of suburban growth: in the first phase,

Figure 16.4. *Changing life styles versus urban built form: stage of desurbanization*

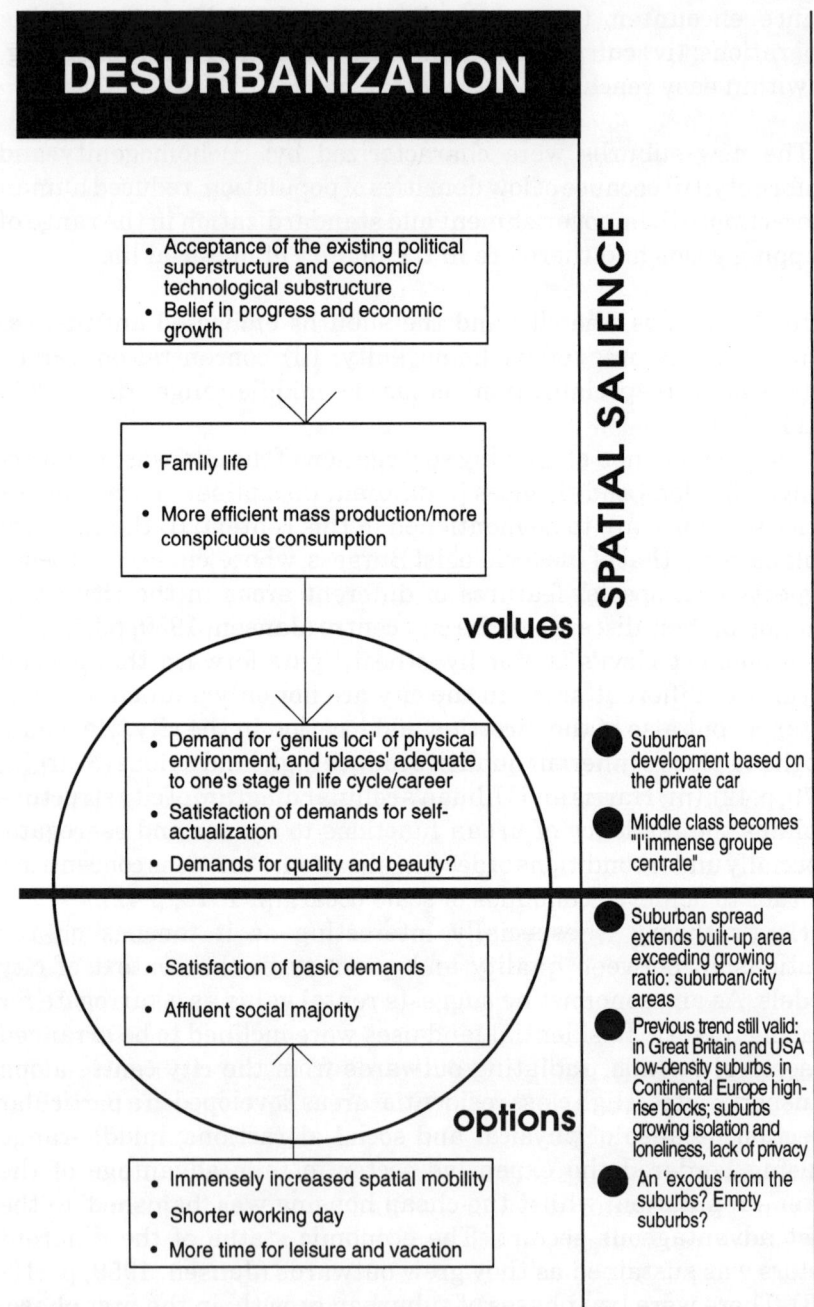

the population of the city increased, the concentration becoming dense in the central areas of the city; in the second phase, the suburban area continues to expand, but their population growth takes place at the expense of the city proper (Sert, 1944, p.52). The pattern of development resembles the general shape of a gigantic octopus. The new consumer's landscape was by some regarded as insufficiently modernist, by others as insolence to traditional tastes (Relph, 1987, p.91).

Desurbanization

The first 20 - 25 years after the Second World War brought growing affluence. The welfare state secured people's basic needs. Social security now encompassed all members of society, both the labour force and the non-productivee part of society. The belief in progress and never-ending growth and prosperity created the *Zeitgeist* of optimism. People thought quantitively in terms of expansion, extension, mechanical multiplication. The pendulum was about to swing back by the symbolic year of 1968. This was a period when life styles were rather homogeneous; the 'first demographic transition' just began to ebb away. The pattern of the normal 'nuclear family', still common for 90% of the adult population, was about to dissolve at the end of the sixties. Leisure was becoming an important element in people's time budget (Figure 16.4.). Millions of families found their homes in the suburbs. While the Anglo-American tradition was a continuation of the garden-city idea with low densities, most European cities planned their garden suburbs with high densities (Barnett, 1986, p.87).

Many sociological works have been written on the quality of life in the new suburbs in North America. The conclusion was rather unanimous: 'urban culture has become a minority culture, practiced by upper-middle-class cosmopolitans in isolated islands of big metropolies' (Hall 1968, p.134 and 1982, p.270). Ordinary people lived in the suburbs and they enjoyed it, ordinary people now being the great majority of the middle class. In the high-density, monotonous suburbs of Europe, on the contrary, dissatisfaction was growing with convenience set against uniformity and loneliness. In literature, the life styles particularly of women living in suburbs were hardly depicted as rosy. Those who could afford it tried to find an *antidotum* against suburban living in the 'second home' for weekend life, far away from suburban uniformity. The car became the symbol of life style. In 1963, Melvin Webber from the University of California at Berkeley presented a new vision of the 'non-place urban realm', which was already a reality in some places in the world: the examples pointed to were Los Angeles, the Bay Area of San Francisco, southeast England, the Dutch Randstad and the Rhine-Ruhr region in western Germany (Webber, 1963). Dynamic surburban growth was increasingly a reality in industrialized countries (Hall, 1982, p.250). Webber maintained that the new system of communication had exploded the old association between propinquity and communication. He predicted a state where the constantly reducing costs of transportation and communication would spread an urban culture freely across the globe, limited only by the surviving need of some activites for proximity and by the fact that some locations are more attractive to live in and to work than others (Hall, 1968, p.250). The concept was not that new. In 1935 Frank Lloyd Wright had already exhibited his model of Broadacres, at the Rockefeller Center. Huge traffic arteries running through the countryside, electric automobiles and helicopters resembling flying saucers emerge on the rural scene, signifying a new spatial organization of life and human settlements (Ciucci, 1980, p.372). Any distinction between urban and rural was finally eradicated and settlement became more and more uniform over regions. The year Hall presented the Berkeleyan theory of the 'nonplace suburban realm' (1968), the first student revolts against such a world as predicted in the theory swept over university cities from Berkeley to Paris.

Re-urbanization?

There are many reasons accounting for the phenomenon of re-urbanization (Figure 16.5) but the change in the prevailing life style which affected most people in the 20-25 years after the Second World War is one of the most important (Wood, 1981, p.257). The period coincided with the 'second demographic transition' which began in the mid-sixties. The high fertility rate of the first post-war years began to fall far below replacement level and was accompanied by a decline in marriage and increasing cohabitation (Strohmeier, 1992). From the end of the sixties onwards, a new wave swept over the western world (Galvany, 1981, p.12). Safe contraception, democratization of education and better prospects for both sexes, have made it possible for the majority to choose their own biographic

Figure 16.5. *Changing life styles versus urban built form: stage of re-urbanization*

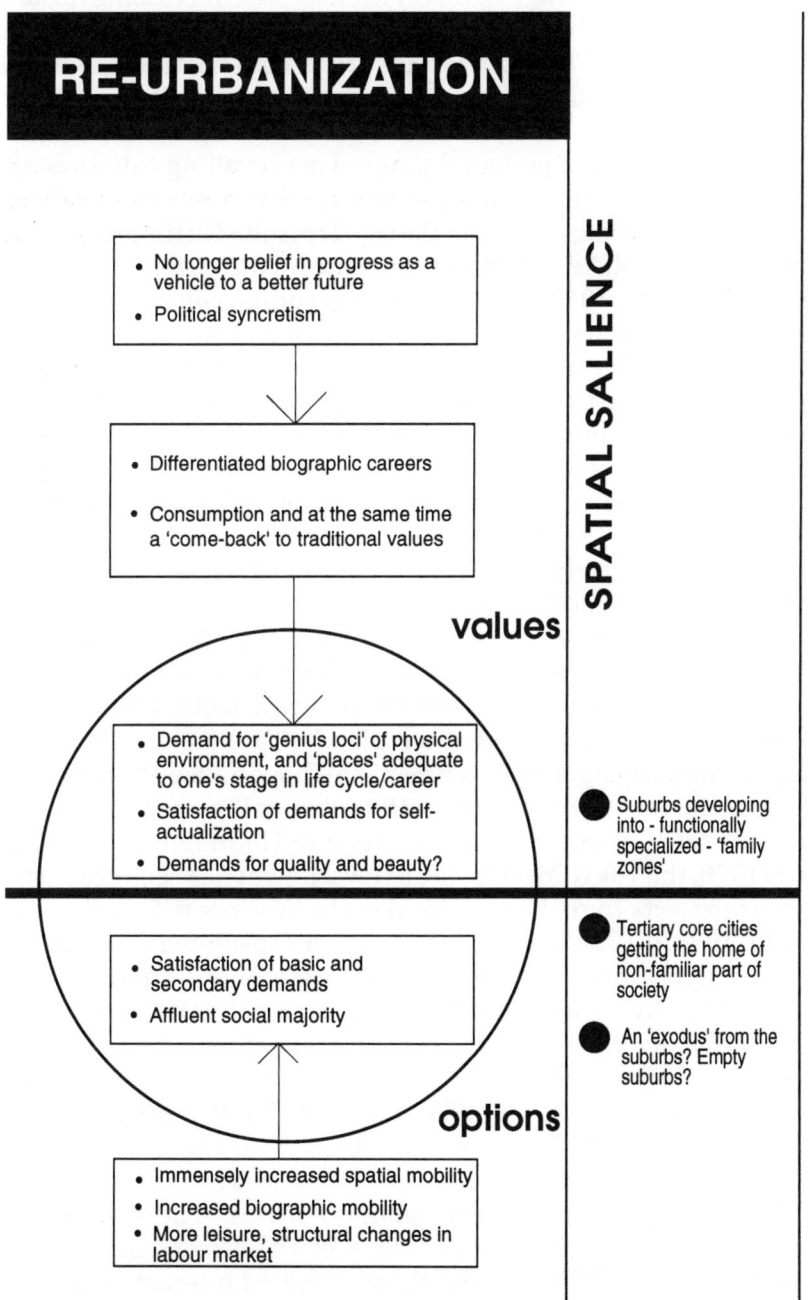

line. Life styles have become more heterogeneous. 'Founding a family is increasingly a biographic option. Where people opt for family, we observe a dissolution of age-norms by a cohort-wise-increasing variance of the ages at first birth or first marriage' (Strohmeier, 1992, and Cazes, 1982, p.64). These tendencies are characteristic of the middle class, but this middle class has become the majority in western societies (Murakami, 1982, p.220). The 'second demographic transition' occurred in a period of profound changes in prevailing value systems with emphasis on quality instead of quantity, a return to primary values and increasing respect for the city. It resulted in the preservation wave in urban policies and it generally meant a 'comeback' to the city (Dunin-Woyseth, 1988a, p.39). The opportunities offered by the city were given more and more priority in preference to those of the suburbs. After 25 years, the predictions about the 'non-place suburban realm' do not seem to have become reality. It is the city as a place, its uniqueness, its *genius loci*, which make the city attractive again (Dunin-Woyseth, 1988a, p.38). From being the abode of 'upper-middle-class cosmopolitans in isolated islands of big metropolies' and of those living on the social fringes, the city has begun to attract a broader part of society who choose to live in cities because they best suit their own life styles. When choosing their 'biographic project', they decide how to place it in the landscape of their future (Earl, 1986, p.116). People... 'stay in places with ecological valencies (infrastructure, housing, jobs etc) which best suit the realization of their biographic project or they move to another place' (Strohmeier, 1992). The suburbs are developing into a functionally-specialized 'family zone', whereas the tertiary core cities will evolve as the abode of the non-familial part of society (Strohmeier, 1992 and Marshall, 1973, p.126). Since 1979, the Tokyo Metropolitan Government has been conducting various projects to realize the ideal of the 'My Town Tokyo' concept (Osaki, 1988, p.126). The image of Tokyo encapsulated in the concept is based upon three basic goals: 'safe metropolitan living', 'active metropolitan living' and 'the hometown metropolis' (Dunin-Woyseth, 1989, p.39). Living, even in the central areas of the city, is one of the objectives of urban policies and plans. (Planning of Tokyo, 1988, p.52; Centenary of Modern City Planning and Its Perspectives, 1988, p.203-265). From 1976 to 1979, the OECD carried out a research project on 'The Future Development of Advanced Industrial Societies in Harmony with that of Developing Countries'. The research was based on several main studies. One of them, conducted by the research team INTERFUTURES, investigated interaction between the superstructure, ie. culture and polity, and the substructure ie. economy and technologies, of societies. Regarding the present situation, the following characteristics have been identified: (i) affluence gives rise to a structural change in demand, ie. from basic to selective demands; (ii) new selective demands are difficult to satisfy through the market.

For development of future life styles, two trends seem to be decisive: (i) macro-homogenization: affluence means high mass consumption, resulting in unprecedented homogenization of life style throughout society. The old concept of class struggle is becoming unacceptable because of this growing homogeneity; (ii) micro-heterogenization: a number of interest groups emerge. Syncretic politics rule (Murakami, 1982, p.221).

These contrasting processes of homogenization and heterogenization proceed together, both within and between societies (The New Encyclopedia Britannica, 1988, p.315). What will be the impact of these processes upon urban form? As an individual's values become more and more varied, so does his/her life styles. The 'city revived' will be in a position to offer more varied opportunities to urban dwellers, the suburbs being a 'harbour' to those who need it within a certain life stage of the individual. A 'megalopolitan' life style will include frequent changes of residence and rented accommodation rather than ownership (including secondary occupational activities) (Cazes, 1982, p.68). Will such an 'exodus' from the suburbs make them empty one day?

Conclusion: possible directions for future work

Changing life styles have made a profound impact on built urban form over the past hundred years. For future life styles, the keywords will be: homogenization and at the same time heterogenization of societies; general affluence; more leisure time; major development within information technologies. Growing suburbs, 'non-place suburban realm', 'cities revived'. Will the next stage be 'abandoned suburbs'?

Only some aspects have been addressed in the paper: (i) The emphasis has been laid on the function of dwelling. The function of leisure has just been touched upon. 'Second homes', the 'cities of leisure/tourism' etc should be investigated in the future; (ii) another important aspect is the role of the human need for beauty. How has the need for beauty formed life styles? Literature on aesthetics such as the philosophy of art should be investigated in their social aspects (ie. Adorno, Bakhtin, Bloch, Bourdieu, Kupfer, Lewis, Sheppard...); (iii) during the last few decades, several ambitious plans for improving the living environment in cities have been implemented, with new forms of living both within the cities and outside them. Have they proved to be successful, have they answered up to expectations? A survey should be done and an assessment made in order to identify the main trends.

This paper is an attempt to form a bridge with other disciplines. The author is, nevertheless, aware of the fact that when she leaves her traditional field, physical planning, her quality control and conceptual clarity decreases. That is why this contribution is mainly a discussion paper.

References

Århundrets Krønike, (1989), *J. W. Cappelens Forlag AS*, Oslo.

Barnett, J., (1986), *The Elusive City, Five Centuries of Design, Ambitions and Miscalculations*, Harper & Row, N Y, London.

Berg v. d., L., Drewett, R., Klassen, L. H., Rossi, A., Vijverberg, H. T., (1982), *Urban Europe. A Study of Growth and Decline*, Pergamon Press, Oxford, New York, Toronto, Sydney, Frankfurt.

Bullock, A. and Stallybrass, O., (1977) *'The Fontana Dictionary of Modern Thought'*, Collins, St. James Place, London.

Carter, H., (1983), *An Introduction to Urban Historical Geography*, Edward Arnold, London.

Cazes, B., (1982), 'Overview of New Demands in 'End-of-Century' France', in Kogane, Y. (ed.), *Changing Value Patterns and Their Impact on Economic Structure,* University of Tokyo Press, Tokyo.

Centenary of modern city planning and its perspectives, (1988), The City Planning Institute of Japan, Tokyo.

Chapin, S. and Logan, T. H., (1970), 'Patterns of Time and Space Use', in Perloff, H. S. (ed.), *The Quality of the Urban Environment*, The Johns Hopkins Press, Baltimore and London.

Ciucci, G., (1989), 'The City in Agrarian Ideology and Frank Lloyd Wright: Origins and Development of Broadacres' in Ciucci, Dal Co, Manieri-Elia, Tafuri, (eds.) *The American City*, Granada, London, New York.

Dahlström, G., (1959) 'Levnadsstandard och Konsumption', in

Dahlström, E., (ed.), *Svensk samhällsstrukur i sociologisk belysning*, Svenska Bokforlaget, Stockholm.

Dahrendorf, R., (1979), *Life Chances. Approaches to Social and Political Theory*, Weidenfeld and Nicholson, London.

Dunin-Woyseth, H., (1988a), Built Form versus Urban Planning Legislation of Last Century: Genius Loci versus International Influences. The Centenary of Modern Urban Planning and Its Future Perspectives. Tokyo Congress, November 1988, Proceedings, Tokyo.

Dunin-Woyseth, H., (1988b), 'Den 'regulerte gestalt'. Lovgivningens tilsiktede og utilsiktede virkning på byformen' in *Arkitektur i Norge. Åarbok 1988,* Norsk Arkitckturmuseum/Bonytt Oslo.

Dunin-Woyseth, H., (1989), 'Planning i Tokyo', Arktektnytt, vol.4, no.28, February.

Earl, P., (1983) *The Economic Imagination: Towards a Behavioural Analysis of Choice*, Armouk, New York.

Fischoff, B. (1981), 'No Man is a Discipline' in Harvey, J. H. (ed.), *Cognition, Social Behaviour and the Environment,* Hillsdale, New Jersey.

Flora, P., (1983), *State, Economy and Society in Western Europe 1815-1975,* Campus Verlag, Frankfurt.

Gallion, A., (1950), *The Urban Pattern,* D. van Norstrand Company, New York.

Galvany, J., and Martin, P., (1981), *Changer le cadre de vie,* Club Socialist du Livre, Paris.

Gebhard, D., and Breton, H. v. (1975), *L. A. in the Thirties 1931-1941,* Peregringe Smith, Inc., New York.

Hall, P., (1982), 'The Urban Culture and the Suburban Culture' in A Symposion of Urban Philosophers: *Man in the City of the Future,* The Macmillan Ltd, London.

Halsey, A. H. (1981), *Change in British society,* Oxford University Press, Oxford.

Janson, C. G., (1959), 'Stadens struktur', in Dahlström, E., (ed.), *Svensk samhallsstruktur i sociologisk belysning.*

Kogane, Y., (ed.), (1982), *Changing Value Patterns and Their Impact on Economic Structure,* A Report to the OECD, University of Tokyo, Tokyo.

Lichtenberg, E., (1986), *Stadtgeographie,* Teubner, Stuttgart.

Marshall, H., (1973), 'Suburban Life Styles: A Contribution to the Debate', in Massotti. L., and Hadden, J. K. (eds.), *The Urbanization of the suburbs*, Sage Publications, London.

Maslow, A. H. (1967), 'A Theory of Motivation: the Biological Rooting of Value-Life', *Journal of Humanistic Psychology*, Vol.7.

Maslow, A. H., (1970) *Motivation and Personality*, Harper & Row, New York.

Mendras, H. and Forsé, M., (1984), *Le Changement social: Tendances et paradigmes*, Armand Colin, Paris.

Murakami, Y., (1982), 'Values, Demands and Scenarios' in Kogane op. cit., Tokyo.

Osaki, M., (1988), 'Urban Improvement Plan. Historical Review since Tokyo Improvement Ordinance', in Tokyo International Symposium: *Centennial Anniversary of Modern Urban Planning Legislation in Japan*, Tokyo, 8-12 November 1988. Proceedings.

Relph, E., (1987) *The Modern Urban Landscape*, Croom Helm, London, Sydney.

Risebero, B., (1985a), *Modern Architecture and Design. An Alternative History*. The MIT Press, Cambridge, Massachussetts.

Risebero, B., (1985b), *The Story of Western Architecture*, The MIT Press, Cambridge, Massachussetts.

Royle, E., (1987), *Modern Britian: A Social History 1750-1985*, Edward Arnold, London.

Scargill, D. J., (1979), *The Form of Cities*, Bell & Hyman, London.

Sert, J. L., (1944), *Can our Cities Survive?*, The Harvard University Press, London.

Singleton, G. H., (1973), 'The Genesis of Suburbia: A Complex of Historical Trends' in Massotti, L. H. and Hadden, J. K., (eds.), *The Urbanization of the Suburbs*, Sage Publications, Beverly Hills, London.

Sobel, M. E., (1981), *Lifestyle and Social Structure: Concepts, Definition Analyses*, Academic Press, New York, London.

Strohmeier, P., (1992), 'Differentiation, Mobility and Urban Change: A Multi-level Perspective', in Drewett, R., and Montanari, A., (eds.), *Innovation and Urban Development: the role of social and technological change*, Avebury, Aldershot, forthcoming.

Sutcliffe, A., (1981), *Towards the Planned City, Germany, Britain, the United States and France, 1750-1914*, Basil Blackwell, Oxford.

Tallman, I. and Morgner, R., (1970), 'Life style differences among urban and suburban blue-collar families', *Social Forces, Vol.48*.

The Bureau of City Planning, The Tokyo Metropolitan Government, (1988), *Planning Of Tokyo*, 1988, Tokyo.

The New Encyclopedia Britannica, (1988), Contrasting process of differentiation and homogenization.

Veblen, T. (1954), 'The Theory of the Leisure Class', in Bendix, R. and Lipset, S. M. (eds.), *Class, Status and Power*, The Free Press, New York.

Veroff, J. Veroff, J. B., (1980), *Social Incentives: A Life-span Developmental Approach,* Academic Press, New York.

Webber, M., (1963), 'Order in Diversity: Community without Propinquity', in Wingo Jr., L. (ed.): *Cities and Space: The Future Use of Urban Form*, Baltimore.

Weber, M., 'Class, Status and Party', in Bendix, R. and Lipseth, S. M. (eds.), *Class, Status and Power,* The Free Press, New York.

Wood, B., (1981), 'Urbanization and Local Government', in Halsey, A. H. (ed.), *Trends in British society since 1900. A guide to the changing social structure of Britain.* Macmillan St. Martin's Press, London.

17. THE DEVELOPMENT OF PUBLIC SPACE AS THE BASIS FOR URBAN LIFE

John Allpass, architect m.a.a., Copenhagen, Denmark*.

Introduction

The phenomenon of towns and cities is indissolubly connected to the polarity of public life and private life, and it is expressed in the polarity of public space and private space in the physical composition of towns and cities. This polarity is specific to the urban situation. Before man started to build towns and cities this division did not exist, and it is still non-existent or weak in the rural areas, e.g. in farming villages. A town, or city, with inadequate public and private spaces cannot be a good place to live. But unfortunately it seems that the modern town and city has not found proper ways of handling the polarity between public and private life, and the polarity of public and private spaces. If modern cities and towns do not function well with regard to their public-private dimensions, how has this happened, and what can be done? One of the things we can do is to learn from historical experience. Old towns and cities have several layers of different public-private relations built into them, as part of their history.

---

* Translated by Bo Grønlund

To study urban development and the public-private relation within it, we have to study the economic, political and social conditions of life in towns and cities. We have to study the relations of production and exchange, family-patterns and the dwelling situation, in short: the daily life of the citizens. At the same time we have to consider the class structure of the society in question, as part of the study.

To understand how towns and cities emerged and developed, we have to study carefully the relationship between the citizens' use of spaces for public life, and the use of space for private life. Usually. maps and drawings give a picture of towns and cities that do not show this. Therefore we can not be satisfied by looking only upon development as a map of the physical structure. 'So strong was this tendency that little attempt was made to examine other implied forces which were operative; nor could it be assumed that because patterns expressed themselves spatially they could be explained spatially' (Pahl, 1975). We have to consider the roles of the citizens, their way of life, and especially the relationship between public life and private life. A great deal of the research on the history of urban development has maps as its major object, but maps are stylised abstractions from the air, which can be highly misleading. E.g. an undulating street can look romantic and self-grown, but reality is often different. Topography and rational thinking may also be forces at play here. Another unsatisfactory approach in studying the development of towns and cities is to focus one-sidedly on the rulers' need for manifestation of power, as the deeper explanation of the physical structure, over-looking the other classes and their use of urban space.

There are some researchers though, who have been interested in the everyday life of towns and cities, (Paulsson 1950). They study how family and housing patterns have developed with changes in the relations of production, and attempt to find a deeper understanding of the polarity of public life and private life. As stated, the notion of the urban is indissolubly connected to the polarity of public life and private life, and it has to be seen as a result of political-economical conditions of life, as expressed in the physical structure of towns and cities. According to Max Weber and many other researchers, the difference between a town and a village is the existence, or non-existence, of public life; public life being impossible in villages, with their web of closely - knit and fully - integrated relationships between the people living there. In 'Die moderne Grosstadt' Hans Paul Bahrdt has given an outstanding analysis of public and private life (Bahrdt, 1961). The development of public space and private space in towns and cities can vary somewhat, depending on the location within Europe. The development of the single town or city is also in many ways unique in its details, depending on local circumstances like topography, political development etc. In spite of this, there are a number of development traits regarding public space and private space that are similar, making it possible to outline a general pattern of development.

The market-place with its public space is the origin of towns and cities

The phenomenon we understand as a town or city, as opposed to a village, emerged around the market-place. The urban market-place, e.g. the urban square, was the first public space, and in northern Europe the market-place was the real reason of existence for towns and cities. When markets were extended from a short season to the whole year, this provided a basis to settle down at the place and develop a market-town. In Scandinavia the market emerged in the Middle Ages (Figure 17.1.) as either a seasonal market, e.g. the large international herring market at Skanør in southern Sweden, or as a more permanent market placed at the crossings of the King's army roads or along the coast protected by fortifications and the King's army. Here both local and international trade could take place. These permanent markets were given the privilege of the Crown and the foundation of towns and cities from the twelfth to the mid-nineteenth century depended on this privilege. The market was placed on the King's land and the Crown therefore became a landowner in the market-towns (Figure 17.2.). These privileges meant a monopoly over trade. No trade was allowed outside the privileged markets and towns. In the beginning the market-place itself - the square - was clearly demarcated and fully controlled and no trade was allowed to take place outside of it. Later trade also spread to the major streets inside the town. The monopoly conferred the sole and exclusive right over trade and crafts within a distance of about 15 kilometres. This kind of monopoly was not abolished in Denmark until 1857.

Figure 17.1. *The market-towns ('købstæder') around 1536 in Denmark*

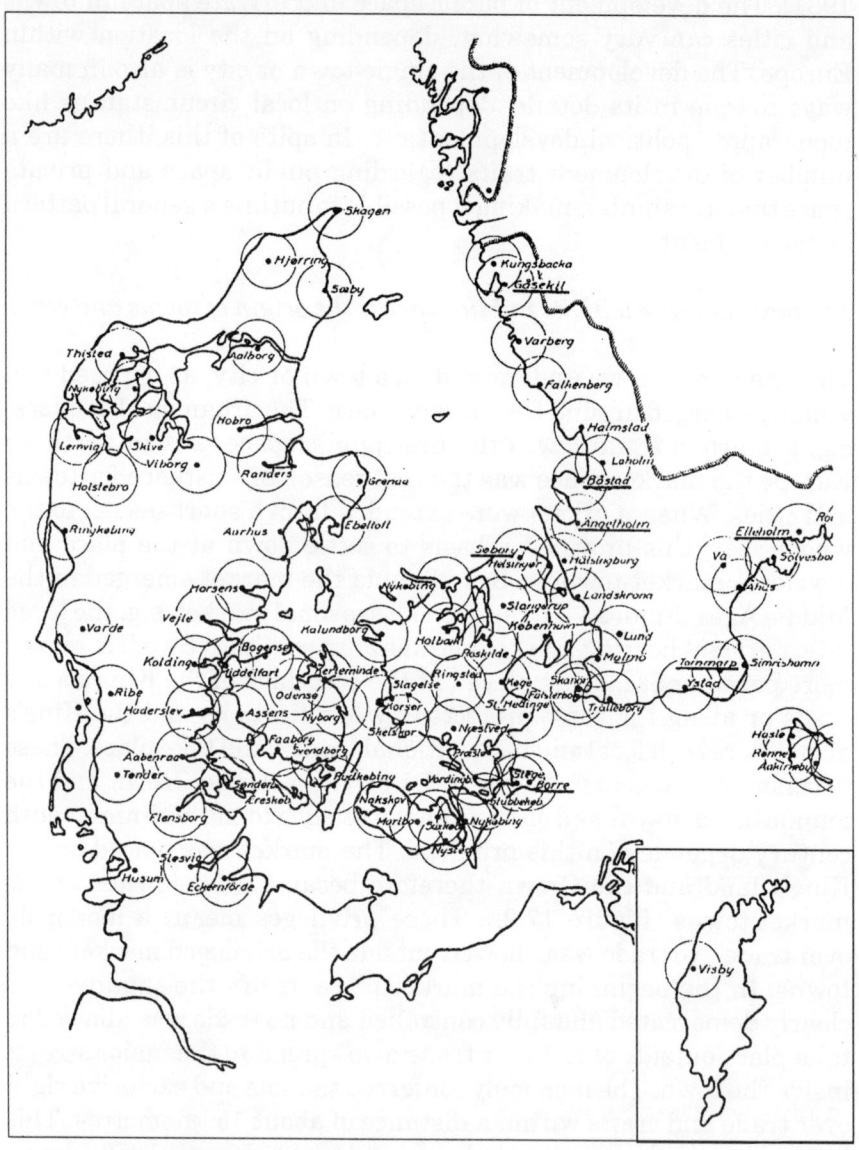

The circles represents a radius of 15 km - the distance a farmer could travel to the market and back in a day. (Lorenzen 1947-58)

Figure 17.2. *The thirteenth century town*

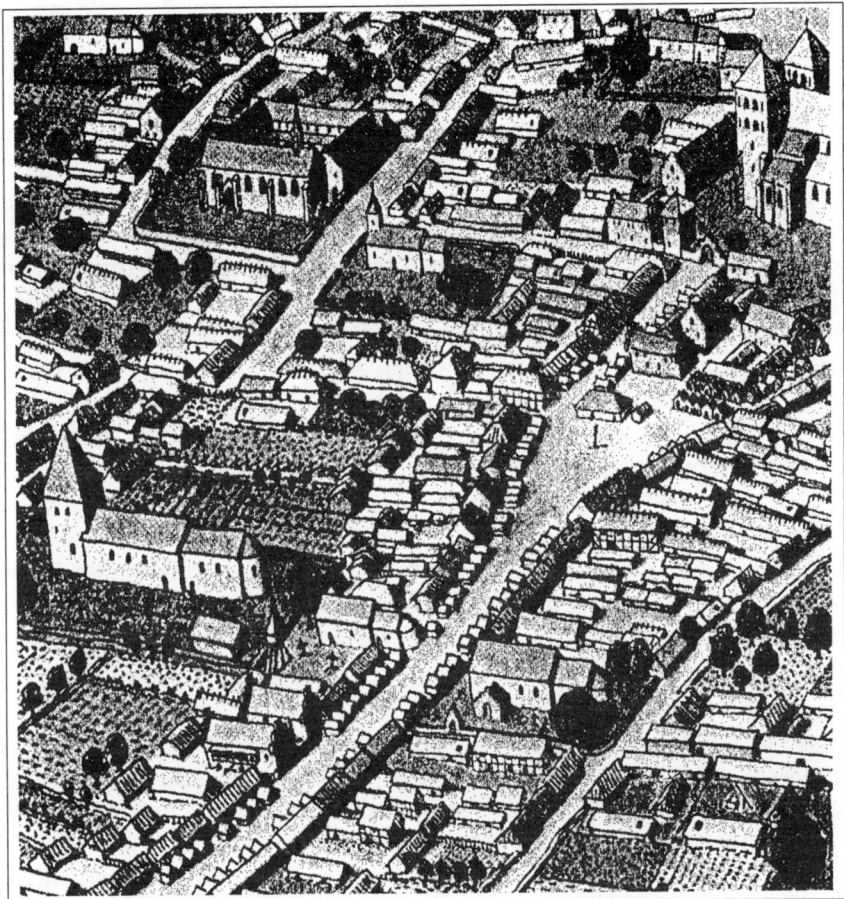

Ragnar Blomqvist's reconstruction of Lund in Sweden. In the top right corner is the dome, in front of it the partly built up square and the long, wide market street with booths in front of the houses. (Blomqvist 1951)

The market-town privileges were given to settlements which had 'rights and freedom' over other settlements. The market-town had its own government with a mayor and a council and it had its own jurisdiction. The price for the privileges were taxes and other duties to the King. The phenomenon is also expressed in our language. The Scandinavian word for this kind of town is *købstad* - a place to buy

341

things - that is a trading town or a market-town. From a linguistic point of view the words market - *marked* in Danish and *købstad* i.e. market-town or trading town originally meant the same. The Latin *mercatus* also means both trade and market-place. The market emerged at a time when society was based on farming villages, where production was organised on the basis of self-sufficient extended households or large families in a closed or rigid feudal system. By closed system, we mean a social order where almost all social relations - including the most intimate - took place within a tight web of personal relations, with a chain of bonds which determined how contacts should take place. In the late Middle Ages, a division of power between the feudal landlords, the Church and the King created possibilities for development. Production and trade grew: the market, serving the surrounding rural areas, started to become important, and the self-sufficient farming households slowly started to change. While the social system in the villages remained feudal, the social system in the market-towns began to break away from feudalism. There has been debate among researchers as to whether market-towns developed from rural villages. This does not seem to be the case - at least not in Denmark. Market-towns developed independently of villages and for external reasons, especially on account of long-distance trade and its demand for protected market-places. (Kjersgaard, 1985).

Max Weber also puts it very distinctly and clearly: the city is a market place. 'Thus, we wish to speak of a 'city' only in cases where the local inhabitants satisfy an economically substantial part of their daily wants in the local market, and to an essential extent by products which the local population and that of the immediate hinterland produced for sale in the market, or acquired in other ways. In the meaning employed here the 'city' is a market place. The local market forms the economic centre of the colony, in which, due to the specialisation in economic products, both the non-urban population and urbanities satisfy their wants for articles of trade and commerce.' (Weber, 1958).

The urban market changes the relations between people and creates an open, dynamic setting, although regulated by law and habits of conduct. Participants in the market-place are free to buy or sell. The market at the same time creates individuals who are disconnected to each other. As individuals they are free to make contacts, agreements and other choices of their own will - even with people of other classes than their own. The contacts are often casual, and the participants often strangers to each other. The result of this activity is not self-evidently predictable - many different things can happen and do happen (Figure 17.3.). A characteristic of the market as such is incomplete social integration. The phenomenon of public life in the market-town is based on this incomplete social integration. Public life emerges not only as an opening in an otherwise closed system. Public life emerges in a situation where communication and agreement exits in spite of differences. Although people are different, communication and agreement is possible in a public setting, with the help of representative manners and behaviour, e.g. social conventions, special clothing and special forms of social gathering. Within certain limits it is possible, with the help of representative behaviour and appearance, to build bridges between different parties. As J. J. Rousseau said: 'Houses make a town, but citizens make a city'. Besides the economic and social aspects of public life - focused around the market and its conventions and social manners - there are also the political aspects - based on participation in the government of the town. The latter has been recently discussed by Habermas, Sennet and others, but has to be left aside in this paper, which concentrates on outdoor public spaces in relation to outdoor and indoor private spaces (Habermas, 1968 and Sennet, 1977).

Public life and public space gives birth to private life and private space

The development of a public sphere creates possibilities for the development of a private life sphere. With the development of public life, it became possible to protect and hide the emotional, intimate aspects of life. These aspects of life also created possibilities of development which did not exist in the closed social order of complete integration. The breaking up of the earlier extended and largely self-sufficient family in favour of smaller, market-based families also have to be seen in this perspective. The foundations for the two-generation bourgeois family were laid. The latter was subsequently transformed into the urban middle-class family, with civil rights and a private life. Private life can be seen as the claim of individuals and groups to determine for themselves when, how, and to what extent information about them is communicated to others. This definition

Figure 17.3. *The development of open air marketplaces into towns*

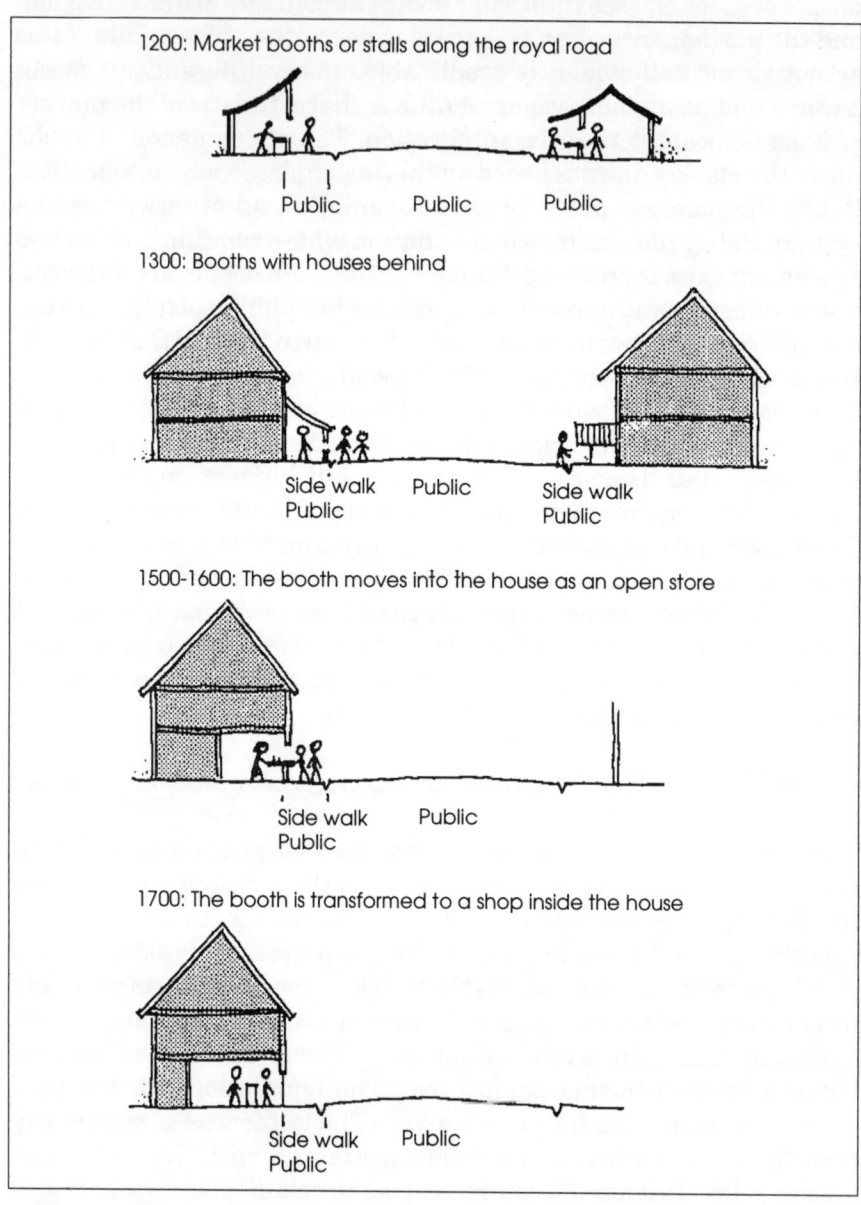

1200: Market booths or stalls along the royal road

Public   Public   Public

1300: Booths with houses behind

Side walk   Public   Side walk
Public   Public

1500-1600: The booth moves into the house as an open store

Side walk   Public
Public

1700: The booth is transformed to a shop inside the house

Side walk   Public
Public

Drawing by the author

344

first developed among the upper bourgeois in the mid-eighteenth century, and private life became almost identical with family-life. Private life takes place within private spaces which are defined by the possibility for a person or a group to decide on the penetrability of the bounds between themselves and others. In this setting it is possible to choose when it is time to be alone, and when it is time to be together with others. Earlier there were no intimate areas in the dwelling; not even the bed was an intimate place. The bed often had to be shared with others, and was also sometimes, for important people, a place to receive guests. As late as the early eighteenth century, not even the upper class, nor the rulers had areas of intimacy inside their dwellings. The dwellings were not subdivided into public and private parts. Life inside the dwelling took place without any spatial separation with regard to intimacy. Even in situations where the household had many rooms, the master of the house lived close together with all members of the family, the servants, the journeymen and other employees. As mentioned, even the most important intimate piece of furniture today - the bed - was not a private intimate thing. It was used by the whole family. In most places the bed was built as an alcove with a curtain. Inside the curtain the master slept with his wife and children. Beside the alcove the employees slept on benches or in settles.

Slowly, a differentiation of the dwelling into more private, intimate rooms and more public, socialising rooms took place. Among the upper class this process started in the mid-eighteenth century, and in the lower middle-class it began in the mid-nineteenth century. At that time private life had a 'facade' of publicly-oriented rooms towards the street, and intimate private rooms towards the courtyard. In the working class this differentiation did not take place until the beginning of the twentieth century. Today among farmers in rural areas, a rigid division of public and private spaces inside the dwelling is still quite usual. The whole family live their daily life together in one room, while they also have a representative room, that is only used for festive occasions, weddings and funerals. The differentiation of public and private spaces within the dwelling happened at different points in time, depending on which class or social group the family belonged to. The divisions of private activity and space into rooms with a different degree of intimacy within the dwelling have had a great influence on the division of public and private life and space in the towns and cities outside the dwelling, and on the urban structure as a whole. The division of activities and space into public and private

345

spheres takes place at several levels in the urban context, and consequently there are public and private spaces at different levels. We can talk of a hierarchy of publicness and privateness. This is valid all the way from the town or city as a whole to the single dwelling.

The urban block is the fundamental unit of towns and cities

The urban block, facing four streets in four directions, is the basic building block of towns and cities. Its primary role is to create two kinds of space, two different worlds, closely connected but clearly separated from each other. On the outside it creates public space in the form of the street and the square (with their public buildings and their facades of town houses). On the inside are the dwellings behind the facades, with their courtyards and gardens. The privacy of the individual unit is protected by individual entrances that can only be reached from public space. In this way the public space of streets and squares functions as a connection between the private units which together make up the urban block.

The importance of intermediary zones or transition zones

The polarity between public and private life does not only take place in clearly defined zones that are either completely public or completely private. There are transition zones or intermediary zones, and these are very important for urban life. It is in the intermediary zones that the individual or group can decide and express the level of publicness they want - on the one hand their wish for contact, communication and social gathering with others, on the other hand their wish to maintain privacy and control. Intermediary zones or transition zones therefore play important roles in the development and maintenance of individual and group identity.

There are different kinds of intermediary zones that are more or less public, more or less private, and with different kinds of control. For practical reasons - to avoid having too many categories and definitions - the intermediary zones can be divided into a single sub-polarity within the greater polarity of public and private zones. This is the polarity of semi-public and semi-private zones. This differentiation of the transition zones into semi-public and semi-private is not absolute, either in time, or at different levels of the urban context. It has to be seen in the light of the specific situation in a temporal and spatial context. What is semi-public and what is semi-private is not always and everywhere the same (and neither is public and private space). However the categories semi-public and semi-private always denote a step-wise transition from public space to private space. The important role of intermediary zones between public space and private space is nearly forgotten in the modern city, and this is one of the reasons for this article. As we have stated above, public life and private life have a dialectical (interactive) relationship with each other, and presuppose each other. As each of them, and the relationships between them, have changed over time, the shape of towns and cities have also changed.

Figure 17.4. *Copenhagen 1674 according to Resen's map*

Although this map is not everywhere true to reality (it shows a Copenhagen that is further developed than was actually the case and it does not have all the details) it is possible to have an impression of a seventeenth-century Danish city. The city was made up of urban blocks with small houses and with the interior of the blocks only developed to a small extent. There were lots of gardens and only a few buildings inside the block. The courtyards were semi-private and the gardens private.

The pre-industrial town and city

From around 1600 to the late eighteenth century, building patterns, including streets, facades and intermediary zones, underwent some very important changes although today they may seem small compared to what happened from the 1850s onwards.

The seventeenth century

a. Economy and work.Small-scale artisan production organised around guilds, sometimes with streets of their own, formed the economic basis of towns and cities together with trade. Merchants with commercial houses did overseas trade besides local trade in the market-place.

b. Family and dwelling. Dwelling and place of work were on the same lot, and not very much differentiated. There were large extended families that also included different kinds of employees.

c. Building pattern. The buildings were placed along the streets, normally taking up the whole width of the building lots. The entrances to the buildings faced the streets but there were also gateways to courtyards behind (Figure 17.4.). Each separate piece of land, that is each lot, was an expression of an economic unit of production consisting of the master, servants and artisans, and besides the dwelling there were also workshops, shops and store rooms.

d. The street. There was a public lane in the middle, with open drains on each side. Between the drains and the facades there was an area with a width of about 2 meters, and by tradition this belonged to the owner of the house. This area was used for sheds, pigsties and garbage. At the entrance each house had a stoop, that is an area in front of the entrance door, where the people living in the house could sit or stand, and be in contact with life in the street (Figure 17.5.). Often there were benches on each side, while a covered stoop or porch was rare.

e. The facade. The facades of the house had openings that could be closed by wooden shutters. Only when opened, it was possible to look out. Glass was very expensive, and when was it introduced only fixed window frames could be afforded. The windows could not be opened and were difficult to look through. Light entered into the dwelling without draft, but contact with the street was lost. This was the reason why the stoop was the only place where it was possible to have contact with the street from the house. The gateway led to a courtyard around which buildings for production, storage and dwellings were erected. Behind these there were private gardens.

f. The intermediary zones. At this time the street was public in the middle, with a semi-public zone on both sides where the public was not supposed to enter but where it could see everything. Through the gateway a semi-private courtyard was reached which the public had access to for errands. Behind the courtyard was the private garden.

Figure 17.5. *Eighteenth-century Dutch single-family houses in an urban setting. (Newman, 1972)*

The picture shows that a street can have different zones. This can be shown with the change in the pavement which indicates otherwise invisible boundaries deciding and controlling peoples behaviour. In the middle there is a public driving and walking zone. The next zone (here to the right only) is also public, but has another covering. If you stand still here, you will be noticed by the residents of the streets. This zone is defined as semi-public. The zone along the facade, the so called 'stoop', is semi-private. If an outsider stops here, the residents in the house will probably come out and ask if they can help him with anything. Quite close to the house is the stepping stone at the entrance door. This is experienced as a part of the house and therefore a private space. On the other side of the house is a private garden.

a. Economy and work. The conditions were basically the same as in the seventeenth century, but some enterprises operated on a larger scale.

Figure 17.6 *Copenhagen 1757 according to Gedde's map*

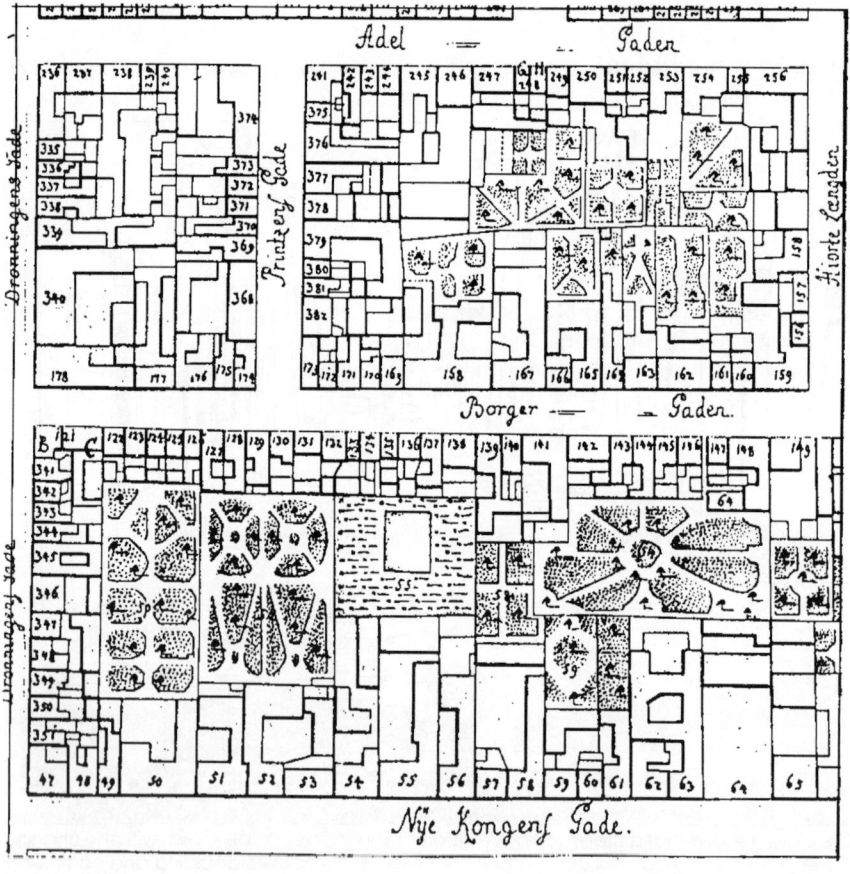

There are still gardens within the urban blocks, but these are disappearing as more buildings are erected on each lot inside the block. The courtyards inside the block were semi-private and the remaining gardens private - as in the seventeenth century.

b. Family and dwelling. With the arrival of the bourgeois family, the dwelling in the late eighteen hundreds started to be divided into a more public part facing the street where guests were received, and a more intimate part facing the back of the building, for love life and other intimate activities. This division of the dwelling into different parts, with varying degrees of intimacy, slowly spread to other classes, but it took more than a hundred years before it become general.

c. The building pattern. The courtyard behind the house along the street was still the frame of the economic unit of production and trade. The private gardens of the seventeenth century were often abolished in order to erect new buildings. The number of buildings on each single building lot in the urban block was increased. In this way each lot came to consist of a network of buildings and courtyards (Figure 17.6.).

d. The street. Traffic grew in the small streets, and it became difficult for pedestrians and carriages to avoid coming into conflict with each other. In the late eighteenth century it was therefore forbidden to use the area between the drain and the facade for private purposes. The *trottoir*, or sidewalk, was introduced. This was reserved for pedestrians. Different kinds of stoops and porches were allowed, though, at entrances to basements and shops as well as stepping-stones in front of the houses - as long as they did not reach too far out to impede pedestrians from passing by. The zone that could be used by the owner of the house was confined to a small strip under the eaves of the roof. The street also changed its character in this century in another way, with the introduction of glass windows that could be opened.

e. The facade. The price of glass fell during the eighteenth century in such a way that by the end of the century it was also used in the more humble houses. It was possible to open the windows and to come into contact with the life in the street. The stoop lost its importance in Scandinavia, while it was kept in Holland all the way up to the twentieth century.

f. The intermediary zones. The street became completely public from facade to facade. The gateway led to a series of semi-private courtyards. The differentiation of outdoor areas inside the block was reduced with the disappearance of private gardens (Figure 17.7.).

Figure 17.7 *Gråbrødretorv in Copenhagen in the early eighteenth century*

The picture shows the semi-private stoop along the facade with porches and stoop (Danish 'bislag') into the basement. These porches disappeared in the eighteenth century because of increased traffic.

The early industrial town and city

The period 1850-1900

a. Economy and work. In northern Europe pre-industrialism changed to industrialism in the period 1840-1870. The market privileges of towns and cities were at the same time abolished (around 1850). Due to industrial production, small-scale artisan production was no longer the dominating form of production in towns and cities. The majority of the population became wage earners who sold their labour on the labour market. This also meant that most people ceased to be members of a joint-production and housing fellowship, and to be subject to the obligations and security this involved. Differing from the market for goods, the market for labour did not have designated public spaces. The labour market was invisible in terms of space (with the exception of a few small offices for employment exchange). But the labour market at the same time created a new kind of space: the commuter area. For the worker, life in the industrial enterprise did not involve any contact with the public life sphere, nor with private life. This was the contrary of the work situation in pre-industrial towns and cities. Here working people, as part of their work, had many direct contacts both with the public world (e.g. the market) and the private world (as a member of a large household based on production and/or exchange of goods). How can we best characterise this new life sphere of wage earner life? It is another life sphere because for the number of hours that the worker has sold his labour, he is no longer free. It is no longer his time, therefore it can neither be public, nor private.

b. Family and dwelling. Industrialisation resulted in the spatial separation of dwelling and place of work. Most of the urban area changed from being a spatial unity of workshops and dwelling rooms to being clearly dominated by dwelling purposes (Figure 17.8.). Life in the home became reproduction of the work force, and leisure-time activities became more and more important as time passed. The urban block changed into a block of the apartment buildings and rented apartments. The density of the built-up fabric increased several times, through speculation for profit, and created housing conditions that were unhealthy and humiliating. As the single unit of the apartment building and the courtyard was still small, and the single lots still clearly separated from each other by walls and fences, there was still a clear division of public and private space (Figure 17.9.). The small scale of the units also meant that everybody knew each other within the unit - and also knew who did not belong there. There was a strong social network formed by common need. Private life within such a confined and walled-in unit gave the residents a sense of belonging, a part of their identity, and the possibility of doing things in an environment with a high degree of security.

c. Use of space in leisure time. Besides the time the wage earner employs for transportation to and from work, where he uses public space, the use of public space is also possible during leisure time. His relationship with public life, and contact with other people in public space, becomes a matter of free personal choice and without obligation (i.e. non-committal). The spatial separation of dwelling and work means therefore that both private life and public life assume a leisure-time character. It has to be remembered, though, that leisure time was very short in the nineteenth century. Not until the victory of the 8-hour work day around World War I, did

Figure 17.8. *Lower middle-class and working-class dwelling at the end of the nineteenth century in Sweden.*

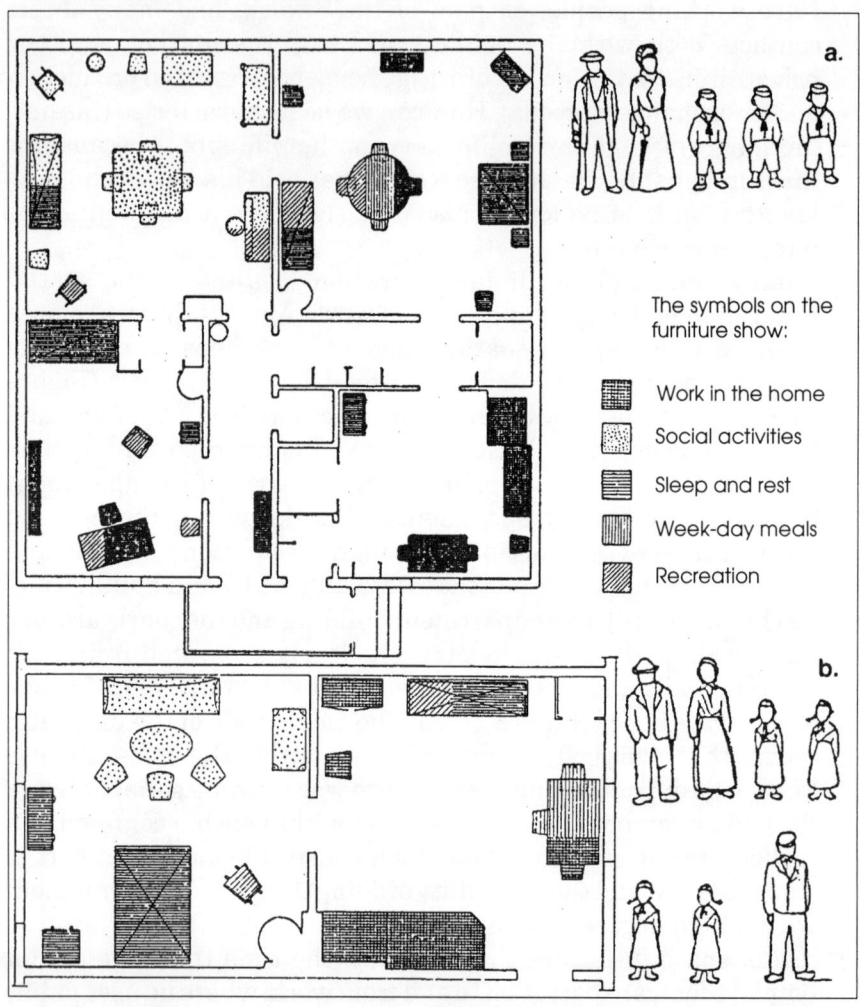

The symbols on the furniture show:

Work in the home

Social activities

Sleep and rest

Week-day meals

Recreation

Source: Paulsson, 1950.
(a) The lower middle-class home with four rooms inhabited by husband, wife and average three children. There was a division of the rooms into social and private spaces, although not as well developed as in the homes of the upper middle-class at that time. (b) The working class home (1876) still had no differentiation of rooms into social and private. The space and number of rooms were quite small - often only one room - so a differentiation was not possible. This flat was inhabited by a husband, wife, four children and a lodger. It was not until the social housing developments in the 1930, that it became possible for the working class in general to divide the dwelling clearly into social and private rooms.

354

the modern concept of leisure time become relevant for the majority of the population.

d. Building pattern. The building pattern was still the urban block subdivided into rather small lots. The built-up density in each lot increased further, with several small courtyards behind each other, often very dark. Apartment buildings were erected on the inside of the lot as well as along the streets.

e. The street. Traffic in the streets grew as a result of the spatial separation of dwellings and places of work, and the number of dwellings and the urban area within the city increased, as a result of the growth of the population. Much of the traffic had no relation to the people living in the particular street. The public space of the street changed further into a traffic channel for vehicles. The street had a public character from one facade to another.

Figure 17.9. *An urban block in Gammelholm, Copenhagen as built around 1870*

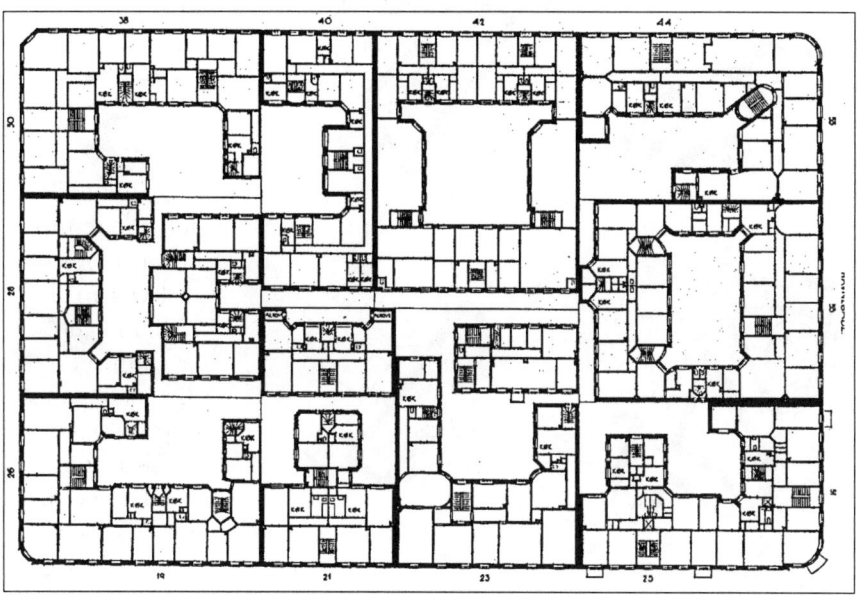

The blocks with apartments for renting built at the beginning of the industrial period were very densely built up with small, dark courtyards and a lot of apartments buildings inside the block. As it was necessary to pass the courtyards to enter the flats inside the block occupied by independent households, the courtyard changed into a semi-public space. The facades towards the street still represented the single lot within the block and there was plenty of architectural detail to give individuality.

f. The facade. Buildings became 2-4 storeys higher, which meant less contact from the windows with life in the street (and vice versa). The facades were still divided into small building units. They corresponded to small separate building lots within the larger urban block, and were built in classical or other historical architectural styles which gave the facades of the different buildings a separate identity. In this way the facades signalled the division of space into public and private.

g. The intermediary zones. The street no longer had any intermediary zones (Figure 17.10.). As the courtyards became a distribution area for visitors coming from outside and going to separate apartments they changed their character from being semi-private to being semi-public.

Figure 17.10. *The development of courtyards in urban blocks in Copenhagen around 1920*

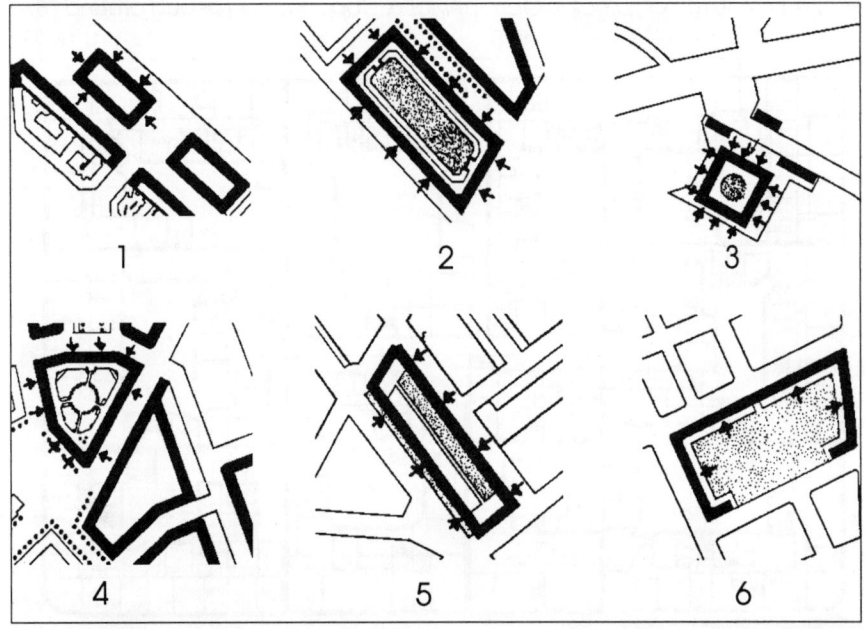

Scale 1:7000

(→) Entrance. (1) Struensegade shows the change from the densely built up urban blocks of the late nineteenth century. Here there are no apartment buildings in the courtyards any more and there is only one courtyard, which is semi-private. (2) Hornbækhus shows a very large block with a garden inside - this block is so big that the courtyard becomes semi-public. (3) Vilh. Thomsen allé is a mixed block with both private, semi-private and semi-public spaces inside it. (4) Borups allé is a large block with a park inside and the entrances from the inside of the block. Here the inside of the block is public and semi-public. (5) Grønnegården. (6) Ved Classens have.

Around 1920 - The development of urban blocks with large courtyards

a. Economy, work and social conditions. Economic growth, workers' building-cooperatives and - at the end of the period - public housing, changed conditions for the building of dwellings.

b. Family and dwelling. The lower classes took over the middle-class habit of having apartments facing the street, although the flats were smaller.

Figure 17.11. Hornbœkhus at Borup's Allé in Copenhagen, built in 1922

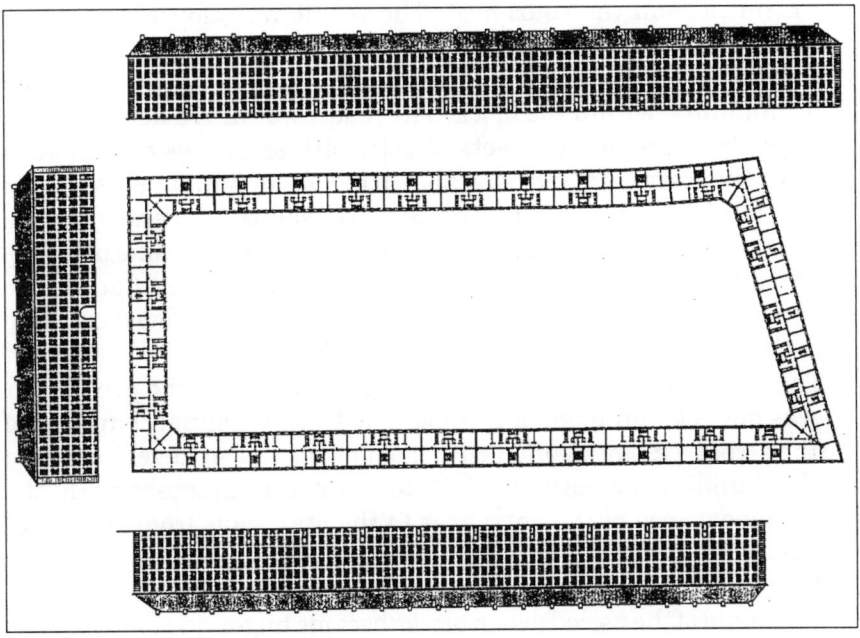

Here the urban block has become extremely large with a fenced park-like garden to look at inside. 300 flats surround the courtyard, and at the beginning more than 1,000 people lived there - many more than it is possible to know. The inside of the block therefore becomes semi-public. The major entrances to the flats are from the street side, but there are also secondary staircases on the courtyard side, giving access to the inside of the block and two gates to the street. The street facades have less ornamentation than before but there is still classical detail, although the general impression starts to get somewhat monotonous.

c. Building pattern. The housing layout continued to be based on the urban block surrounded by streets until the end of the early industrial period in the 1920s. But there was no construction of apartment-buildings in the courtyards inside the block any more, and no fences dividing the inner part of the block into separate courtyards. Light, air and sun were the objective. The courtyard became a common area serving the whole block (Figure 17.11.). At first this development took place in urban blocks of about the same size as before, and the relation between the perimeter-buildings and the street was also the same. Around 1920 the courtyard in the new developments became a garden and - as the size of the urban blocks grew - a park. At first, the relation between the perimeter buildings and the streets did not change. But very soon the entrance to the stairway was moved to the inside of the block, with entrance from the courtyard. The result was that the building turned its backs onto the rest of the city. The purpose was to create a quiet and peaceful environment, an oasis with focus on the 'community' within the block. The result was also a deterioration of public space in the streets. Blocks with entrances to stairways from the courtyard had been known before for buildings that were erected inside the block. The novelty was the inversion of the entrances in the buildings along the streets. This became possible, when the earlier system with two staircases to each flat for fire protection reasons was set aside as a result of improved building technology.

d. The street. The traffic continued to grow although the private car was not yet a problem in Europe. The traffic channel character of the street was intensified with each new stage of development of the building pattern until it took over completely with the disappearance of the entrances to the stairways from the street side of the buildings.

e. The facades. The facades still had ornaments and a personality, but at the end of the period urban blocks became huge and monotonously repetitive. The building of medium-height buildings (4-6 storeys) continued. At the upper levels the contact with the street through windows were weak. Balconies and terraces were still quite rare.

f. The intermediary zones. The new 'common' courtyard for the block was at first semi-private, as the courtyards and the number of dwellings in the blocks were rather small, and the entrances to the stairways were from the outside. When the urban block grew in size and the courtyard became a park, it was possible to know only a small proportion of the people living there. As a result the inside of the block became semi-public. Finally, when the entrances were moved to the inside of the block, the inner space became almost completely public, although the degree of publicness was to some extent modified by the size of the block (Figure 17.12.).

Figure 17.12. *The urban block Solgården (The Sun Yard), Copenhagen, built 1929-31*

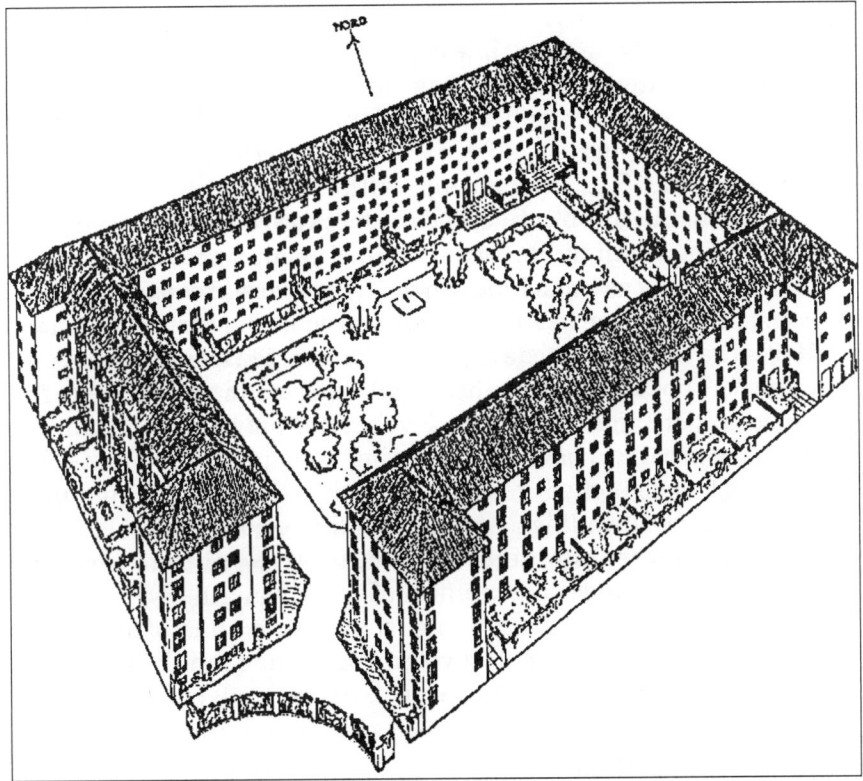

As a result of the criticism against densely built-up urban blocks, the large courtyard type of urban block developed with a single courtyard without apartment buildings inside the block. As the main entrances were to the courtyard, this acquired a semi-public character. There were about 150 flats around this courtyard built as a single development.

The modern industrial town and city 1920-70

a. Economy, work, social and and political conditions. With the 1930s societies in Scandinavia began to change into welfare states, where governments started to guarantee a minimum level of social security and other welfare benefits, e.g. the supply of housing for the lower classes. The co-operative and public housing companies grew in strength as did the centralisation of decisions and administration of housing. Mass production of housing was an important political goal, but it was with the rapid economic growth, starting in the late 1950s, that mass production of housing first became a reality, industrialised building methods being part of the picture from that time on.

b. Family and dwelling. Dwelling units became standardised for reasons of equality and rational production. The provision of privacy in the dwellings for each nuclear family was the major housing goal. Healthy housing, with a lot of air and sunlight, continued to be an objective, and modern conveniences in the home became increasingly important. Functionalist architects looked upon the home as a machine to live in, the mass-produced machine being an ideal even before it became possible in practice. From the 1950s the capitalist mode of production created hitherto unknown wealth. The majority of the population could afford means of consumption high above the level of minimum existence. The dwelling became a part of leisure-time consumption benefits. The economic development made it possible to have larger and more well-equipped dwellings. At the end of the period, around 1970, the average flat was quite large compared to 50 years before, about 80-100 sq. meters, and every member of a family normally had a room of their own and a separate living room. A larger part of private life took place inside the single dwelling units, and family life started to become isolated from other aspects of life. Modern affluence and consumption goods increased the isolation of individual families; when each dwelling had its own water, bathroom, refrigerator, TV etc, it was no longer necessary to have contact with neighbours. Leisure time outside the dwelling became institutionalised in the form of clubs and other organisations. Social life in the neighbourhood was normally confined with only a few contacts to other households living in the same stairway.

c. Building pattern. The urban block, with buildings around the perimeter disappeared giving way to linear blocks of flats laid out in a parallel pattern (Figure 17.13.). This happened all over northern Europe. The linear block was theoretically of infinite length. There were two reasons for this building pattern: equality, especially concerning sunlight, and rational building construction suitable for industrial building methods, e.g. the use of building cranes on linear tracks at the building site. The major change in the building pattern from the beginning of the period around 1930 to the end of the period around 1970 was that the single building unit grew in size. Both the length and the height of the single block, and the size of the housing estate, increased to a large extent.

d. The street. The traditional street between urban blocks disappeared (Figure 17.14.). The 'street' even disappeared as a word in the new developments and was replaced by the word 'way', meaning 'road'. This expressed the desire to get rid of the concept of the traditional urban street. At the same time the open housing estates were typically called 'park-developments'. The transformation of the street into a traffic channel was completed around 1960, with the advent of mass ownership of private cars in northern Europe.

e. The facades. Facades had balconies - one for each housing unit. At the same time they were stripped of the traditional historical ornamental clothing and became more anonymous, especially in the 1960s, as estates and single blocks grew in size. With high-rise apartment blocks and towers, contact through the windows from the apartment to the ground level disappeared almost completely.

f. Transition zones. With modern urban planning residential areas and apartment blocks now had vast areas of free space around the building in the form of lawns. In northern Europe, this space is also called 'common space' or 'community space'. But these spaces were neither public nor private from the point of view of their use; rather they have to be characterised as 'no man's land'. The spatial result of the building pattern was a uniform area of 'public' outdoor space. There was no semi-private or semi-public outdoor space any more. On one side of the block (to the east or sometimes to the north) was the entrance road or walkway, with entrances to the stairways. On the other facade of the block, the living-rooms - normally with balconies - faced the sun and an outdoor common recreational space towards the west (or sometimes the south). This space could only be reached by walking around the single block. As a result of this uniform repetitive approach all of the outdoor space from the

Figure 17.13. *Ryparken, Copenhagen, 1931. Three storey public housing.*

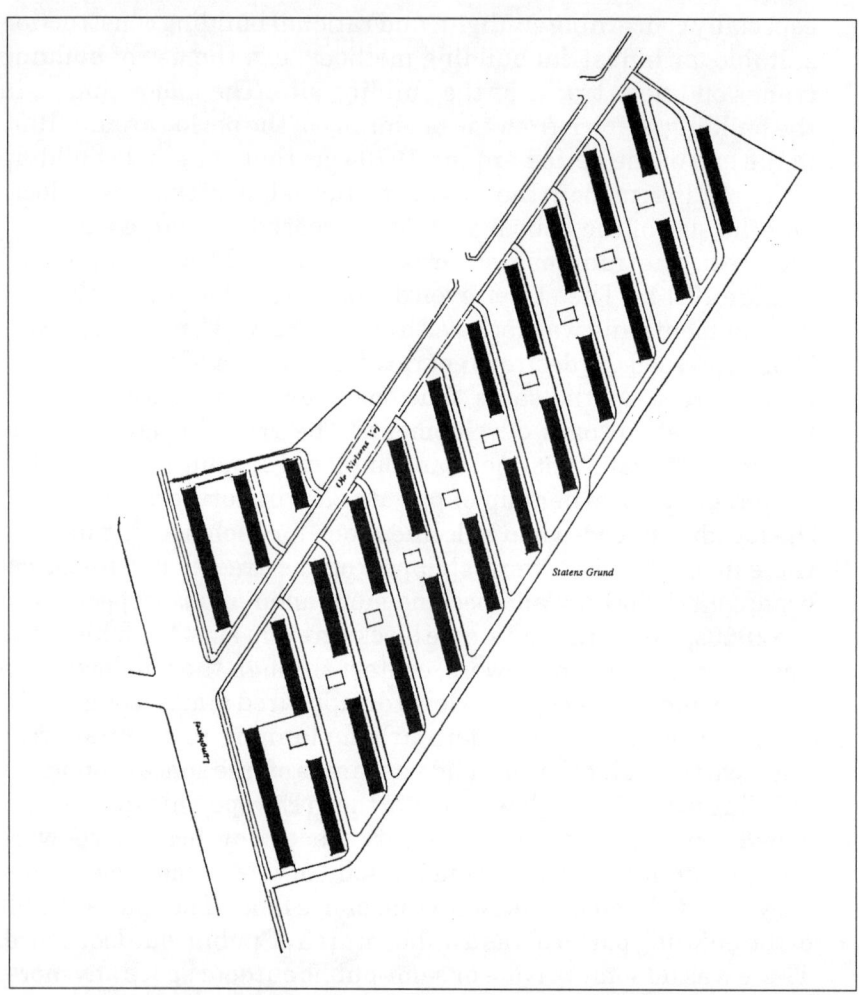

The estate is built up of parallel apartment blocks in a park-like setting. Entrance road and entrances are to the east of each building, lawns to the west. The free distance between the buildings is a minimum of 26,5 meters. The plan reflects the improvements in the single dwelling, all having sunlight and fresh air. The outdoor area was one great common green space. The earlier neighbourhood around a semi-private courtyard was looked upon as petty-bourgeois and outdated. In reality a private dwelling of a high standard was created, while the outdoor areas became too public and anonymous.

Figure 17.14. *The development of the street in the period 1600-1970*

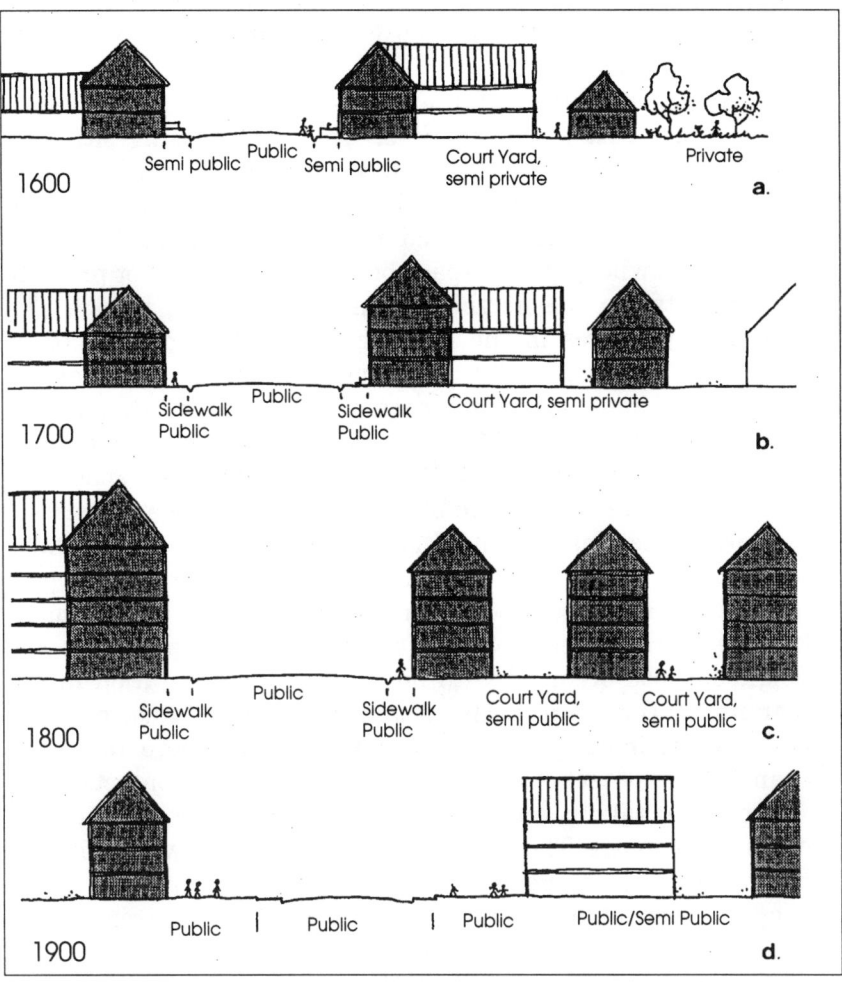

(a) Around 1600. The street is not very wide. There is a public lane in the middle, an open drain on each side and semi-public stoops along the facades. Courtyards are semi-private and gardens private. (b) At the end of the eighteenth century. The street becomes a bit wider and completely public with sidewalks along the facades. Courtyards are still semi-private, a system with courtyards behind courtyards is developing. (c) Around 1870. The all-public street gets wider still and the amount of traffic grows. The buildings become higher and apartment buildings for individual families are erected in the courtyards, which became semi-public. (d) Modern housing development in the twentieth century. The street is very wide and there is a green zone in front of and around the buildings. All this outdoor space, both on the street side and inside the estates, is public or semi-public. (Drawing by the author) point of view of access became 'public space'. Inside the single block the stairways became semi-public. As there were no semi-private space outside the private dwelling units, the entrance door to the single dwelling became the division line between private and public space. There was no intermediary zone at all, or a very weak one (Figures 17.15. and 17.16.).

g. Social consequences of housing development in modern industrialism. At the same time, while the housing situation improved from some points of view, it worsened from other points of view. Above all people lost identification with their housing area, and the building they lived in. The layout of parallel linear blocks, with wholly public space between them, did not promote contact between the neighbours in the adjoining blocks. The 'community' spaces in the housing areas were too vast to be populated, and too empty to be meaningful for a variety of activities. Community space became so 'common', that it was regarded as space owned by the municipality, and it therefore had to be regarded as public space. The so called 'community' space also had to be regarded as a public space, because there were no community feelings about it. This being the real situation, no rules of conduct for the vast public open spaces in the suburbs developed, and with the absence of real use of these spaces, the environment became more insecure. Social contact between people in the neighbourhood occurred with no, or little, consideration in the planning of these areas. This happened partly because equality and freedom in obligations were looked upon as more important than social contacts, and partly because social contacts were supposed to take place in an institutionalised way in a local community centre. The planners had invented a new meaning for neighbourhood - the neighbourhood unit equalling a school and a shopping district. As a result of the development of a tremendous amount of no man's land within the modern city, the concept of public space became severely diluted.

Post-industrialism: new directions since the early 1970s

a. Economy, political and social conditions. Grass-roots have been demanding participation in and the exertion of influence upon a multitude of issues. A desire for more personal involvement has

Figure 17.15. *A suburban public housing area in Copenhagen as built in the 1960s and a proposal for improvements from the late 1980s*

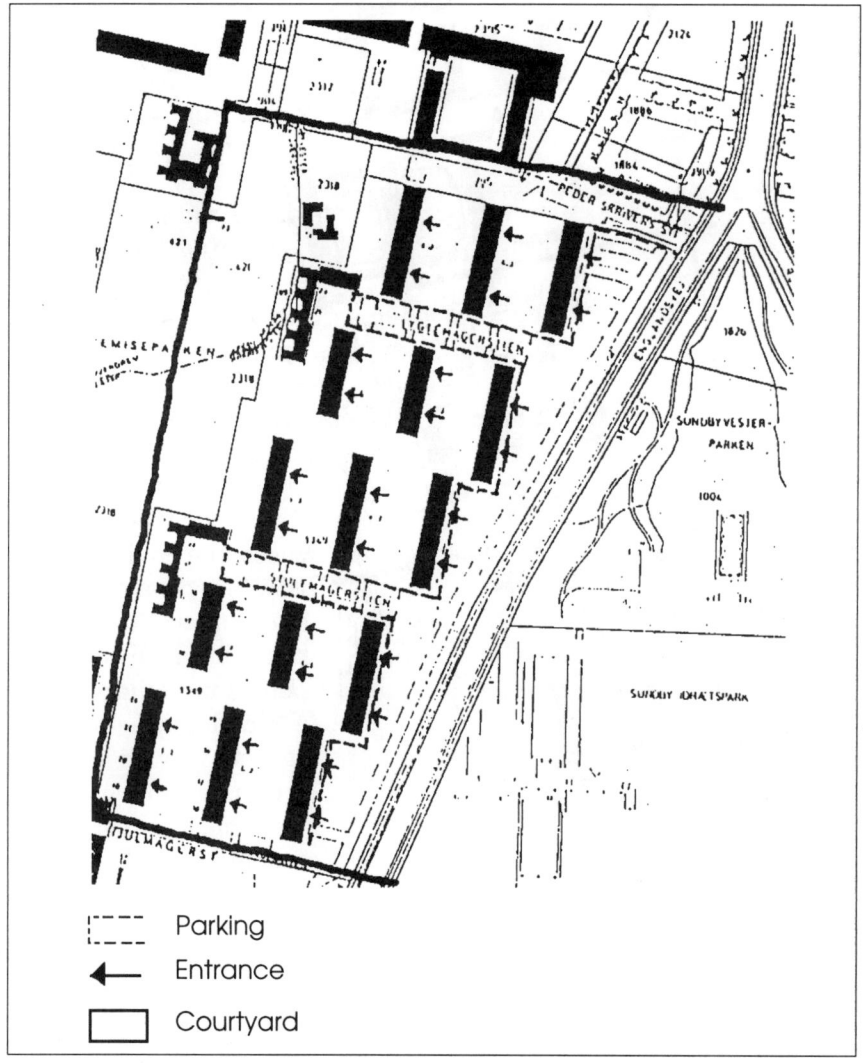

In the 1960s many public housing areas were planned in a stereotype way with parallel blocks place in an almost limitless public space which had the character of a 'no man's land'. The system of entrances were often manifold and unclear.

Figure 17.16. *A suburban public housing area in Copenhagen, proposal for improvements from the late 1980s*

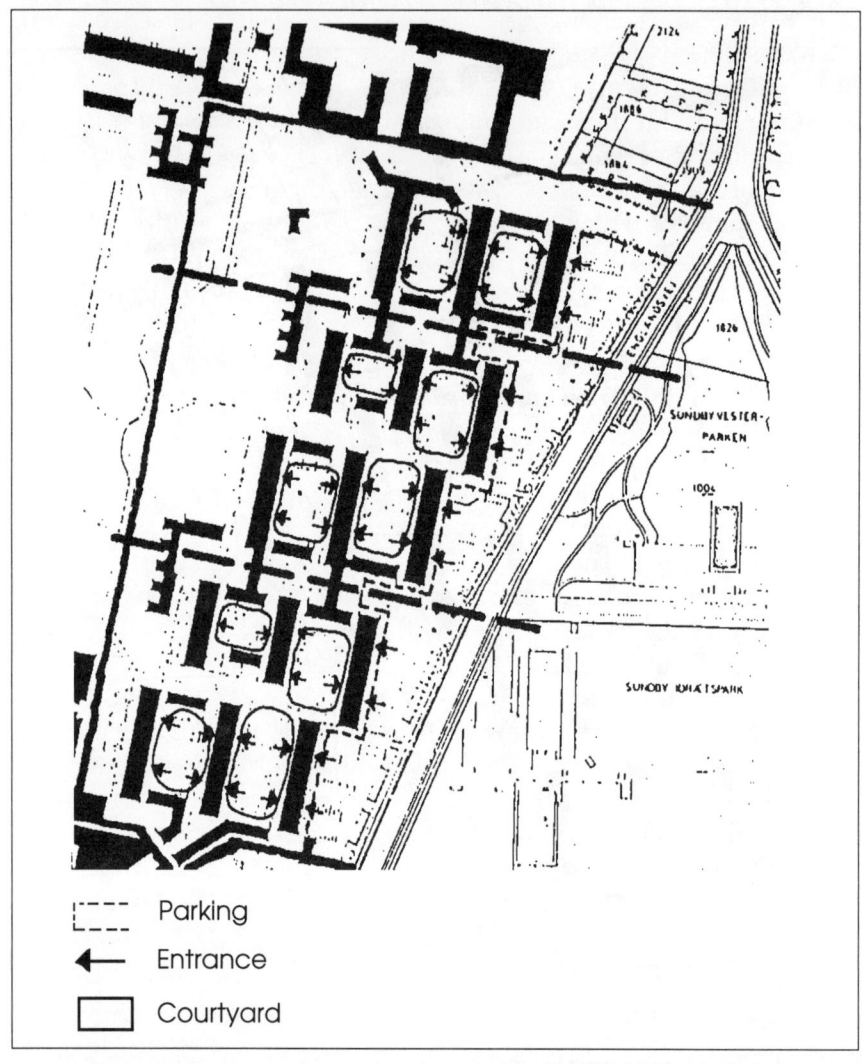

[----] Parking

⟵ Entrance

[ ] Courtyard

The plan shows a proposed reconstruction of the area with a subdivided space, some new supplementary buildings and a rearrangement of the entrances. Here the urban block is recreated. The purpose of the plan is primarily to create outdoor spaces with different qualities and clear transition zones. In this way the area will have both private, semi-private, semi-public and public open spaces. (The plan is prepared by the author) arisen, and a reaction against the institutionalisation of life has occurred. The welfare state has been getting into economic trouble and it becomes difficult to raise more government funds, as the level of taxation is already very high. The welfare state, based on institutions, is changing into a welfare society based on self-management.

b. Family and dwelling. People living in rented dwellings have demanded and have had more influence on the management of their housing areas ('self-management'). There is a growing wish for more and closer contacts with neighbours and for housing developments which strengthen the possibility of identification with the place of dwelling (personal responsibility and self-esteem). Today there also seems to be a change in private life within the dwelling and differentiation between intimate and more public parts is tending to fade away. The dwelling as a whole is now accessible to guests to show who You are, and how You want to present all aspects of Yourself. Even sexual life is not a tabu-subject any more, nor the naked body. The public sphere penetrates into the intimate sphere of private life, as the outer world enters through the media. (The deeper meaning of this, and the possible consequences at the urban level, have yet to be explored.)

c. Building pattern. The traditional urban block is coming back in slightly modified forms, and with them the historically known pattern of public space in the form of streets surrounding small housing estates, groups of apartment buildings or row houses. Inside the block there is one or a couple of courtyards and gardens of a semi-private character. The number of dwelling units in the block and around each courtyard or garden is rather small.

d. The streets. There has been new interest in the street as a public space. Pedestrian streets and pedestrianised areas, which had been introduced at the end of the modern period, have become even more important. But since the mid-1970s the mixed street, with both driving and pedestrian traffic, has again become acceptable, as elaborate street design and regulations of car traffic make it possible to reduce the environmental disadvantages and dangers of cars.

e. The facades. There is a desire to express personality in the facade. The estates are divided into smaller units with variations between them, etc. Small size and individuality is coming back, but in an industrialised way.

f. The intermediary zones. There is a clear intermediary zone in the form of semi-private courtyards or gardens that belong to the people living close to them. The division of outdoor space into more public and more private parts has been reintroduced. In this way people have to pass through public space on the way to other blocks (Figure 17.16.). At the same time they have an outdoor semi-private space close to the dwelling that promotes the feeling of belonging to the place, the use of outdoor space for daily activities and contacts with the neighbours.

g. Social effects. The individual really feels a greater identity with the place, the outdoor spaces are used, the social situation has become comprehensible and the number of contacts with the neighbours has grown.

Public life in town and city centres

Public life in towns and cities started, as we have mentioned, with the market-place. The market-place in the form of a public square still played the most important role in public life all the way up to the nineteenth century, but new forms and spaces for public life were added. In the eighteenth century public life was extended with the emergence of coffee-houses and theatres. Newspapers were allowed and became common at the the beginning of the eighteenth century. With them 'public opinion' was formalised. In England pubs started to play an important role, pub meaning a public place. In the nineteenth century outdoor public space in the city centre started to become relatively less important, as both trade and production changed their character. Workers only had a leisure-time relationship with the city centre, and trade was transferred to new types of buildings instead of taking place outside: the department store and the covered market hall. A lot of other kinds of public spaces also became available: schools, libraries, museums, railway stations, hotels, sports facilities, exhibition areas, etc. In the 1920s, with the diffusion of the radio, it was no longer necessary to go to the city centre to listen to the news and to music, but at the same time new entertainment in the form of the cinema gave new life to the city centre. In both cases public life became still more dominated by one-way mass media.

In the modern industrial period, especially after World War II, urban centres were transformed because of large-scale suburban development, cars, external shopping-centres, chain-stores and TV. The old central urban areas became relatively less important, and the new suburban centres almost completely focused on trade with standardised things in a standardised way. Public life became poor. In the 1950s and 1960s - and to some extent also in our post-industrial period urban centres were also drained of many public functions, as town halls and other public services grew in size and moved out of the city centres. In the late 1960s a reverse trend started to develop that strengthened public life in the central urban areas again. Streets and squares became pedestrianised, and there were restrictions on the use of cars in central areas. In the 1970s and 1980s the number of visitors to the old central urban areas has grown several times, and participation in all kinds of public leisure-time activities has increased, as has the number of more serious cultural activities. Retail trade is now assuming some of the forms of the old mixed urban market-place, with new kinds of food markets differing from supermarkets in the suburbs.

So the last 10 to 20 years have seen a great demand for revitalisation of the use of public spaces in the old urban cores - both of outdoor and indoor public spaces (in streets, squares, restaurants, entertainment, pubs). People from all over the urban and metropolitan areas use these central public urban spaces more than before. In a way it is the old urban market function coming back, with partly a new content and in partly a different form. Some of the reasons for the new popularity of the old urban centres have to do with conduct. Public streets and places and public parks in the old pre-modern city have well-defined rules for conduct, especially in the city centre. There are also forms of public protection here. The rules are rooted in history, tradition and environment. These rules contribute to the attractiveness of these spaces and areas, which are open to anyone, as long as the rules are respected. It is still possible to improve public spaces in the central urban areas in many ways. And so we should, as the urban population in many towns and cities is still deprived of good urban spaces - especially in suburban areas.

Conclusions

The foundation of towns gave origin to the creation of public and private space, but public space today often means a piece of no man's land. At the same time the need for well-functioning public and private spaces has been further strengthened with the development of a work-sphere that is neither public nor private. The vast public spaces in suburban housing areas that have been built since the 1920s have diluted the meaning and use of public, semi-public and semi-private spaces. There are often no rules of conduct in these areas, as they are regarded not as publicly or privately owned, but as a no man's land. What is the reason behind the creation of this no man's land? Why have the planners promoted 'common space', when most people actually want smaller spaces, where groups can create its own rules. There are normally no fixed rules for the use of and behaviour in semi-public and semi-private spaces defined by society at large. It is up to the neighbourhood with its social and community life to define rules for these spaces in a self-management setting. This gives freedom to form different habits in different housing areas and different spatial units that can suit different groups of people, as well as giving towns and cities an increased variety, which in itself can be a quality.

One of the reasons why outdoor spaces became 'public' and anonymous was that the administration of the estates were centralised and distant. People were not asked what they really wanted. Another reason can be found in the political ideologies of the planners who wanted to reduce 'privatization', or even worse, of petty-bourgeois ideas and ways of living. Anyhow, 'community' space was laid out without any real thinking of what makes a community work - of what make people gather around common tasks and activities. Therefore much was built in the period 1920-70 which is not functioning very well.

Today large housing areas have to be rebuilt, the public spaces in these areas have to be reorganised, and rules of behaviour have to be developed by the residents, in order that these areas can become more attractive, safe and usable. The historical overview shows that our understanding and use of public and private spaces have changed with time and with the change of modes of production and reproduction. Every period has reconsidered its perceptions about and its use of public and private spaces and the polarity between them.

References

Allpass, J. (1983), *Demokrati og Planlægning*, København - Oslo.

Allpass, J. (1983), *Miljøforbedring*, København.

Allpass, J. (1984), *Om livet i boligområder, Frygten for vold og hærværk*, København

Bahrdt, H. P. (1961), *Die moderne Grosstadt*, Hamburg.

Blomqvist, R. (1951), *Lunds historie*, Lund.

Habermas, J. (1968), *Strukturwandel der Öffentlichkeit*, Luchterland Verlag.

Kjersgaard, E. (ed.) (1985), *Markedsgade, Markedsbod, Gadebod; Den gamle by, årbog 1985*, Århus.

Lorenzen, V. (1947-58), *Vore byer, studier i bybygning 1-5*, København.

Newman, O. (1972), *Defensible Space*, New York - London.

Pahl, R.E. (1975), *Whose City?*, London.

Paulsson, G. (1950), *Svensk stad, liv och stil i svenska städer under 1800-talet, 1-3*, Stockholm.

Sennet, R. (1977), *The Fall of Public Man*, New York.

Weber, M. (1958), *The City*, New York.

18. INNOVATION AND ECONOMIC TRANSFORMATION OF THE URBAN WATERFRONT: RECENT DEVELOPMENTS AND PROSPECTS FOR THE FUTURE

Armando Montanari, National Research Council (CNR), Naples, Italy.

The relationship between waterfronts, the urban areas of which they are part, and the hinterland upon which they depend, is rather complex and varies according to different cities. It lies in forms of mutual influence which are the result of several geographic and historical as well as economic and social factors. Such a relationship has continually changed over time, according to the production systems and the economic setting of each individual country, and especially in relation to the world economic scenario, the evolution of navigation techniques and ground transportation systems, and the spreading of new cultures and lifestyles. This confined and therefore extremely valuable strip of land lying, and at times compressed, between a metropolitan area and the water is a particularly important region. It has always served as a magnet, absorbing the effects of technological, cultural, economic and social innovations which have been produced worldwide. Identifying the most important development trends of metropolitan port areas will make it possible to anticipate the changes that will characterize the nineties. It will also provide an opportunity to devise forms of greater scientific collaboration among research institutes in the industrialized countries that belong to the OECD, that is those of Europe, the United States and Japan.

Battery Park City in New York, Rokko Island in Kobe, Minato Miray 21 in Yokohama, Ohkawabata Redevelopment Plan in Tokyo, the Docklands in London, the Management Area in Naples and the Olympic Village in Barcelona are among the major redevelopment plans recently implemented in port areas or in connected industrial areas. All of the industrial waterfront areas of the major port cities throughout the world are undergoing redevelopment, but the plans mentioned above envisage huge operations requiring considerable grants, great managerial skills, wholly original cultural approach, distructive social groups and different historical conditions (Ley, 1985, p. 419). It is a matter of understanding whether these examples can provide specific indications as to future developments (Knox, 1987), or whether instead, they are bound to remain isolated experiences. The aim of this article, however, is not to describe a set of case studies. Rather, the intention is to place these experiences within the framework of a critical analysis of the historical invention, innovation and dissemination processes that have characterized port cities over the past century.

Innovative processes in port cities

The different functions performed by port facilities and the evolution of maritime technologies in the last decade have determined the transfer of port activities to areas often located at some distance from towns (Hayut, 1982). This phenomenon has entailed the gradual functional and morphological conversion of the urban waterfront, and is considerably affecting the economic setting and the possibilities for development of port cities and related metropolitan areas (Montanari, 1989). It has thus become essential all over the world to study the ongoing transformations and analyze the possible operating strategies. In any event, it may be stated that the service area of a port is much more restricted than its area of regional influence, within the framework of which it is an element of paramount importance for land planning and urban area management (Norcliffe, 1981). Port cities do remain urban areas and have been affected by the same changes which have influenced the cities of industrialized countries in the past two decades. Recent studies have shown that economic phases are especially essential to urban development (Berg et al., 1982). On the other hand, the size, shape, population density, and location of these towns, along with the date marking the onset of industrialization and therefore urbanization, are less important, as is the economic system of the countries within which these metropolitan areas develop (Muscarà, Soricillo and Vallega, 1982).

The waves of technological invention-innovation processes have surely had an impact in the sector of maritime transportation. Their much-travelled international routes and their dependence on maritime traffic (Mayer, 1973) have caused port cities to be affected by invention-innovation flows to a greater degree and immediately, at a technological (Slack, 1980), cultural and social level. The innovations characterizing the railway system, however, have evolved differently. They were introduced at different times in the various countries, for interconnecting the various national systems was less of a priority. High-speed railway systems were introduced in Japan in the sixties, in France in the eighties, and they are likely to be established in the rest of Europe only after the year 2000. The spreading of innovations in urban transportation was even more differentiated a process (Marchetti, 1988). In the area of transportation by sea, reference may be made to the innovations introduced in the propulsion systems and ship-building materials, which have therefore more directly influenced the development and transformation of port cities (Vigariè, 1981).

In the United States, steamships, initially driven by coal and later by oil, began replacing sailing ships during the nineteenth century and were outnumbering them altogether by the end of the century (U.S. Dept. of Commerce, 1970). At the beginning of the twentieth century, diesel engine-driven ships in turn began taking the place of steam-driven ships, although they are expected to outnumber the latter only after 2010 (Nakicenovic, 1985). This constant transformation trend was also confirmed by several studies investigating the processes replacing the technology applied to naval propulsion, considering the tonnage of the ships registered under the English flag: steamships replaced sailing ships around 1890, and were in turn substituted by diesel-driven ships around 1950, right after the end of the Second World War (Marchetti, 1986, p. 28). Another innovation certainly consisted in the change in the material used to build ships, particularly in view of its impact on the characteristics and dislocation of dockyards. In the second half of the nineteenth century, metal was introduced. Iron was first used, followed by steel, as a new material for ship building, following the new metallurgical technologies ushered in by Bessemer in 1857. In the United States, the ships built with metal outnumbered those built in wood as of the beginning of the twentieth century. Such innovations led to the rapid and indiscriminate growth of port facilities which took up what was left of the unoccupied waterfront regions in proximity of previously builtup areas, to the detriment of the existing urban structures.

Additional technological changes came about in the second half of the twentieth century, which influenced cargo loading and unloading techniques in ports. The introduction of containers changed the ways and time required to handle cargo, and had an impact on all of the activities that were once carried out on the waterfront. By the same token, new industrial development approaches caused productive infrastructures to move away from urban areas, and consequently from traditional port areas. During the same period of time, the advent of commercial aviation virtually eliminated passenger maritime traffic over long routes. As a result, this process led to the physical, functional and even cultural separation of port facilities from the urban areas of which they were once part.

Finally, over the past decade other innovations have come into play in the port-city relationship, including greater investments, the internationalization and globalization of the economy, the concomitant reduction of jobs, the need for greater space to be allotted to port activities, and their absolute independence with respect to previosuly occupied areas. Moreover, new tertiary activities bearing no relationship at all with port activities, have developed in the areas situated between cities and the waterfront. The existing port structures, which are currently experiencing an absolute decline, are being converted into homes, tourist resorts and leisure-time facilities. Finally, urban populations and therefore city port dwellers too, have placed greater emphasis on the quality of life, on the reduction of atmospheric and water pollution, and on the possibility of having access to the waterfront. At the beginning of the sixties, port change and evolution dynamics were examined, as well as the effects of technological changes which have occurred over time in the surrounding areas (Bird, 1963, p.21-36). In his study, Bird examined six phases of port development directly related to the same number of technological innovations: primitive port, limited wharf extension, limited wharf reorganization, dock implementation, non-specialized wharf construction, and special wharf construction. If the six development phases are examined in terms of the effects that they caused on the urban structure, it may be observed that the first three refer to a port-city system that is still integrated. The fourth and the fifth are characterized by the interrelation between the city and the port, two different dimensions which are physically separate, and yet are still related. Finally, the sixth refers to two completely independent functional systems.

More recently another scholar studied the city-port relationship, and singled out five development phases (Hoyle, 1988, p.7): original port city, expansion of the port city, industrialization of the port city, withdrawal from the waterfront, waterfront revitalization. In this case too, three different situations come to the fore. The first refers to an integrated port-city system, the second and the third refer to two autonomous and yet interrelated elements, the city and the port, while the fourth and fifth refer to two self-sufficient functional systems that develop autonomously.

These two studies had the merit of recognizing and thus describing a series of situations and development processes that have affected port cities since ancient times. They have, however, failed to analyze the relations that might exist between the different types of innovation and the evolutionary process of the port-city system. Gaining an empirical insight into the metropolitan areas where port functions were once performed will make it possible to relate the evolutionary phases of the port-city interfaces with the phases of urban development (Montanari, 1988). It has been proven that the primary port city matches the preurbanization phase, the expanding port city matches the phase of urbanization, the industrialization of the port city matches the phase of suburbanization, the withdrawal from the waterfront matches the process of deurbanization, and the revitalization of the waterfront matches the phase of reurbanization.

Innovative examples for waterfront redevelopment

Battery Park City

Battery Park City (BPC) extends over an area which in 1976 was filled with earth, between pier A and pier 25, along the Hudson river in the southwestern part of Manhattan, New York City (Figure 18.1.). Battery Park City is known as the most extensive urban restructuring plan in the United States, as it covers an area of 36.8 ha. Once it is completed, it will comprise five towers providing office space (World Financial Center), various residential buildings for a total of 12,000

Figure 18.1. *New York, Battery Park City: location of the project in the metropolitan area and its planimetry*

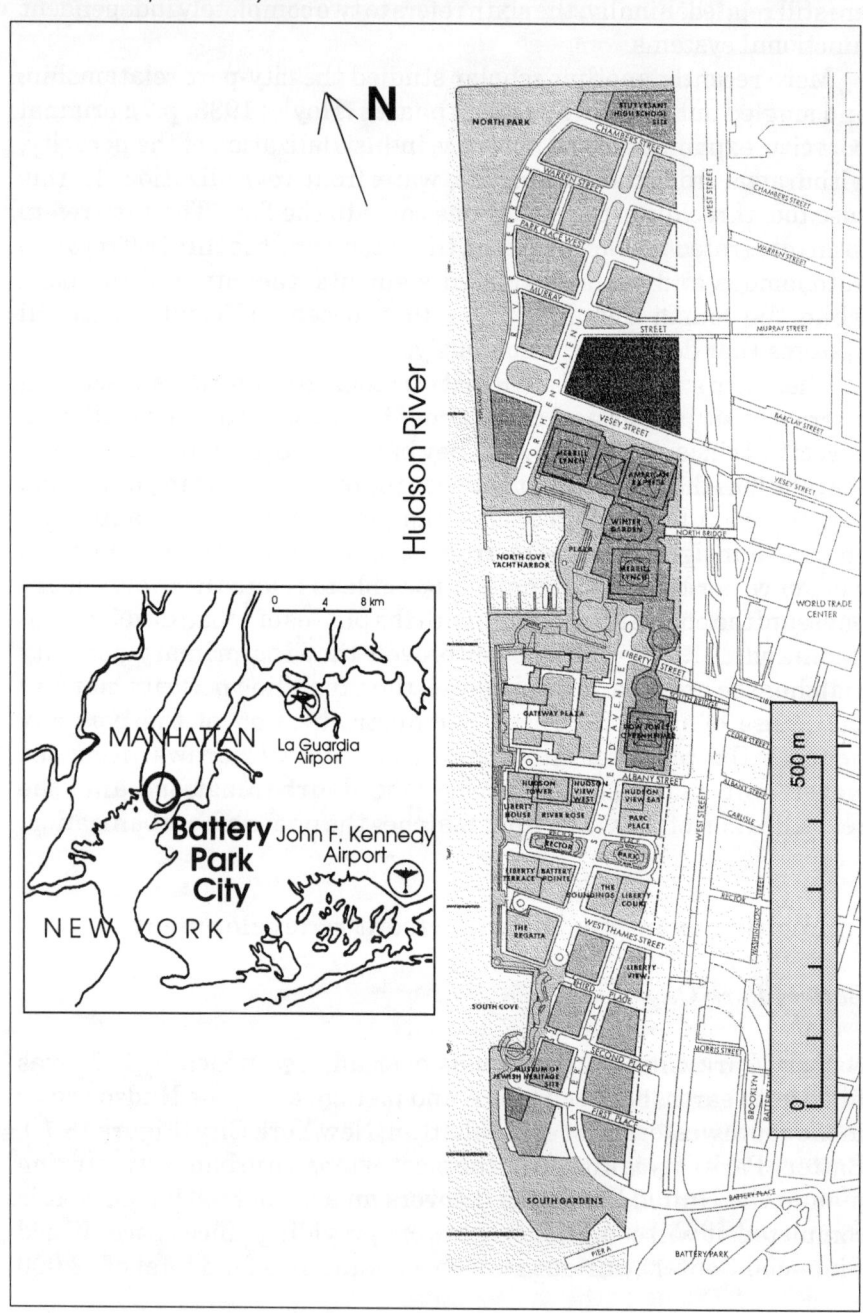

flats, a hotel and a museum (the Jewish Heritage Museum). It is envisaged that twelve hectares of land, accounting for about 30% of the total, will be available for park space and outdoor public recreation, along with a two-kilometer promenade along the Hudson, a 14,000 square meter outdoor plaza, and a 1,700 square meter indoor plaza in the shape of a large greenhouse with 16 palm trees almost as high as a five-story building. In the summer of 1989, a tourist harbor was also completed, which can accomodate up to 26 yachts. The urban planning, building design, architecture, promenades and public recreational areas have made BPC a much acclaimed project. The estimated cost for land management operations amounts to about 200 million dollars, while about 4 billion dollars were allocated for the implementation of the plan, including a 1.5 billion dollar private grant for the construction of the World Financial Center.

The history of Battery Park goes back to 1968, when the Sate of New York decided to set up a company, the Battery Park City Authority (BPCA), which was assigned ownership of the area, along with the task of building a shopping center and a set of residential homes. For this purpose, in 1972 the company issued 200 million dollars worth of treasury bonds. BPCA has retained ownership of the area and is responsible for the management of the public recreational areas, while the grounds were leased for the construction of the buildings. These activities generated a 60 million dollar income for BPCA in 1988. The operation on the whole is expected to yield a profit of about 15 billion dollars for the New York Administration within the next 35 years. This would allow for the implementation of the 'Housing New York' low-cost housing project, which is already in progress and entails the construction of 60,000 homes, without adding to public expenditure.

The only activities undertaken up to 1979 consisted in land management, due to the general economic slump, the financial problems faced by the New York City Administration, and the inability of the local public body to attract private financing. In 1979 a new agreement was set up between the City and State of New York. BPCA was reorganized, refinanced, and administrative procedures for the issue of construction licenses were streamlined. In 1980 construction of the first residential homes got underway. The World Financial Center (Figure 18.2.) was financed and construction was begun in 1981 by Olympia and York Battery Park Company, a Canadian enterprise. The first tower was completed in 1985 and was occupied by some of the major financial corporations including Dow

379

Jones, Daiwa, Nippon and Nikko Securities. In the same year, American Express began settling into the second tower, which it has taken over completely, while in 1986 Merrill Lynch began taking over the remaining two towers. Olympia and York was thus able to complete the program in five years, building on average over 100,000 square meters of office space annually, a sort of record in this field. A fifth tower is currently being planned.

Figure 18.2. *New York, Battery Park City: axonometrical view of the World Financial Center and Plaza*

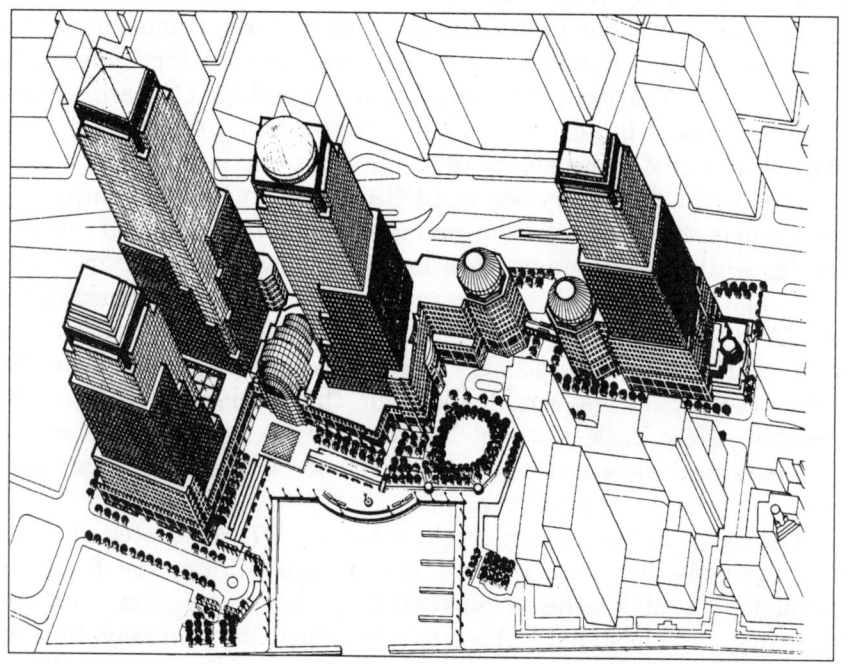

When in 1981 Olympia and York revealed for the first time their intention to build 560,000 square meters worth of new office space, many real estate brokers in New York were doubtful as to the outcome of the operation. With approximately 25 million square meters, Manhattan presented the highest office density worldwide, and it seemed highly unlikely that it could suddenly absorb another half a million. The doubts were dissipated only in 1984, when negotiations began with Merrill Lynch for the purchase of over 300,000 square meters.

London Docklands

The London Docklands were also defined as the biggest and most important urban renewal plan[1]. The plan involved an area of approximately 22 square kilometers (Figure 18.3.), divided into four main areas (Wapping, Surrey Docks, Isle of Dogs, Royal Docks) on which the public administration spent about 317 million pounds between 1981-87. The major British newspapers have already moved their offices from Fleet Street to this part of London: the Financial Times, The Daily Telegraph, The Guardian, The Mail Newspapers, and The News International where The Times, The Sunday Times, The Sun, The News of the World and Today are published.

The pier and deposit system which developed after 1800[2] made the London port one of the major harbors of the world. In 1964 the Docks again registered a movement of 61 million tons of cargo, and the Port of London Authority (PLA) continued to submit projects for their expansion. In reality, however, the Docks had started slowly and inexorably to decline long before, especially following the introduction of new cargo handling technologies. The East India Docks were the first to shut down in 1967, shortly followed by all the others. In October 1981 the news reported the mooring of the last ship[3]: an area measuring about 2,025 ha had been abandoned in less than 15 years, and almost 150,000 people had lost their jobs.

The first attempt to reconvert the Docklands dates back to 1973, when a hotel was opened in the St. Katharine Docks[4]. In 1981 the Government decided to set up the London Docklands Development Corporation (LDDC) for the purpose of fostering a new physical development in the area, but also in order to spur the economic and social development of one of the most depressed areas in London. The focal point of the whole operation was an area defined as the Enterprise Zone, covering about 203 ha, including 53 ha of bodies of water. Numerous administrative, financial and urban planning allowances are envisioned here to attract private enterprises. Between

1981 and 1987, 11,975 new homes were built, and just as many are expected to be completed by 1991. Private enterprises invested a total of 2,200 million pounds, and about 10,000 new jobs were created, which made it possible to partly make up for the jobs that had been lost in the industrial sector, and even generate new ones. It is estimated, however, that 80,000 new jobs will be created by 1991, and 200,000 by the year 2000. Some scholars (Church, 1988) consider such projections to be excessive, especially in light of the results obtained thus far.

Figure 18.3. *London, Canary Wharf: location of the project in the metropolitan area and within the context of the docks redevelopment plan*

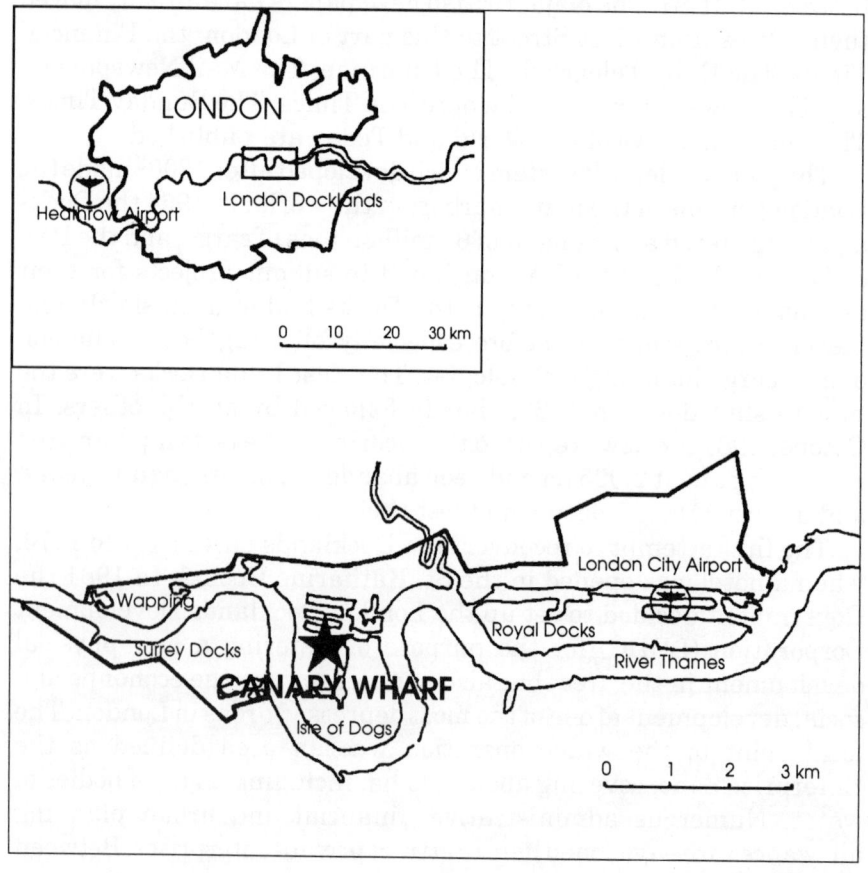

But the most significant part of the Enterprise Zone consists of the Canary Wharf, an area of 28.4 ha, on which Olympia and York, the same company operating in New York, has planned to invest between 3,000 amd 4,000 million pounds for the implementation of 22 building blocks (Figure 18.4.) which will provide 930,000 square meters worth of office space. A total of 46,000 jobs are expected to be created, thus making this project certainly one of the most extensive building operations in Europe. This operation has sparked many controversies in London, since the building of a 260 meter tower is envisaged in the middle of the area. This tower would be the tallest building in Europe, and would surely form a striking contrast with its surroundings. But concern, both in England and in the nearby regions of the mainland, is fuelled by the fact that this area will become London's third financial center, after the City and the West End, and even now it is referred to as Wall Street on the water. Some doubts have also been voiced as to whether the London real estate market will be able to absorb this additional mass of office space, especially after the 1987 recession. Olympia and York, however, took over the project in July 1987 from two previous partners, two banks[5]. The company probably felt that Canary Wharf would offer the possibility to create a great hub, a pole of attraction for large international financial corporations which would become a point of reference for the London metropolitan area as well as all of Southeastern England. But it would extend even beyond that, to other parts of Europe now within reach thanks to the tunnel under construction below the Channel. The great concern that this aroused at an international level, especially in France, was countered by the attitude in Britain, where the initiative met with the approval of the Government, and the Premier attended the cerimony for the laying of the foundation stone[6].

The Naples Management Centre

The Management Centre in Naples (CDN) is currently the biggest example of urban renewal in Italy; it extends over an area not far from the sea, in one of the industrial regions connected to the port (Figure 18.5.). It covers an area of about 110 ha, of which 52 will be set aside for parks, recreational activities, etc. The buildings will have a total volume of over 6 million cubic meters; public offices will account for 21% of the total, private offices for 63%, while homes will account for the rest. The CDN will house as many as 12,000 residents, and will provide work for 45,000 people. The overall cost of the infrastructure is estimated at around 500 billion lire; investments in the order of 5 trillion have thus been fostered. The implementation of the CDN was promoted by MEDEDIL, an IRI group company, in 1978, but only towards the end of 1982 did the plan take on its current features (Figure 18.6.).

Minato Mirai 21 in Yokohama and Rokko Island in Kobe

In Japan, the most extensive urban renewal effort is known by the name of Minato Mirai 21 (MM 21), which was initiated in 1983 and is under implementation in Yokohama (Figures 18.7. and 18.8.). The area covers 186 ha (76 ha were obtained by filling with earth the port docks), on which buildings will be constructed for about 10,000 residents and 190,000 employees. Of the 186 ha, 87 ha will be used for commercial and residential purposes, 46 ha will be set aside for public parks, 11 ha for port activities, while roads and railway lines

Figure 18.5. *Naples, Management Center: location of the project in the metropolitan area and its planimetry*

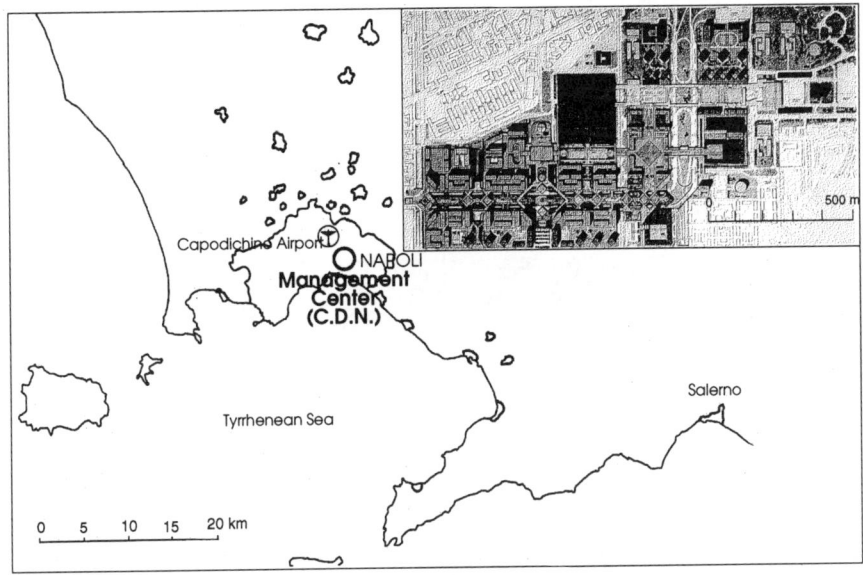

Figure 18.6. *Naples, Management Center: plan*

will be built in the remaining 42 ha. The allocation of 330 billion yen is envisaged for land management and for the implementation of the infrastructure; investments for approximately 1,670 billion yen are expected to follow.

Figure 18.7. *Yokohama, Minato Mirai 21: location of the project in the Tokyo Bay*

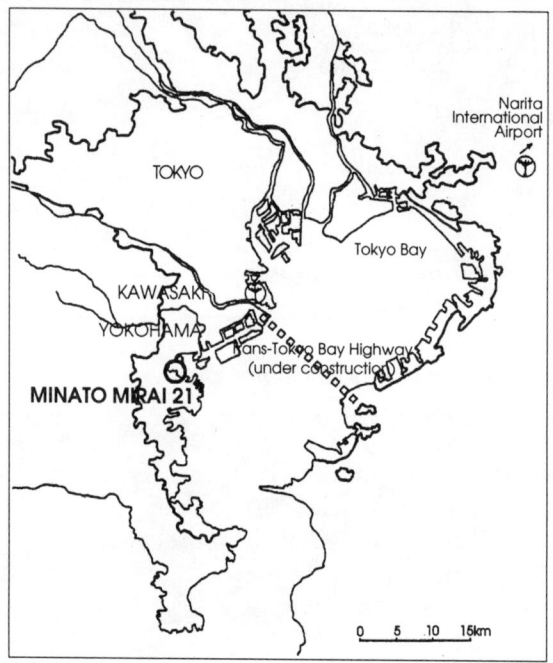

Over recent years, Yokohama has been increasingly converted into a region where Tokyo dwellers simply sleep. MM 21 was thus planned to provide the city with a new service center (Figure 18.9.), capable of enhancing the role of the city, the country's second largest in statistical terms. The settlement is the result of a joint effort between the public administration and private enterpreneurs (Mitsubishi Estate Company[7], Mitsubishi Heavy Industries, and the Bank of Yokohama).

Figure 18.8. *Yokohama, Minato Mirai 21: project planimetry*

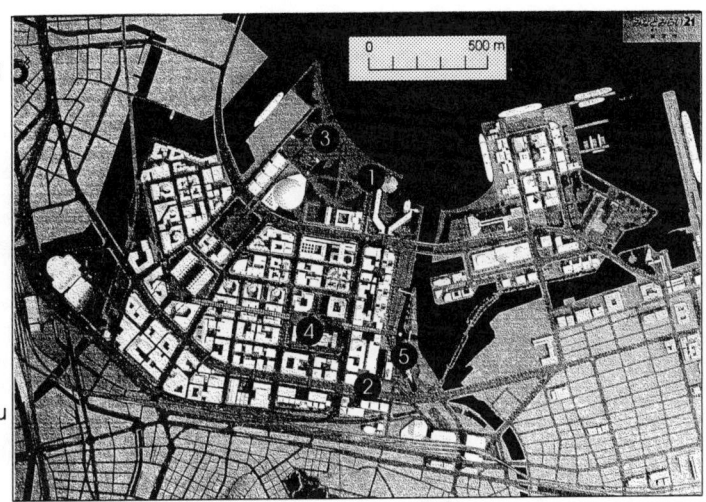

1) International congress center, international exhibit center and hotel
2) Maritime Museum
3) Waterfront park
4) Art museum
5) Nippon Maru Memorial Park

The Rokko Island project is currently under way in Kobe, the number one port in Japan. Rokko Island is a 580 ha totally man-made island, located in front of the existing urban area (Figure 18.10.) and built in 1972, using 120 million cubic meters of backfill. It is a multi-purpose facility which is expected to be completed by the beginning of the nineties, and in which port, tertiary and residential activities will be integrated. In fact 345 ha, accounting for 60% of the total area, is to be devoted to manufacturing activities and to cargo transportation, 35 ha (6%) will be covered by parks, and 131 (23%) will be set aside for homes and cultural as well as commercial activities (Figure 18.11.). The construction of 8,000 dwellings should allow for the settlement of at least 30,000 people. The Kobe Administration considers Rokko Island as a tool to trigger the necessary changes and integrations for a twenty first century metropolitan area. The total cost for the implementation of the plan is estimated at 1,240 billion yen.

Figure 18.9. *Yokohama, Minato Mirai 21: plan of the complex comprising an international congress center, an international exhibit center and a hotel*

Figure 18.10. *Kobe, Rokko Island: location of the project in the Osaka Bay*

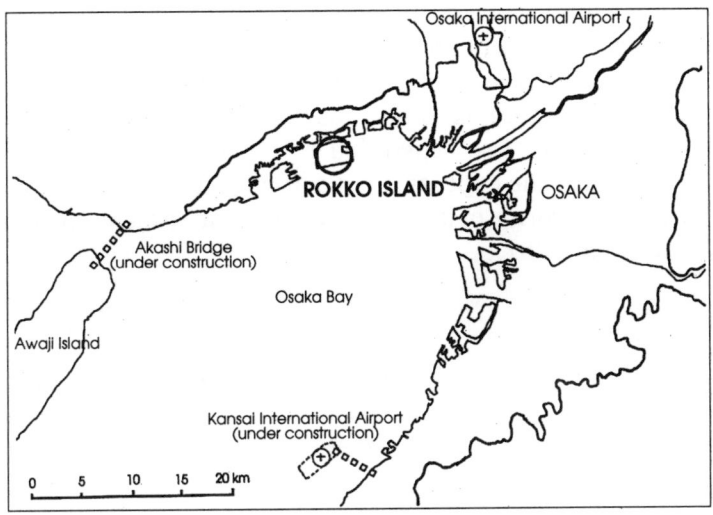

Figure 18.11. *Kobe, Rokko Island: plan of the project, with part of the Kobe urban area in the background, squeezed in between the mountains and the sea*

Innovative elements for the revitalization of the waterfront.

The examples described in this paper in an obviously concise form, have several points in common, although they have been implemented in rather different social and economic settings:

1 - they are concrete examples, part of which have already been implemented, which are scheduled to be completed by the early nineties. They are the expression of a society undergoing transition from Fordist-type forms of industrial capitalism to forms of advanced capitalism (Soja, 1987);

Table 18.1. *Battery Park City (BPC), Canary Wharf (CW), Naples Management Centre (CDN), Minato Mirai 21 (MM21), Rokko Island (ROKKO): cost (billions ECUS), size (ha), land use (ha), employees (x 1,000 people), residents (x 1,000)*

	BP	CW	CDN	MM21	ROKKO
Total project cost (1)	4	6	3.7	13.3	8
Total surface	36.	28.4	110	186	580
Homes and offices (2)	24.	-	58	87	131
Parks (2)	12	-	52	46	35
Roads and railways lines (2)	-	-	-	42	69
Port activities and related functions (2)	-	-	-	11	301
Estimated jobs	40	46	45	190	63
Number of estimated residents	25	-	12	10	30

The investment ranges from a little over 90 billion per ha of homes and office space, in Kobe and Naples, to over 200 billion in Yokohama and New York City, to 300 billion in London. The cost necessary to generate each job is more balanced and at the same time more significant; indeed, it ranges from 105 million in Yokohama, to 122 in Naples, 150 in New York City and 190 in Kobe and London. The number of planned residents appears to be especially dictated by the need to create an integration between work site and place of residence. This figure varies according to the supply of luxury homes offered in the areas surrounding the site of the project. (1) rough estimates expressed in billions of ECUs. The figures provided by the administrations responsible for each individual project are reported in the text and are expressed in local currencies. (2) the sub-areas indicated by the various administrations responsible are not perfectly comparable, as they were evaluated according to different criteria. However, port activities and related functions are envisaged only in Yokohama and Kobe.

2 - the plans envisage operations on a very large scale (Table 18.1.). They entail operations over areas covering tenths of hectares, located near the center of metropolitan areas, where initiatives of this sort and scope had never been planned before. The sums envisaged are considerable, comparable to those needed for the building of a large-scale infrastructure. The quality and cost of such operations obviously determine particularly high rents or sales costs for the individual office areas or homes. In Battery Park City, a studio apartment was rented for about 1,600 dollars a month in 1988; it may thus be estimated that an annual income of at least 80,000 dollars is necessary to live in the area. In London, the Docks have been invaded by the nouveaux riches, who move there from the City into modern homes or refurbished port facilities affording a view over the Thames. The Docks are in fact also commonly known as the yuppie towers, much like the Cascades district in the Isle of Dogs (Short, 1989). These homes, however, are still very much in demand. In Kobe the demand for apartments, which in 1988 were still under construction, was so high that they were assigned by picking names out of a hat;

3 - regardless of the architectural style and the chosen urban form, the projects are characterized by high quality work, for which massive investments have been made to improve the quality of life, and to create welcoming and culturally significant public areas. The handling of buidings volumes, the organisation of public and private spaces, and a more extensive planning of care services (culture, amusement, recreation and entertainment) (Schwanke, 1987) show ludic combinatorial, localistic and anti-narrative characteristics which are typical of post-modern culture;

4 - all of the projects require the investment of large sums by private enterpreneurs, who ask for effective administrative coordination by the public body, along with continual attention in the bureaucratic handling of affairs, and the prompt fulfillment of the obligations envisaged by the law, in order to save time and cut costs. These features of the projects, together with those already mentioned, show the attempt to deconstruct modernist organisation which is typical of the 'postmodernism of resistance' in anthithesis to an aestheticising and superficial form defined 'postmodernism of reaction' (Foster, 1985). Since local administrations had been previously unable to respond to the needs of the private enterpreneurs, specific agencies were established (Battery Park City Authority, London

Docklands Development Corporation, etc.) which followed the project from the very beginning, in its technical, financial and administrative aspects. Such agencies have recognized the needs of private enterpreneurs, but have also defended the interests of the community;

5 - the private companies that undertook this type of operation are mostly multinationals of great standing. Their investments to date have been remarkably successful; it is thus likely that they will continue to engage in these projects, which will be further facilitated by the growing internationalization and globalization of the world economy;

6 - all of these projects were carried out in areas of particular significance in terms of the landscape and environment. Moreover, all of the districts were built directly on the waterfront or facing it. Their success stems from their having taken advantage of a natural resort's power of attraction, which in the case of port cities has led to their enhancement. The offered natural beauty and quality of life have proven to be essential elements for the success of an initiative implemented in a typical phase of reurbanization;

7 - at the beginning of the eighties our cities, at an advanced stage of deindustrialization, exhibited numerous voids, and the most traditional industrial areas had been abandoned and were in a state of obsolescence. They were considered to be 'weak' and 'soft' as compared to the vitality and development capacity of service centers. The most significant of such areas include the Docks in London, the Citroen factory in Paris, the Rotterdam port, and the industrial areas in Turin and Milan (Montanari, 1992). A few years have elapsed since then, and it may be stated that the industrial waterfront, precisely due to its position and characteristics, is the one area to have been renewed and revitalized before all others;

8 - renewal initiatives in urban areas on the water have shown an innovative characteristic compared to the past even in the field of physical planning. Indeed, the once dispersed, fragmented, heterogeneous plans, sometimes drawn up according to no pre-established standards, have been replaced by a unitary management and implementation plan. Therefore it is precisely in their phase of reurbanization that these urban areas have gone back to those design methods which have been seldom implemented, although they are at the origin of modern urban planning.

9 - the recent large-scale operations described here propose alternative developments as compared to those which have taken

place in contemporary cities. Indeed the latter were essentially built according to lots, which are dispersed in a continuous and unwarranted sprawl, and confusedly take in the free areas still surrounding them. Actually, the most recent studies and cultural trends confirm that the urban areas of the next decades already exist: it is a matter of renewing, modifying and revitalizing the areas built in the last century, in the course of the different phases of the industrialization and urbanization process.

Notes

1. The following statement was made in an article published in 'The Times': 'the regeneration of London's Docklands will undoubtedly be seen as one of the greatest successes in urban renewal anywhere in the world'.
2. In 1799 the West India Dock Company was authorized to implement its own Docks in the then uninhabited Isle of Dogs. The Docks were thus opened in 1802. This surely represented a remarkable innovation, and the traffic in the port of London went up from 14,000 ships in 1794 to 23,600 ships in 1824.
3. Given the exceptional nature of the event, the news also reported the name of the last ship that unloaded its cargo in the Docklands, the Chinese ship Xingfeng.
4. When the Tower Hotel was inaugurated in 1973, it sparked a wave of protest by the trade unions, which claimed that homes should be built before hotels, with banners reading 'Homes before Hotels'.
5. Credit Suisse First Boston and Morgan Stanley were the two banks.
6. In attending the cerimony for the laying of the foundation stone (May 11, 1988), Margaret Thatcher, the British Premier, congratulated Olympia & York for having placed their trust in London and in Great Britain, and for the fact that the initiative was totally financed by the private sector 'I congratulate Olympia & York on their vision for Canary Wharf and I welcome the project. It is a vote of confidence in Britain, in London and in the future. I particularly welcome the fact that this development is being driven and financed by private enterprise'.
7. The Mitsubishi Real Estate Company owns 33 large office buildings in Tokyo, for a total of 1.77 million square meters, estimated at about 1 trillion yen.

References

Berg, v. d., L., Drewett, R., Klassen, L. H., Rossi, A., Vijverberg, H., T. (1982), *Urban Europe. A Study of Growth and Decline*, Pergamon Press, Oxford.

Bird, J. H. (1963), *The Major Seaport of the United Kingdom*, Hutchinson, London.

Bird, J. H. (1984), 'Seaport development: some questions of scale', in Hoyle, B. S. and Hilling, D. (eds.), *Seaport Systems and spatial change: technology, industry and development strategies*, Wiley, Chichester, p. 21-41.

Church, A. (1988), 'Demand-led Planning, the Inner-City Crisis and the Labour Market: London Docklands Evaluated' in Hoyle, B.S., Pinder, D.A., and Husain, M.S. (eds.), *Revitalising the Waterfront, International Dimensions of Docklands Redevelopment*, Belhaven, London, p. 199-221.

Foster, H. (ed.) (1985), *Postmodern Culture*, Pluto Press, London.

Hayut, Y. (1982), 'The Port-urban Interface: an area in transition', *Area*, 14 (3), p. 219-24.

Hoyle, B. S. (1988), 'Development Dynamics at the Port-city Interface', in Hoyle, B. S., Pinder, D. A. and Husain, M. S. (eds.), Revitalising the Waterfront, International Dimensions of Dockland Redevelopment, Belhaven, London, p. 3-19.

Knox, P. L. (1987), 'The Social Production of the built Environment: Architects, Architecture and the Post-modern City', *Progress in Human Geography*, 11, p. 354-78.

Ley, D. (1985), 'Cultural/Humanistic Geography', *Progress in Human Geography*, 9, p. 415-23.

Marchetti, C. (1986), *Stable Rules in Social Behaviour*, Academia Brasileira de Ciências/IBM Conference, Rio de Janeiro.

Marchetti, C. (1988), 'Infrastructures for Movement: Past and Future', in Ausubel, J. H. and Herman, R. (eds.), 'Cities and their Vital Systems: Infrastructure, Past, Present and Future', National Academy Press, Washington, D. C., p. 146-74

Mayer, H. M. (1973), 'Some Geographical Aspects of Technological Changes in Maritime Transportation', Economic Geography, 49, p. 145-55.

Mensch, G. (1975), *Das Technologische Patt*, Umschau, Frankfurt am Main.

Montanari, A. (1988), 'A Modern Perspective: the Recent Development of Port Cities in Southern Europe', in Malkin, I. and Hohlfelder, R.L. (eds.), *Mediterranean Cities, Historical Perspectives*, Cass, London, p. 166-85.

Montanari, A. (1989), 'Barcelona and Glasgow: the Similarities and Differences in the History of Two Port Cities', *The Journal of European Economic History*, 18 (1), p.171-89.

Montanari, A. (1992), 'Toward a Change in the Urban Design Paradigms?', in v. Böwenter, E., Koll, R. and Schubert, U. (eds.), *Innovation and Urban Development: A multi-discipline perspective*, Avebury, Aldershot, forthcoming.

Muscarà, C., Soricillo, M., Vallega, A., (eds.) (1982), *Changing Maritime Transport*, I.U.N. - Istituto di Geografia Economica, Naples.

Nakicenovic, N. (1985), 'The Automotive Road to a Technological Change: the Diffusion of the Automobile as a Process of Technological Substitution', *IIASA WP*, IIASA, Laxenburg.

Norcliffe, G. B. (1981), 'Industrial Change in Old port Areas: the Case of the Port of Toronto', *Cahiers de Géographie du Québec*, 25, p. 237-54.

Schwanke, D. (1987), '*Mixed-use Development Handbook*', Urban Land Institute, Washington.

Slack, B. (1980), 'Technology and Seaports in the 1980s', *Tijdschrift voor Economische en Sociale Geografie*, 71, p. 108-13.

Soja, E. (1980), 'The Postmodernization of Geography: a Review', *Annals of the Association of American Geographers*, 77, p. 289-94.

U.S. Dept. of Commerce (1970), 'Historical Statistics of the United States, colonial times to 1970', part 1 & 2 (Washington D.C.:Bureau of the Census).

Vigariè, A. (1981). 'Maritime Industrial Development Areas: Structural Evolution and Implications for Regional Development', in Hoyle, B. S. and Pinder, D. A. (eds.), 'Cityport Industrialization and Regional Development: Spatial Analysis and PLanning Strategies', Pergamon Press, Oxford, p. 23-36.

Mcaleese, A. (1988), "A Modern Perspective on the Recent Development of Iron Ochre in Southern China," in Mathias, T. and Habib, N. or R.L. (eds.) *New Internment Cities, Hong and Recreation on China*, London, b.166-95.

Moulanet, A. (1993), *Aerospace and Glasgow: the Similarities and Differences in the History of two World Port Cities*, The Journal of European Economic History, 151, pp. 17-43.

Mulrenin, A. (1999), "Towards a Change in the Urban Process," in Williams, K., Burgess, B., Roll, E. and Scott, M. (eds.), *Journal on Modern Development: a Sustainable Urban Form?*, E. & F.N. Spon, London.

Chukola, T., Cannillo, M., Valero, A., (eds.) (1992), "Turin: Maritime Economy," UN., Istituto di Geografia Economica, 42-pla?.

Muamede, K. (1988), "The Automotive Road to a Technological Change and Diffusion of the Automobile as a Process of Technological Development," Ph.D., U.S.A., Laxenburg.

Nowak, K., F.(1981), "Industrial change in Old port Areas: the Case of the Lower Texaco," *International Inequalities in Quality*, 82-a, 388-84.

Seward, T. (1983), *Mixed-use Development*, Broadside, Urban Land Institute, Washington.

Susi, R. (1980), "Technology and Reaction in the 1980s," *Journal of Economic and Social Geography*, 4, pp.105-15.

Sojka, K. (1980), "The Predemocratization of the American Frontier," *Annals of the Association of American Geographers," 77, p. 589-94.

U.S. Dept. of Commerce (1979), *Historical Statistics of the United States: Colonial Times to 1970*, part 1 & 2, Washington, D.C. Bureau of the Census.

Vicente, A. (1981), "Maritime Industrial Development Areas: Theoretical and Practical Implications for Regional Development," in Hoyle, B. and Pinder, D. A. (eds.), *Cityport Industrialization and Regional Development: Spatial Analysis and Planning Strategies*, Pergamon Press, Oxford, pp. 23-36.